Advances in Corrosion Science and Surface Engineering

Principles, Mechanisms, and Emerging Technologies

Edited by

Rajender Boddula[1]

Ramyakrishna Pothu[2,3]

G. Kausalya Sasikumar[4]

Noora Al-Qahtani[5]

[1] School of Sciences, Woxsen University, Hyderabad - 502345, Telangana, India.

[2] Center for Innovation and Inclusive Research, Sharda University, Greater Noida – 201310, India

[3] School of Physics and Electronics, College of Chemistry and Chemical Engineering, Hunan University, Changsha 410082, P.R. China.

[4] Centre of Research and Development, KPR Institute of Engineering and Technology, Coimbatore 641407, Tamil Nadu, India.

[5] Center for Advanced Materials, Qatar University, Doha, Qatar.

Published by **Materials Research Forum LLC**
Millersville, PA 17551, USA

Published as part of the book series
Materials Research Foundations
Volume 188 (2026)
ISSN 2471-8890 (Print)
ISSN 2471-8904 (Online)

Print ISBN 978-1-64490-390-2
eBook ISBN 978-1-64490-391-9

Distributed worldwide by

Materials Research Forum LLC
105 Springdale Lane
Millersville, PA 17551
USA
https://mrforum.com

Manufactured in the United States of America
10 9 8 7 6 5 4 3 2 1

Table of Contents

Preface

Corrosion has remained one of the most persistent problems in materials science and engineering, causing losses in the form of billions of dollars every year and giving rise to the deterioration of the most important infrastructure across the globe. The increased growth rate of modern industries, renewable technologies and green materials has changed the field of research on corrosion making it necessary to have an integrated knowledge that balances chemistry, physics, and engineering practice.

The book, Advances in Corrosion Science and Surface Engineering, is a collection of extensive knowledge in the basics of corrosion, electrochemical process, metal degradation, protective coating, corrosion inhibitors and advances in mitigating corrosion. All chapters present both traditional principles and the latest advances, including nano-engineered surfaces, bio-inspired inhibitors, and data-driven prediction models.

The book aims to close the gap between theoretical work and practical technology to empower researchers, engineers, and graduate students with the instruments needed to develop sustainable corrosion-control strategies that can resolve some of the current industrial and environmental concerns.

I would like to offer my very special thanks to the scientists who contributed, as well as my peer reviewers and colleagues whose critical examination and technical assistance has enhanced this piece of work. The compilation is an outcome of the work of a world community that has committed itself to the spread of corrosion science to make the world a safer and more sustainable place.

Advances in Corrosion Science and Surface Engineering   Materials Research Forum LLC
Materials Research Foundations 188 (2026) 1-21   https://doi.org/10.21741/9781644903919-1

Chapter 1

Types of Coating on Metallic Materials

K.V. Satheesh Kumar[1]*, M. Bhuvanesh Kumar[1], V.N. Kowshalya[2], G. Kausalya Sasikumar[3], Ramyakrishna Pothu[4,5]

[1]Department of Mechanical Engineering, Kongu Engineering College, Perundurai-638060, Erode District, Tamilnadu, India

[2]Department of Chemistry, Kongu Engineering College, Perundurai-638060, Erode District, Tamilnadu, India

[3]Centre for Research and Development, KPR Institute of Engineering and Technology, Coimbatore 641407, Tamil Nadu, India

[4]Center for Innovation and Inclusive Research, Sharda University, Greater Noida – 201310, India

[5]School of Physics and Electronics, College of Chemistry and Chemical Engineering, Hunan University, Changsha 410082, P.R. China

kvs@kongu.ac.in

Abstract

Corrosion is a significant challenge for industries and economies worldwide, contributing to substantial material degradation, safety concerns, and financial losses. It is estimated that corrosion accounts for a considerable percentage of many countries' GDP, reflecting the urgency to address this pervasive issue. This study investigates the fundamental mechanisms of corrosion, emphasizing its electrochemical nature and the influence of environmental and material-specific factors. The roles of corrosive agents, including oxygen, moisture, and industrial chemicals, are examined to provide a comprehensive understanding of their impact on metallic surfaces. This chapter also highlights the critical importance of corrosion-resistant materials, such as stainless steel, titanium alloys, etc., which offer inherent durability in aggressive environments. Protective coatings, as a primary defense mechanism, are explored in detail, with a focus on recent advancements in organic, inorganic, and hybrid coatings designed to enhance longevity and inhibit corrosion. The study underscores the economic and industrial implications of corrosion, emphasizing the need for innovative materials and protective strategies to mitigate these losses. This study provides valuable insights into understanding and combating corrosion, supporting global efforts to enhance industrial reliability and reduce its economic impact.

Keywords

Corrosion, Corrosion Resistance Materials, Coatings, Metallic Materials

Contents

1.   Introduction to Corrosion Resistance Coating

Corrosion is a natural electrochemical process that occurs when metals undergo chemical or electrochemical reactions with their environment, leading to degradation. This process typically involves the oxidation of the metal, resulting in the formation of compounds such as rust (iron oxide) on steel or aluminum oxide on aluminum. The consequences of corrosion are significant, leading to material degradation, failure of structural components, and substantial economic losses in industries such as construction, aerospace, automotive, and marine engineering. Understanding the underlying mechanisms of corrosion is essential for developing effective prevention strategies

Advances in Corrosion Science and Surface Engineering   Materials Research Forum LLC
Materials Research Foundations 188 (2026) 1-21   https://doi.org/10.21741/9781644903919-1

and improving the longevity of metallic structures. This article explores the basic principles of corrosion, the electrochemical mechanisms involved, and the factors that influence its rate and severity, drawing on recent literature for a comprehensive discussion. Corrosion can take various forms, each driven by different environmental conditions. The primary types include Uniform Corrosion, Galvanic Corrosion, Pitting Corrosion, Crevice Corrosion, and Stress Corrosion Cracking (SCC). Uniform Corrosion is the most common and widespread form of corrosion, where the metal deteriorates uniformly across its surface. It typically occurs due to the direct exposure of a metal to moisture or acidic conditions. For instance, iron subjected to oxygen and water forms iron oxide uniformly on its surface [1]. When two different metals are in electrical contact in a corrosive environment, one metal becomes anodic (sacrificially corroding), while the other becomes cathodic (protected). This type of corrosion is often referred as Galvanic Corrosion and observed when metals like steel and copper are in contact, leading to accelerated corrosion of the anode [2].

Pitting Corrosion occurs when localized corrosion leads to the formation of small holes or pits on the surface of the metal. This type of corrosion is particularly dangerous because it is not always visible, and the damage can progress rapidly once initiated. Pitting is common in stainless steel when chloride ions penetrate the passive oxide layer, leading to localized breakdown [3]. Crevice Corrosion is Similar to pitting corrosion, and it occurs in confined spaces or crevices, such as beneath washers, seals, or rivets. These areas often experience localized depletion of oxygen and increased acidity, creating ideal conditions for the breakdown of protective oxide films and the initiation of corrosion [4]. Stress Corrosion Cracking (SCC) is a type of corrosion results from the combination of tensile stress and a corrosive environment, which leads to the formation of cracks in the metal. SCC is most commonly observed in high-strength alloys used in aerospace and marine applications, where mechanical stresses and chloride-rich environments interact [5].

1.1 Electrochemical Mechanism of Corrosion

The corrosion of metals is fundamentally an electrochemical process, which involves the transfer of electrons between the metal and its environment. This process can be broken down into several steps as follows:

Electrochemical Cells: Corrosion generally occurs in electrochemical cells, where a metal (the anode) is in contact with an electrolyte (typically water or an aqueous solution containing salts or acids). The metal acts as the anode, undergoing oxidation reactions that release electrons. The cathode, often a different part of the metal or surface, consumes these electrons in reduction reactions. For example, in the corrosion of iron (rusting), the overall reaction can be written as:

$$Fe \rightarrow Fe_2^+ + 2e \qquad (1)$$

The Fe^{2+} ions then react with oxygen and water to form iron oxide (rust):

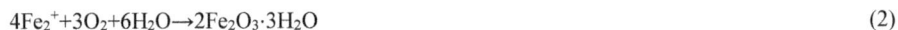

$$4Fe_2^+ + 3O_2 + 6H_2O \rightarrow 2Fe_2O_3 \cdot 3H_2O \qquad (2)$$

Anodic and Cathodic Reactions: In a typical corrosion scenario, the metal surface forms an electrochemical cell with anodic and cathodic regions. The anode undergoes oxidation, releasing electrons, which flow through the metal to the cathode. The cathodic reaction typically involves the reduction of oxygen or hydrogen ions, such as:

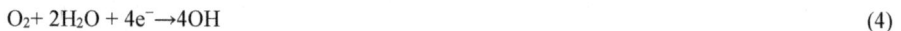

$$2H^+ + 2e^- \rightarrow H_2 \tag{3}$$

$$O_2 + 2H_2O + 4e^- \rightarrow 4OH \tag{4}$$

Or

Formation of Corrosion Products: As the metal loses electrons at the anode, corrosion products such as metal oxides or hydroxides form. For example, in the case of steel, iron oxide (rust) forms as a product of the reaction between iron and oxygen in the presence of water or moisture.

Electrochemical Potential Difference: The corrosion rate is influenced by the electrochemical potential difference between the anodic and cathodic regions. A greater difference leads to faster electron transfer and, consequently, higher corrosion rates [6].

1.2 Factors Influencing Corrosion

Several factors can influence the rate and type of corrosion that occurs. These factors include the environmental conditions, temperature, material composition, pH of the environment and mechanical stress. The presence of moisture, salts, acids, and other chemicals in the environment plays a crucial role in determining the corrosion rate. Chlorides, for example, accelerate corrosion by breaking down protective oxide films on metals such as stainless steel [1]. High humidity levels and exposure to seawater are particularly corrosive for many metals. Similarly, temperature affects the corrosion rate by increasing the rate of electrochemical reactions. For example, in marine environments, increased temperatures accelerate the dissolution of metal ions and the formation of corrosion products [7].

The type of metal or alloy significantly influences its susceptibility to corrosion. Materials with a passive oxide layer, such as stainless steel, are generally more resistant to corrosion. However, defects in the oxide layer can lead to localized corrosion [8]. On the other hand, metals like iron and copper are more prone to uniform and galvanic corrosion in aggressive environments. Likely, Acidic conditions (low pH) tend to increase the corrosion rate, as they facilitate the dissolution of metal ions. For instance, in acidic conditions, the electrochemical reactions at the anode and cathode become more pronounced, leading to faster degradation of the material. The presence of mechanical stress can also influence corrosion behavior, especially in the case of stress corrosion cracking (SCC). Metals subjected to tensile stress in a corrosive environment are more likely to develop cracks that propagate over time, leading to material failure [9].

1.3 Corrosion Protection Strategies

Given the significant impact of corrosion on materials and structures, various corrosion protection strategies have been developed. These include:

Coatings: Protective coatings such as paint, galvanization (zinc coating), and anodizing (for aluminum) are widely used to prevent the direct contact of metals with corrosive environments. Coatings act as physical barriers that reduce the exposure of the metal to oxygen and moisture.

Corrosion Inhibitors: Chemical inhibitors are added to environments to reduce the rate of corrosion. These inhibitors can work by forming protective films on the metal surface or by scavenging corrosive agents such as chloride ions.

Sacrificial Anodes: In galvanic protection, sacrificial anodes made of a more electrochemically active metal (such as zinc or magnesium) are attached to the structure. These anodes corrode preferentially, protecting the underlying metal.

Material Selection: Choosing materials that are inherently resistant to corrosion, such as stainless steel, titanium, and certain alloys, can significantly reduce the risk of corrosion in specific environments.

Corrosion is a complex process driven by electrochemical reactions that degrade materials, primarily metals, when they interact with their environment. The process is influenced by various factors such as environmental conditions, material properties, temperature, and mechanical stress. A clear understanding of the electrochemical mechanisms behind corrosion can help in the design of effective prevention and mitigation strategies. The development of corrosion-resistant materials, protective coatings, and inhibitors continues to evolve, offering promising solutions for extending the service life of metallic structures and reducing the economic costs associated with corrosion.

2. Corrosive Materials

Corrosive materials are substances that can cause degradation or destruction of materials, typically metals, through chemical or electrochemical reactions. Corrosion is an important phenomenon in a wide range of industries, including construction, transportation, energy production, and manufacturing, where the integrity of materials is crucial for safety, performance, and durability. Corrosive materials can be found in natural environments, such as seawater, and in industrial settings, such as acids, alkalis, and salts. The impact of corrosive materials on infrastructure and industrial assets is substantial, often leading to failure or significant damage if not adequately managed.

The process of corrosion is highly dependent on the type of corrosive material involved and the interactions it has with the materials in question. This article aims to provide a detailed discussion on the different types of corrosive materials, how they interact with metals and other materials, and the consequences of their actions, supported by insights from recent journal articles.

2.1 Types of Corrosive Materials

Acids are among the most common corrosive materials, capable of causing severe corrosion on many metals. Their corrosive effects are primarily due to their ability to donate protons (H^+ ions), which accelerate the oxidation process of metals. Sulfuric Acid is one of the most aggressive industrial acids widely used in processes like metal cleaning, etching, and battery production. A study by [10] discussed how sulfuric acid can significantly degrade the surface of steel and iron, leading to the formation of iron sulfate. This degradation can be particularly problematic in the petroleum and chemical industries, where sulfuric acid is used extensively. Another common corrosive material, hydrochloric acid is used in metal pickling, oil well acidizing, and the production of chlorides. Its interaction with metals such as steel, copper, and aluminum is well-documented, with the acid accelerating the metal dissolution process by reacting with the surface to form soluble metal chlorides [11]. The corrosion rate increases significantly in the presence of impurities like iron oxide, making it particularly challenging to control in certain applications.

Alkalis, such as sodium hydroxide (NaOH) and potassium hydroxide (KOH), are also highly corrosive, particularly to metals such as aluminum and zinc. Alkalis can break down the protective oxide layers on metals, allowing further corrosion to occur. Similarly, Sodium hydroxide is commonly used in cleaning, etching, and as a catalyst in various chemical reactions. Research by [12] highlighted the aggressive nature of sodium hydroxide towards aluminum alloys, where the alkali reacts with the oxide film on aluminum, causing rapid dissolution and forming soluble aluminum hydroxide. This is especially problematic in the chemical processing industry, where sodium hydroxide is used in cleaning reactors and pipelines.

Salts, especially chlorides, are major contributors to corrosion, particularly in environments such as marine and industrial settings. The chloride ions (Cl^-) are particularly corrosive because they break down passive oxide layers on metals, leading to localized corrosion such as pitting or crevice corrosion. Sodium Chloride (NaCl) is commonly found in seawater, causing corrosion in metal structures, including pipelines, ships, and offshore platforms. A study by [13] emphasized that chloride ions penetrate the passive films on stainless steel and aluminum alloys, initiating localized corrosion that is often difficult to detect until significant damage has occurred. In addition, the combination of moisture and chloride ions accelerates the breakdown of protective films, making metal structures more susceptible to rapid deterioration.

Calcium Chloride ($CaCl_2$) also widely used as a de-icing agent on roadways, also promotes corrosion, particularly in automotive and transportation infrastructure. Research by [14] highlighted that calcium chloride can interact with the surface of steel and other metals, leading to increased metal loss, particularly in cold weather environments where salt application is frequent. Certain gases, particularly oxygen, hydrogen sulfide, and carbon dioxide, are potent corrosive agents when present in the environment, especially in confined spaces or industrial facilities. Oxygen while not traditionally classified as a corrosive material, can contribute to corrosion, especially in the presence of water. Oxygen promotes the oxidation of metals, such as iron and steel, leading to the formation of rust. Research by [15] discussed how the presence of oxygen, combined with moisture, accelerates the corrosion of carbon steel in marine environments, where the rates of corrosion are often exponential.

Hydrogen sulfide (H_2S) is a corrosive gas found in natural gas and petroleum industries. It reacts with metals like carbon steel, causing hydrogen embrittlement and the formation of iron sulfide scales on the metal surface. This can significantly reduce the material's strength and integrity. A previous study highlighted the effects of H_2S on pipeline steels, noting that exposure to H_2S can lead to serious cracking and failure if not properly mitigated [16].

Microbial corrosion, also known as microbiologically influenced corrosion (MIC), occurs when microorganisms, such as bacteria and fungi, produce acids or other corrosive by-products that lead to metal degradation. This type of corrosion is particularly relevant in water systems, oil and gas pipelines, and marine environments. Sulfate-Reducing Bacteria (SRB) is known to produce hydrogen sulfide (H_2S), which is corrosive to metals. SRB are commonly found in anaerobic environments, such as pipelines and storage tanks. Another researcher showed that the biofilm formed by SRB on metal surfaces accelerates the corrosion process, especially in the presence of moisture and organic matter [17]. This phenomenon is a particular concern in the oil and gas industry, where SRB can cause significant damage to infrastructure.

Corrosive materials, ranging from acids and alkalis to salts and gases, play a significant role in the degradation of materials, especially metals. The impact of these materials is felt across various industries, from construction and transportation to oil and gas production. Understanding how these materials interact with metals is essential for developing effective corrosion prevention and mitigation strategies. Advances in material science, including the development of corrosion-resistant alloys, coatings, and inhibitors, continue to improve the longevity of structures exposed to corrosive environments. The following sections discuss corrosion resistance materials, and the types of coating methodologies in detail.

3.   Corrosion Resistance materials

Corrosion-resistant materials are critical in various industries, as they enhance durability, reliability, and safety while minimizing maintenance costs. The selection of corrosion-resistant materials is crucial for prolonging service life and maintaining the integrity of structures and components. Advances in alloy design, surface engineering, and coatings continue to improve the efficiency and lifespan of these materials, addressing industrial challenges effectively. These materials are selected based on their chemical composition, mechanical properties, and resistance to environmental factors.

3.1   Necessities of Corrosion Resistance Materials

Economic Impact - Material loss is largely caused by corrosion, which is projected to cost the world more than $2.5 trillion year, or 3–4% of GDP. Utilizing materials that are resistant to corrosion lowers maintenance expenses, increases equipment longevity, and lessens the need for replacements.

Safety and Reliability - Corrosion-related failures can have disastrous outcomes, such as fatalities, environmental catastrophes, and infrastructure destruction. Hence the use of corrosion resistant materials enhance the durability of metals, ensuring the safety of structures in aggressive environments.

Application in Extreme Environments - Equipment is subjected to high pressures, corrosive media, and extremely high temperatures in sectors like chemical processing, aerospace, and marine. Under these demanding circumstances, corrosion-resistant materials guarantee performance and dependability.

Enhancing Longevity and Performance - The development of corrosion-resistant materials contributes significantly to the longevity and efficiency of systems and structures.

3.2   Stainless Steels

Stainless steels are widely used across various industries, owing to its excellent corrosion resistance, physical properties, mechanical properties, ease of manufacture and cost effcetiveness. Alloying composition of stainless steel is responsible for its stainless atrribute and the chromium content (typically above 10.5%) forms a adhering passive layer and protects the underlying metal from corrosion. Other alloying element including Ni, Mo, N, and Mn are also vital to control the microstructure, stability, thickness of passive layer etc [18]. A wide range of stainless steel grades are manufactured by varying the chemical composition for various applications including nuclear power plants, food industries and biomedical implants [19]. Austenitic grades (e.g., 304 and 316) have excellent toughness and ductility, while ferritic and martensitic grades provide moderate resistance.

3.3   Aluminum Alloys

Aluminium alloys, due to its light weight, durability, low maintenance requirements, and resistance to corrosion, utilized in a wide range of industrial applications, including the automotive, marine, and aerospace industries. In recent years, aluminum alloys have gained popularity in structural, bridge, and offshore engineering. [20]. Structural aluminum is often referred to as the "green metal" as it can be recycled completely without losing quality, making it a highly desirable sustainable construction component in a world where sustainability is increasingly crucial. It is anticipated that the need for structural aluminum alloys will only increase in the future due to the ongoing emphasis on sustainability and the circular economy.

3.4   Titanium and Titanium alloys

Titanium and its alloys are utilized extensively in many engineering applications because of its great strength, resistance to corrosion, and ability to function well at temperature as high as about 600 °C. The densities of titanium and its alloys are roughly 50% and 60% higher, respectively, than those of nickel and steel alloys [21]. Despite issues with supply and pricing, titanium and its alloys are used in many industries, including the automotive, aerospace, and naval sectors. Titanium and its alloys are also presently employed as metallic structural biomaterials in implants, including artificial hip joints and dental roots [22]. They are predominantly utilized in hard tissue replacements due to their superior biocompatibility, specific strength, extended fatigue life, and corrosion resistance.

3.5   Nickel based Alloys

Nickel alloys can withstand a wide range of corrosive, aqueous conditions that are commonly seen in environmental technologies and industrial processes. By alloying nickel with cobalt, chromium, aluminium, titanium, and other refractory elements, nickel-based superalloys are created [23], and they have exceptional high temperature properties. Nickel-based alloys are widely employed in the nuclear power, oil and gas, and chemical processes industries. In the oil and gas sector, nickel-based alloys are the material of choice for high-temperature, high-pressure conditions with high CO_2 and H_2S concentrations. The addition of Mo and Cr provides better corrosion resistance in these circumstances. Even though these alloys have a high resistance to corrosion, the presence of sulfur and halide ions can still cause pitting, crevice corrosion, and hydrogen embrittlement. Nickel-based alloys are also a viable option for acid industries because of their great resistance to stress corrosion cracking, particularly at high temperatures [24].

The necessity of corrosion-resistant materials spans economic, safety, environmental, and performance considerations. Their importance is highlighted in critical applications across various industries, ensuring sustainability, safety, and efficiency. Advances in material science, such as coatings and alloy developments, continue to address the challenges posed by corrosion.

3.6   Selection Criteria for Corrosion Resistance Materials

For metallic materials, choosing appropriate corrosion-resistant coatings requires a thorough assessment of the material's environment, intended performance attributes, and viability from an economic standpoint. This procedure guarantees the coated material's best possible protection and durability. Crucial elements to take into account include:

Material Compatibility

In order to avoid problems like adhesion failure or galvanic corrosion, the coating must be compatible with the base metal. Research shows that in order to prevent fast degradation, coatings need to be compatible with the substrate both chemically and electrochemically. Zinc-rich primers, for example, are frequently applied to steel because of their sacrificial anode effect, which uses cathodic protection mechanisms.

Environmental Exposure

The choice of coating is greatly influenced by the surrounding environment's chemical composition. Coatings, for example, need to withstand particular chemicals in situations that are acidic, alkaline, or chloride-rich. Coatings with excellent chloride resistance, such as zinc-rich [25] and epoxy-based coatings, are preferred for maritime applications. Coatings that are subjected to high humidity or temperatures must continue to provide protection from condensation and heat stress. For settings that need abrasion resistance and UV stability, polyurethane coatings are recommended

3.6.1   Properties of Coatings

A coatings capacity to stop the infiltration of oxygen and moisture is a basic requirement. Due to their impermeability, epoxy coatings are frequently utilized in this context [26]. Coatings ought to be able to withstand mechanical stresses like impact and abrasion. In these circumstances, thermoplastic coatings frequently provide outstanding durability. To demonstrate good corrosion inhibition, the coating should have a high impedance and a low corrosion current density.

3.6.2   Application Method

Application techniques, such as brushing, electroplating, dipping, or spray coating, should be in line with the geometry of the material and its intended purpose. The thickness and homogeneity attained during application appear to have a direct bearing on corrosion resistance [27], according to research. In high-performance applications and intricate geometries, for example, thermal spray coatings are perfect because they produce a thick, sticky film.

3.6.3   Economic Considerations

In industrial applications, cost-effectiveness is a crucial consideration. In addition to the material cost, the application procedure, maintenance, and lifespan costs are all included in the overall cost of a coating system. Recent research recommends high-performance coatings in situations where maintenance is difficult, including drilling rigs, and calls for striking a compromise between initial expenditure and long-term performance.

3.6.4 Sustainability and Environmental Regulations

Increasing emphasis on eco-friendly coatings has driven research into waterborne and bio-based coatings (reduce VOC emissions). A thorough assessment of material compatibility, environmental factors, coating characteristics, application methods, and cost is necessary to choose the best corrosion-resistant coating. The lifespan and performance of metallic materials can be considerably increased by matching the coating selection to the particular needs of the application.

The necessity of corrosion-resistant materials is highlighted in critical applications across various industries, ensuring sustainability, safety, and efficiency. Advances in material science, such as coatings and alloy developments, continue to address the challenges posed by corrosion.

4. Types of coating for Metallic Materials

Coatings for metallic materials serve to protect against corrosion, wear, and environmental degradation, while also enhancing appearance. Common types include paint coatings, Powder Coatings, Metallic Coatings, ceramic coatings, organic coatings, conversion coatings, thermal spray coatings, Physical Vapor Deposition (PVD) Coatings and Electrophoretic Deposition (E-Coating).

4.1 Paint Coatings

Paint coatings on metallic materials primarily function as a protective barrier that prevents corrosive elements like moisture, oxygen, and chemicals from reaching the metal surface. By isolating the metal, these coatings inhibit the electrochemical reactions that cause corrosion. In some cases, coatings also offer sacrificial protection; for example, zinc-rich paints corrode preferentially, protecting the underlying metal [28]. It is compatible with a range of metallic substrates, including steel, aluminium, and copper alloys. Each type of metal may require specific types of coatings to ensure optimal adhesion and protection. For example, epoxy coatings are commonly used on steel for their excellent adhesion and chemical resistance, while polyurethane coatings are favoured for aluminium due to their superior UV resistance and flexibility [29]. Paint coatings provide an effective barrier that prevents rust and degradation of metal surfaces [30]. It also offer various colours and finishes, enhancing the visual appeal of metallic structures [31]. Compared to other protective methods, paint coatings are relatively inexpensive and easy to apply [32]. It is always Suitable for complex shapes and sizes, ensuring comprehensive protection [27]

Paint coatings are prone to scratches and impact damage, which can compromise their protective capability [33]. The prolonged exposure to UV light and harsh chemicals can lead to degradation of certain coatings [34]. Regular inspection and reapplication may be necessary to maintain the integrity of the coating [35] . Some paint coatings may not withstand high-temperature environments unless specially formulated [36].

Paint coatings protect vehicle bodies from environmental damage while enhancing aesthetics [37]. It is most widely used on structural steel to prevent rust in buildings, bridges, and other infrastructure [38]. To resist saltwater corrosion paint coatings are applied to ships and offshore structures to resist saltwater corrosion [39]. Paint coatings extend the service life of machinery exposed to harsh conditions [40]. It also enhances the durability and appearance of everyday items such as refrigerators and washing machines [41].

4.2   Powder Coatings

Powder coatings are applied to metallic materials through an electrostatic process where finely ground particles of pigment and resin are charged and sprayed onto the surface of the metal as shown in Figure 1. The charged powder adheres to the electrically grounded metal substrate. The coated metal is then heated in a curing oven, where the powder melts and forms a continuous, uniform film that is both protective and decorative. This solvent-free process minimizes the environmental impact associated with traditional liquid coatings [42].

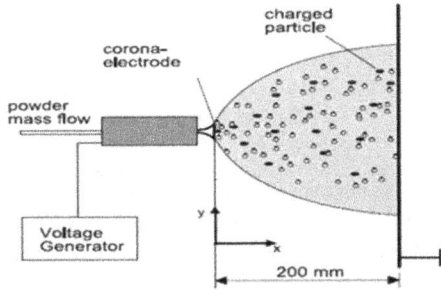

Figure 1. Powder coating principle (Courtesy of Ye and Domnick [47]).

Powder coatings are compatible with various metallic substrates, including steel, aluminum, and zinc-coated metals. The surface of the metal typically requires preparation, such as cleaning and phosphating, to enhance the adhesion and performance of the powder coating [43]. Powder coatings are free of solvents and emit negligible volatile organic compounds (VOCs), making them environmentally preferable compared to liquid coatings [44]. It also offers excellent resistance to impact, abrasion, and corrosion, which extends the service life of the coated products [45]. Mainly it provides a consistent finish with minimal runs or drips, ensuring an aesthetically pleasing appearance [46].

Powder coatings are compatible with various metallic substrates, including steel, aluminum, and zinc-coated metals. The surface of the metal typically requires preparation, such as cleaning and phosphating, to enhance the adhesion and performance of the powder coating [43]. Powder coatings are free of solvents and emit negligible volatile organic compounds (VOCs), making them environmentally preferable compared to liquid coatings [44]. It also offers excellent resistance to impact, abrasion, and corrosion, which extends the service life of the coated products [45]. Mainly it provides a consistent finish with minimal runs or drips, ensuring an aesthetically pleasing appearance [46].

The setup costs for powder coating equipment and curing ovens can be significant, which may be a barrier for small enterprises. The high temperatures required for curing limit the use of powder coatings to substrates that can withstand such conditions without damage [48]. Changing colors in the coating process requires thorough cleaning of equipment to prevent contamination, which can be time-consuming [49]. Powder coatings are extensively used for wheels, body panels, and engine parts, providing a durable finish that resists environmental wear [50]. It can also be applied to aluminum extrusions used in windows, doors, and building facades, offering excellent weather resistance and aesthetic options. And can also be used on appliances like refrigerators and washing machines, where durability and appearance are key considerations [51]. Coatings protect machinery and tools from corrosion and mechanical damage in demanding environments[52].

4.3   Metallic Coatings

Metallic coatings are applied to metallic substrates to enhance their surface properties, including corrosion resistance, wear resistance, and aesthetic appeal. These coatings can be deposited through various methods, such as electroplating, thermal spraying, chemical vapor deposition (CVD), and physical vapor deposition (PVD). In electroplating, a metal is deposited on a substrate by passing an electric current through a solution containing the metal ions, which are reduced and form a coating on the cathode as shown in Figure 2. Thermal spraying involves melting the coating material and spraying it onto the substrate to form a layer that adheres through mechanical interlocking. Both CVD and PVD involve vapor-phase deposition, where the coating material is vaporized and then deposited on the substrate under vacuum conditions [53]. Metallic coatings are compatible with a wide range of metals and alloys, including steel, aluminum, copper, and titanium. Steel is commonly coated with zinc (galvanization) to prevent rusting. Aluminum and titanium are often coated with other metals to improve their surface hardness and corrosion resistance. The compatibility depends on the properties of both the substrate and the coating material, as well as the intended application.

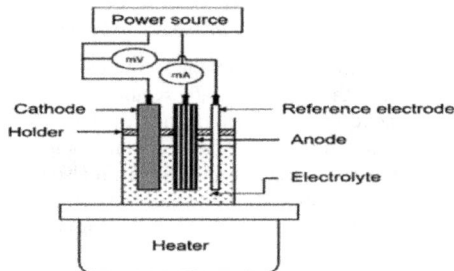

Figure 2. Metallic coating principle (Courtesy of Shaikh, Mane [54]).

Metallic coatings protect substrates from corrosive environments, significantly extending their service life. Zinc and nickel coatings are widely used for this purpose [55]. Coatings like chromium or tungsten carbide enhance the wear resistance of metal parts, making them suitable for high-stress applications [56]. Decorative coatings such as gold or silver plating improve the visual appearance of consumer products [57]. Advanced techniques like PVD and CVD require sophisticated equipment and controlled environments, which can be expensive [58].

It is used as on aircraft components to protect them from oxidation and wear in extreme conditions [59] . Zinc and nickel coatings are applied to car parts to enhance corrosion resistance and extend their lifespan [60]. Metallic coatings are crucial in the manufacturing of electronic devices to ensure reliable electrical connections and protect against corrosion [61].

Advances in Corrosion Science and Surface Engineering
Materials Research Foundations 188 (2026) 1-21

Materials Research Forum LLC
https://doi.org/10.21741/9781644903919-1

4.4   Ceramic Coatings

Ceramic coatings are applied to metallic materials to enhance their surface properties, such as thermal resistance, corrosion resistance, and mechanical strength. These coatings are typically deposited using methods like plasma spraying, chemical vapor deposition (CVD), and physical vapor deposition (PVD). In plasma spraying, the ceramic material is heated to a molten state and sprayed onto the metallic surface, where it solidifies to form a protective layer. CVD and PVD involve the deposition of ceramic materials in vapor form onto the substrate, creating a uniform and dense coating that adheres strongly to the metal surface.Ceramic coatings are compatible with a variety of metallic substrates, including steel, aluminum, titanium, and nickel-based alloys. The selection of the ceramic material depends on the application requirements.

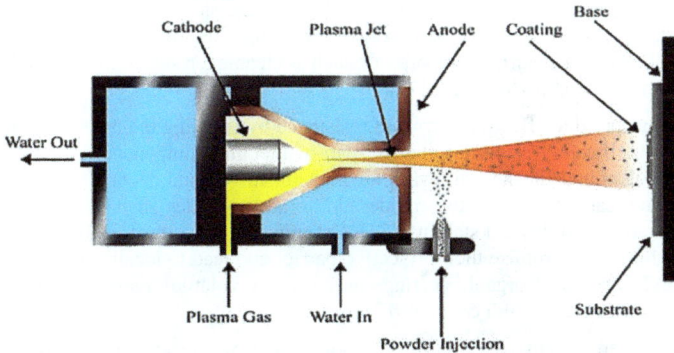

Figure 3. Plasma spraying principle (Courtesy of Bosco, Van Den Beucken [63]).

Common ceramic materials used for coatings include alumina (Al_2O_3), zirconia (ZrO_2), an d silicon carbide (SiC). These materials are chosen for their exceptional hardness, thermal stability, and resistance to chemical attack, making them ideal for protecting metallic components in harsh environments [62].

Ceramic coatings, such as yttria-stabilized zirconia (YSZ), are used as thermal barrier coatings (TBCs) to protect engine components from high temperatures, enhancing efficiency and lifespan [64]. It offers superior resistance to oxidation and chemical corrosion, making them suitable for applications in aggressive environments, such as marine and chemical industries [65]. The hardness of ceramic materials provides excellent wear resistance, reducing maintenance and extending the service life of components [66]. Certain ceramic coatings serve as insulators, preventing electrical conductivity and protecting components from electrical hazards [67].

Ceramic materials are inherently brittle, which can lead to cracking or chipping under mechanical stress or impact [68]. Ceramic coatings are used on turbine blades and other engine components to withstand high temperatures and reduce heat transfer, enhancing fuel efficiency. Engine parts, such as pistons and exhaust systems, are coated with ceramics to improve thermal resistance and reduce wear [69]. Biocompatible ceramic coatings, such as hydroxyapatite, are used on metallic implants to enhance osseointegration and reduce wear in joint replacements [70].

Advances in Corrosion Science and Surface Engineering   Materials Research Forum LLC
Materials Research Foundations 188 (2026) 1-21   https://doi.org/10.21741/9781644903919-1

4.5   Organic Coatings

Organic coatings on metallic materials function primarily as a barrier to environmental factors that cause corrosion, such as moisture, oxygen, and chemicals. These coatings form a continuous film over the metal surface, preventing contact with corrosive agents. The performance of organic coatings depends on their adhesion to the substrate, which is influenced by surface preparation, the nature of the coating material, and the application method. Modern advancements in organic coatings include the incorporation of functional additives that enhance properties like UV resistance, hydrophobicity, and self-healing capabilities [71].

Organic coatings are compatible with a wide range of metallic substrates, including steel, aluminum, copper, and zinc. Compatibility depends on the adhesion properties of the coating material and the substrate's surface condition. For instance, steel and aluminum are commonly coated with epoxy or polyurethane-based organic coatings due to their excellent adhesion and protective qualities. Proper surface preparation, such as cleaning, roughening, or priming, is crucial to ensure optimal coating performance [72].

Organic coatings provide effective corrosion resistance by isolating the metal from the corrosive environment, significantly extending the lifespan of metallic components [45]. These coatings offer various colors and finishes, allowing for both functional and decorative applications [73]. Organic coatings can be tailored to provide additional properties such as abrasion resistance, chemical resistance, and thermal stability [74]. Organic coatings are susceptible to scratches and impacts, which can compromise their protective barrier and lead to localized corrosion [45]. The production and disposal of organic coatings may involve volatile organic compounds (VOCs), raising environmental and health concerns [75].

Organic coatings are extensively used on car bodies and components to prevent corrosion and provide aesthetic finishes [76]. Structural metals in buildings and infrastructure are coated to protect against environmental damage and improve longevity [52]. Organic coatings are used to protect electronic components from moisture and dust, ensuring reliable performance [27].

4.6   Other Coatings

Conversion coatings form a protective layer on metal surfaces through chemical or electrochemical reactions. The base metal is transformed into a surface film, such as phosphates or oxides, which enhances corrosion resistance and adhesion of subsequent coatings. These coatings are commonly used on aluminum, magnesium, zinc, and their alloys [77]. Conversion coatings are cost-effective with simple processing [32]. Automotive and aerospace industries effectively utilize conversion coatings for corrosion protection and it can be used in electronic components for reliability enhancement [78].

Thermal spray coatings involve melting or heating a material and then spraying it onto a surface. The material cools to form a solid coating, providing wear resistance and corrosion protection [79]. It is suitable for repairing worn components [80]. PVD involves vaporizing a solid material in a vacuum and depositing it onto a substrate to form a thin, durable coating. Most compatible materials are metals, ceramics, and certain polymers [81]. E-Coating uses an electric field to deposit charged particles from a suspension onto a conductive substrate, creating a uniform coating layer [82]. It is primarily used on metals like steel and aluminum [83].

Conclusion

Corrosion presents a persistent challenge to industries worldwide, causing significant material degradation and economic losses. This study has explored the mechanisms of corrosion and the key factors influencing its progression, providing a foundation for understanding its complexities. While the role of corrosive agents and environmental conditions is crucial, the development and application of corrosion-resistant materials and protective coatings emerge as the most effective strategies for mitigation. Innovations in material science have led to the creation of alloys such as stainless steel, titanium alloys, nickel alloys, etc with enhanced properties, tailored to withstand specific corrosive conditions. Equally important are protective coatings, which serve as an external barrier against corrosive agents providing improved adhesion, chemical stability, and longevity, ensuring superior performance even in the harshest conditions. Integrating protective coatings with corrosion-resistant materials is essential for securing machinery and infrastructure. Future studies should focus on creating self-healing materials, eco-friendly coatings, and nanotechnology-driven solutions to improve corrosion resistance even more. Industries may overcome corrosion issues with increased operational reliability, cost effectiveness, and sustainability by concentrating on these developments.

References

[1] Lee, S., et al., Oxygen isotope labeling experiments reveal different reaction sites for the oxygen evolution reaction on nickel and nickel iron oxides. Angewandte Chemie, 2019. 131(30): p. 10401-10405. https://doi.org/10.1002/ange.201903200

[2] Zhang, Y., et al., A comparative study between the mechanical and microstructural properties of resistance spot welding joints among ferritic AISI 430 and austenitic AISI 304 stainless steel. J. Mater. Res. Technol., 2020. 9(1): p. 574-583. https://doi.org/10.1016/j.jmrt.2019.10.086

[3] Choudhary, S., R. Kelly, and N. Birbilis, On the origin of passive film breakdown and metastable pitting for stainless steel 316L. Corrosion Science, 2024. 230: p. 111911. https://doi.org/10.1016/j.corsci.2024.111911

[4] Liu, Y., et al., Synergistic damage mechanisms of high-temperature metal corrosion in marine environments: A review. Progress in Organic Coatings, 2024. 197: p. 108765. https://doi.org/10.1016/j.porgcoat.2024.108765

[5] Willis, M.M., Localised Corrosion of Ni-base Superalloys in Seawater. 2019: The University of Manchester (United Kingdom).

[6] Tang, H.-Y., et al., Iron corrosion via direct metal-microbe electron transfer. MBio, 2019. 10(3): p. 10.1128/mbio. 00303-19. https://doi.org/10.1128/mBio.00303-19

[7] Refait, P., et al., Corrosion of carbon steel in marine environments: role of the corrosion product layer. Corrosion and Materials Degradation, 2020. 1(1): p. 10. https://doi.org/10.3390/cmd1010010

[8] Böhni, H., Localized corrosion, in Corrosion mechanisms. 2020, CRC Press. p. 285-327.

[9] Alqahtani, I.M., A. Starr, and M. Khan, Experimental and theoretical aspects of crack assisted failures of metallic alloys in corrosive environments-A review. Materials Today: Proceedings, 2022. 66: p. 2530-2535. https://doi.org/10.1016/j.matpr.2022.07.075

[10] Marcos-Meson, V., et al., Durability of Steel Fibre Reinforced Concrete (SFRC) exposed to acid attack-A literature review. Construction and Building Materials, 2019. 200: p. 490-501. https://doi.org/10.1016/j.conbuildmat.2018.12.051

[11] Anderez, A., F.J. Alguacil, and F.A. López, Acid pickling of carbon steel. Revista De Metalurgia, 2022. 58(3): p. e226-e226. https://doi.org/10.3989/revmetalm.226

[12] Hassan, N., et al., Development of sustainable superhydrophobic coatings on aluminum substrate using magnesium nanoparticles for enhanced catalytic activity, self-cleaning, and corrosion resistance. Journal of Molecular Liquids, 2023. 383: p. 122085. https://doi.org/10.1016/j.molliq.2023.122085

[13] Parangusan, H., J. Bhadra, and N. Al-Thani, A review of passivity breakdown on metal surfaces: influence of chloride-and sulfide-ion concentrations, temperature, and pH. Emergent Materials, 2021. 4(5): p. 1187-1203. https://doi.org/10.1007/s42247-021-00194-6

[14] Xu, X., et al., Corrosion of stainless steel valves in a reverse osmosis system: analysis of corrosion products and metal loss. Engineering Failure Analysis, 2019. 105: p. 40-51. https://doi.org/10.1016/j.engfailanal.2019.06.026

[15] Kumar, V., et al., Atmospheric corrosion of materials and their effects on mechanical properties: A brief review. Materials Today: Proceedings, 2021. 44: p. 4677-4681. https://doi.org/10.1016/j.matpr.2020.10.939

[16] Li, K., Y. Zeng, and J.-L. Luo, Influence of H2S on the general corrosion and sulfide stress cracking of pipelines steels for supercritical CO2 transportation. Corrosion Science, 2021. 190: p. 109639. https://doi.org/10.1016/j.corsci.2021.109639

[17] Sheng, M.-Q., et al., Fabrication of medetomidines/epoxy coatings for marine anticorrosion and antifouling. Progress in Organic Coatings, 2025. 200: p. 109033. https://doi.org/10.1016/j.porgcoat.2024.109033

[18] Sun, Y., et al., Mechanisms of inclusion-induced pitting of stainless steels: A review. Journal of Materials Science & Technology, 2024. 168: p. 143-156. https://doi.org/10.1016/j.jmst.2023.06.008

[19] Hakimian, S., et al., Application of machine learning for the classification of corrosion behavior in different environments for material selection of stainless steels. Computational Materials Science, 2023. 228: p. 112352. https://doi.org/10.1016/j.commatsci.2023.112352

[20] Sun, Y. The use of aluminum alloys in structures: Review and outlook. in Structures. 2023. Elsevier. https://doi.org/10.1016/j.istruc.2023.105290

[21] Najafizadeh, M., et al., Classification and applications of titanium and its alloys: A review. Journal of Alloys and Compounds Communications, 2024: p. 100019. https://doi.org/10.1016/j.jacomc.2024.100019

[22] Rack, H. and J. Qazi, Titanium alloys for biomedical applications. Materials Science and Engineering: C, 2006. 26(8): p. 1269-1277. https://doi.org/10.1016/j.msec.2005.08.032

[23] Weber, J.H. and M.K. Banerjee, Nickel and Nickel Alloys: An Overview☆, in Reference Module in Materials Science and Materials Engineering. 2019, Elsevier. https://doi.org/10.1016/B978-0-12-803581-8.02572-8

[24] Karimihaghighi, R. and M. Naghizadeh, Effect of alloying elements on aqueous corrosion of nickel-based alloys at high temperatures: A review. Materials and Corrosion, 2023. 74(8): p. 1246-1255. https://doi.org/10.1002/maco.202213705

[25] Hussain, A.K., et al., Research progress in organic zinc rich primer coatings for cathodic protection of metals - A comprehensive review. Progress in Organic Coatings, 2021. 153: p. 106040. https://doi.org/10.1016/j.porgcoat.2020.106040

[26] Zhang, B., et al., Enhancement of barrier and corrosion protection performance of epoxy coatings through adding eco-friendly lamellar biochar. Materials and Corrosion, 2022. 73(5): p. 720-732. https://doi.org/10.1002/maco.202112697

[27] Fotovvati, B., N. Namdari, and A. Dehghanghadikolaei, On coating techniques for surface protection: A review. Journal of Manufacturing and Materials processing, 2019. 3(1): p. 28. https://doi.org/10.3390/jmmp3010028

[28] Takahashi, M., et al., Corrosion behavior of carbon steel coated with a zinc-rich paint containing metallic compounds under wet and dry cyclic conditions. Materials and Corrosion, 2021. 72(11): p. 1787-1795. https://doi.org/10.1002/maco.202112465

[29] Gomez-Lopez, A., et al., Poly (hydroxyurethane) adhesives and coatings: state-of-the-art and future directions. ACS Sustainable Chemistry & Engineering, 2021. 9(29): p. 9541-9562. https://doi.org/10.1021/acssuschemeng.1c02558

[30] Croll, S., Surface roughness profile and its effect on coating adhesion and corrosion protection: A review. Progress in organic Coatings, 2020. 148: p. 105847. https://doi.org/10.1016/j.porgcoat.2020.105847

[31] Puthran, D. and D. Patil, Usage of heavy metal-free compounds in surface coatings. Journal of Coatings Technology and Research, 2023. 20(1): p. 87-112. https://doi.org/10.1007/s11998-022-00648-4

[32] Butt, M.A., Thin-film coating methods: a successful marriage of high-quality and cost-effectiveness-a brief exploration. Coatings, 2022. 12(8): p. 1115. https://doi.org/10.3390/coatings12081115

[33] Kumar, S., S. Kumar, and E.P. Namburi, Functional Paints and Coatings, in Novel Defence Functional and Engineering Materials (NDFEM) Volume 1: Functional Materials for Defence Applications. 2024, Springer. p. 219-246. https://doi.org/10.1007/978-981-99-9791-6_8

[34] Sørensen, P.A., et al., Anticorrosive coatings: a review. Journal of coatings technology and research, 2009. 6: p. 135-176. https://doi.org/10.1007/s11998-008-9144-2

[35] De Landtsheer, G., Corrosion under insulation (CUI) guidelines: technical guide for managing CUI. Vol. 55. 2020: Woodhead Publishing.

[36] Mondal, K., et al., Thermal barrier coatings overview: Design, manufacturing, and applications in high-temperature industries. Industrial & Engineering Chemistry Research, 2021. 60(17): p. 6061-6077. https://doi.org/10.1021/acs.iecr.1c00788

[37] Kanchana, R., et al., Coatings in the Automobile Application. Functional Coatings for Biomedical, Energy, and Environmental Applications, 2024: p. 343-361. https://doi.org/10.1002/9781394263172.ch15

[38] Dohare, S., Corrosion protection and modern infrastructure, in Introduction to Corrosion-Basics and Advances. 2023, IntechOpen. https://doi.org/10.5772/intechopen.111547

[39] Munger, C.G., L. Vincent, and D.A. Shifler, Marine coatings. LaQue's Handbook of Marine Corrosion, 2022: p. 527-571. https://doi.org/10.1002/9781119788867.ch19

[40] Sharun, V., et al., Study on developments in protection coating techniques for steel. Advances in Materials Science and Engineering, 2022. 2022. https://doi.org/10.1155/2022/2843043

[41] Kustiawan, H., S.K. Boontanon, and N. Boontanon, Utilization of sanitaryware waste product (SWP) as an admixture ingredient for eco-cooling paint. Waste Management, 2024. 190: p. 1-11. https://doi.org/10.1016/j.wasman.2024.08.033

[42] Faccini, M., et al., Environmentally friendly anticorrosive polymeric coatings. Applied Sciences, 2021. 11(8): p. 3446. https://doi.org/10.3390/app11083446

[43] Zhang, J., et al., Enhancing adhesion and anti-corrosion performance of hot-dip galvanized steels by sandblasting/phosphating co-treatment. Surface Topography: Metrology and Properties, 2021. 9(4): p. 045037. https://doi.org/10.1088/2051-672X/ac3c9d

[44] Kargarfard, N., et al., Self-stratifying powder coatings based on eco-friendly, solvent-free epoxy/silicone technology for simultaneous corrosion and weather protection. Progress in Organic Coatings, 2021. 161: p. 106443. https://doi.org/10.1016/j.porgcoat.2021.106443

[45] Aljibori, H., A. Alamiery, and A. Kadhum, Advances in corrosion protection coatings: A comprehensive review. Int. J. Corros. Scale Inhib, 2023. 12(4): p. 1476-1520. https://doi.org/10.17675/2305-6894-2023-12-4-6

[46] Pélissier, K. and D. Thierry, Powder and high-solid coatings as anticorrosive solutions for marine and offshore applications? A review. Coatings, 2020. 10(10): p. 916. https://doi.org/10.3390/coatings10100916

[47] Ye, Q. and J. Domnick, On the simulation of space charge in electrostatic powder coating with a corona spray gun. Powder technology, 2003. 135: p. 250-260. https://doi.org/10.1016/j.powtec.2003.08.019

[48] Czachor-Jadacka, D. and B. Pilch-Pitera, Progress in development of UV curable powder coatings. Progress in Organic Coatings, 2021. 158: p. 106355. https://doi.org/10.1016/j.porgcoat.2021.106355

[49] Prashar, G., H. Vasudev, and L. Thakur, Influence of heat treatment on surface properties of HVOF deposited WC and Ni-based powder coatings: a review. Surface Topography: Metrology and Properties, 2021. 9(4): p. 043002. https://doi.org/10.1088/2051-672X/ac3a52

[50] Vasiljević, S., et al., Review of the coatings used for brake discs regarding their wear resistance and environmental effect. Proceedings of the Institution of Mechanical Engineers, Part J: Journal of Engineering Tribology, 2022. 236(10): p. 1932-1949. https://doi.org/10.1177/13506501211070654

[51] GmbH, N.D. and B.E.E. SA, Coating Refrigerators Fully Automatically and Flexibly. IST International Surface Technology, 2019. 12(4): p. 24-27. https://doi.org/10.1007/s35724-019-0072-8

[52] Lazorenko, G., A. Kasprzhitskii, and T. Nazdracheva, Anti-corrosion coatings for protection of steel railway structures exposed to atmospheric environments: A review. Construction and Building Materials, 2021. 288: p. 123115. https://doi.org/10.1016/j.conbuildmat.2021.123115

[53] Mehta, N., Overview of Coating Deposition Techniques. Tribology and Characterization of Surface Coatings, 2022: p. 1-32. https://doi.org/10.1002/9781119818878.ch1

[54] Shaikh, A.V., et al., Electrochemical deposition of cadmium selenide films and their properties: a review. Journal of Solid State Electrochemistry, 2017. 21: p. 2517-2530. https://doi.org/10.1007/s10008-017-3552-0

[55] Farag, A.A., Applications of nanomaterials in corrosion protection coatings and inhibitors. Corrosion Reviews, 2020. 38(1): p. 67-86. https://doi.org/10.1515/corrrev-2019-0011

[56] Burkov, A. and M. Kulik, Wear-resistant and anticorrosive coatings based on chrome carbide Cr 7 C 3 obtained by electric spark deposition. Protection of Metals and Physical Chemistry of Surfaces, 2020. 56: p. 1217-1221. https://doi.org/10.1134/S2070205120060064

[57] Vorobyova, M., et al., PVD for Decorative Applications: A Review. Materials, 2023. 16(14): p. 4919. https://doi.org/10.3390/ma16144919

[58] Adeoye, A.E., O. Adeaga, and K. Ukoba, Chemical Vapour Deposition (CVD) and Physical Vapour Deposition (PVD) techniques: Advances in thin film solar cells. Nigerian Journal of Technology, 2024. 43(3): p. 479-489.

[59] Bonu, V. and H.C. Barshilia, High-temperature solid particle erosion of aerospace components: its mitigation using advanced nanostructured coating technologies. Coatings, 2022. 12(12): p. 1979. https://doi.org/10.3390/coatings12121979

[60] Chakraborty, A., et al., Evolution of microstructure of zinc-nickel alloy coating during hot stamping of boron added steels. Journal of Alloys and Compounds, 2019. 794: p. 672-682. https://doi.org/10.1016/j.jallcom.2019.04.164

[61] Tushinsky, L., Coated metal: structure and properties of metal-coating compositions. 2002: Springer Science & Business Media.

[62] Chate, G.R., et al., Ceramic material coatings: emerging future applications, in Advanced Ceramic Coatings for Emerging Applications. 2023, Elsevier. p. 3-17. https://doi.org/10.1016/B978-0-323-99624-2.00007-3

[63] Bosco, R., et al., Surface engineering for bone implants: a trend from passive to active surfaces. Coatings, 2012. 2(3): p. 95-119. https://doi.org/10.3390/coatings2030095

[64] Liu, Q., S. Huang, and A. He, Composite ceramics thermal barrier coatings of yttria stabilized zirconia for aero-engines. Journal of materials science & technology, 2019. 35(12): p. 2814-2823. https://doi.org/10.1016/j.jmst.2019.08.003

[65] Wei, T., F. Yan, and J. Tian, Characterization and wear-and corrosion-resistance of microarc oxidation ceramic coatings on aluminum alloy. Journal of Alloys and Compounds, 2005. 389(1-2): p. 169-176. https://doi.org/10.1016/j.jallcom.2004.05.084

[66] Hu, H., et al., Wear-resistant ceramic coatings deposited by liquid thermal spraying. Ceramics International, 2022. 48(22): p. 33245-33255. https://doi.org/10.1016/j.ceramint.2022.07.267

[67] Contreras, J.E. and E.A. Rodriguez, Nanostructured insulators-A review of nanotechnology concepts for outdoor ceramic insulators. Ceramics International, 2017. 43(12): p. 8545-8550. https://doi.org/10.1016/j.ceramint.2017.04.105

[68] Rubeša, D., B. Smoljan, and R. Danzer, Main features of designing with brittle materials. Journal of materials engineering and performance, 2003. 12: p. 220-228. https://doi.org/10.1361/105994903770343385

[69] Weng, H., et al., Electrical insulation improvements of ceramic coating for high temperature sensors embedded on aeroengine turbine blade. Ceramics international, 2020. 46(3): p. 3600-3605. https://doi.org/10.1016/j.ceramint.2019.10.078

[70] Huang, C.-H. and M. Yoshimura, Biocompatible hydroxyapatite ceramic coating on titanium alloys by electrochemical methods via Growing Integration Layers [GIL] strategy: A review. Ceramics International, 2023. 49(14): p. 24532-24540. https://doi.org/10.1016/j.ceramint.2022.12.248

[71] Deflorian, F., "Advances in Organic Coatings 2018". 2020, MDPI. p. 555. https://doi.org/10.3390/coatings10060555

[72] Okokpujie, I.P., et al., Effect of coatings on mechanical, corrosion and tribological properties of industrial materials: a comprehensive review. Journal of Bio-and Tribo-Corrosion, 2024. 10(1): p. 2. https://doi.org/10.1007/s40735-023-00805-1

[73] Ielo, I., et al., Nanostructured surface finishing and coatings: Functional properties and applications. Materials, 2021. 14(11): p. 2733. https://doi.org/10.3390/ma14112733

[74] Nazari, M.H., et al., Nanocomposite organic coatings for corrosion protection of metals: A review of recent advances. Progress in Organic Coatings, 2022. 162: p. 106573. https://doi.org/10.1016/j.porgcoat.2021.106573

[75] Wang, D., et al., Increasing volatile organic compounds emission from massive industrial coating consumption require more comprehensive prevention. Journal of Cleaner Production, 2023. 414: p. 137459. https://doi.org/10.1016/j.jclepro.2023.137459

[76] Thomas, J., et al., A comprehensive outlook of scope within exterior automotive plastic substrates and its coatings. Coatings, 2023. 13(9): p. 1569. https://doi.org/10.3390/coatings13091569

[77] Hafeez, M., et al., Phosphate chemical conversion coatings for magnesium alloys: a review. Journal of Coatings Technology and Research, 2020. 17: p. 827-849. https://doi.org/10.1007/s11998-020-00335-2

[78] Zhu, H. and J. Li, Advancements in corrosion protection for aerospace aluminum alloys through surface treatment. International Journal of Electrochemical Science, 2024: p. 100487. https://doi.org/10.1016/j.ijoes.2024.100487

[79] Tejero-Martin, D., et al., Beyond traditional coatings: a review on thermal-sprayed functional and smart coatings. Journal of thermal spray technology, 2019. 28: p. 598-644. https://doi.org/10.1007/s11666-019-00857-1

Advances in Corrosion Science and Surface Engineering   Materials Research Forum LLC
Materials Research Foundations 188 (2026) 1-21   https://doi.org/10.21741/9781644903919-1

[80] Ashokkumar, M., et al., An overview of cold spray coating in additive manufacturing, component repairing and other engineering applications. Journal of the Mechanical Behavior of Materials, 2022. 31(1): p. 514-534. https://doi.org/10.1515/jmbm-2022-0056

[81] Saad, K.S.K., T. Saba, and A.B. Rashid, Application of PVD coatings in medical implantology for enhanced performance, biocompatibility, and quality of life. Heliyon, 2024. 10(16). https://doi.org/10.1016/j.heliyon.2024.e35541

[82] Saji, V.S., Electrophoretic-deposited superhydrophobic coatings. Chemistry-An Asian Journal, 2021. 16(5): p. 474-491. https://doi.org/10.1002/asia.202001425

[83] Verma, K., H. Cao, and P. Mandapalli, A Study of a Fast and Precise Thickness Simulation of an Industrial e-Coating Process. 2020, SAE Technical Paper. https://doi.org/10.4271/2020-01-0898

Advances in Corrosion Science and Surface Engineering
Materials Research Foundations 188 (2026) 22-34

Materials Research Forum LLC
https://doi.org/10.21741/9781644903919-2

Chapter 2

Corrosion Mechanisms

Anusree S. Gangadharan[1]*, G. Kausalya Sasikumar[2], A.S. Shilpa[2], Rajender Boddula[3]

[1]Department of Chemistry and Centre for Nanoscience and Technology, KPR Institute of Engineering and Technology, Coimbatore 641407, Tamilnadu, India

[2]Centre for Research and Development, KPR Institute of Engineering and Technology, Coimbatore 641407, Tamilnadu, India

[3]School of Sciences, Woxsen University, Hyderabad - 502345, Telangana, India

anusreegangadharan95@gmail.com

Abstract

Corrosion is a complex phenomenon affecting various materials, with significant economic, environmental, and safety implications. This review delves into the underlying principles and mechanisms of various corrosion types, encompassing uniform corrosion, pitting corrosion, crevice corrosion, galvanic corrosion, and microbiologically influenced corrosion, to provide a comprehensive understanding of the complex processes driving corrosion. Environmental factors, material properties, and microorganisms are discussed as contributors to corrosion. Understanding the interactions between materials, environments, and microorganisms is crucial for developing effective corrosion prevention and control strategies, ensuring the reliability and safety of infrastructure, equipment, and materials.

Keywords

Uniform Corrosion, Pitting Corrosion, Crevice Corrosion, Galvanic Corrosion, Microbiologically Influenced Corrosion

Contents

1.   Introduction

Corrosion is a natural process that causes materials to degrade over time due to environmental factors, affecting not just metals, but all types of materials [1]. Metal surfaces undergo permanent damage through corrosion, a chemical reaction-driven process that transforms pure metals into more stable substances, such as oxides, hydroxides, and sulfides, when exposed to corrosive environments, including solids, liquids, and gases [2, 3]. Corrosive environments, also known as electrolytes, facilitate the transfer of ions (cations and anions) and enable two primary reactions: anodic and cathodic. When two dissimilar metals are immersed in an electrolyte, a galvanic cell form. In this scenario, the less noble metal becomes the anode, undergoing corrosion, while the more noble metal acts as the cathode, remaining protected [4]. Corrosion is a destructive electrochemical process where electrons flow from a less noble metal to a more noble one, triggering a catastrophic reaction that ravages the former [5]. This phenomenon is starkly illustrated by the corrosion of zinc in a copper-containing electrolyte, where zinc's inherent vulnerability, as dictated by its reduction potential, seals its fate. As corrosion takes hold, electrons are wrenched from the anode, while the cathode remains intact, shielded from the corrosive onslaught. This insidious process can assume various guises, including uniform surface degradation, localized attacks, and intergranular corrosion, which exploits the metal's inherent weaknesses, laying bare the complex, multifaceted nature of corrosion [6]. Most naturally occurring metals exist as compounds, with the notable exception of noble metals like gold and platinum. This phenomenon stems from the fact that metals in their compound form tend to be more thermodynamically stable than in their elemental state [7].

Corrosion manifests in various forms, including uniform corrosion, which results in homogeneous metal loss across a surface, as seen in rusting and tarnishing [8]. Localized corrosion types include pitting, which creates patches or pits, often in stainless steels exposed to halide-containing solutions, and crevice corrosion, which occurs in stagnant solutions within metal surface crevices [9]. Other forms of corrosion include galvanic corrosion, which arises from electrical contact between dissimilar materials in a corrosive electrolyte, stress corrosion cracking, which combines tensile stress and corrosive environments, and intergranular corrosion, which affects metal grain boundaries [10]. Additionally, dealloying, erosion corrosion, and other forms of corrosion can occur, each with distinct mechanisms and consequences, highlighting the complexity and diversity of corrosion phenomena [11]. The economic toll of corrosion is substantial, with estimated annual losses ranging from 3.5% of a nation's GDP to a staggering Rs. 2.0 lakh crores (approximately $27 billion USD) in India, and $300 billion in the United States equivalent to 3.2% of the domestic product [12]. Corrosion's far-reaching impact extends beyond economic losses, affecting national infrastructure, industrial operations, environmental sustainability, and human safety, leading to premature deterioration and failure of materials, disruption of industrial operations, environmental hazards, and compromised human safety [13]. Therefore, raising awareness about corrosion, the mechanisms of corrosion and implementing timely, effective control measures are crucial to mitigating these risks and preventing corrosion-related failures.

This chapter provides an in-depth examination of the complex and multifaceted corrosion mechanisms that underlie various forms of corrosion, which can have devastating consequences on materials and infrastructure. The discussion explores the underlying mechanisms of uniform corrosion, pitting corrosion, crevice corrosion, galvanic corrosion, stress corrosion cracking, and microbiologically influenced corrosion. Specifically, the chapter delves into the mechanisms by which uniform corrosion leads to homogeneous metal loss, pitting corrosion creates localized patches or pits, and crevice corrosion occurs in stagnant solutions within metal surface crevices. Additionally, the chapter examines the mechanisms of galvanic corrosion, which arises from electrical contact between dissimilar materials, stress corrosion cracking, which combines tensile stress and corrosive environments, and microbiologically influenced corrosion, which involves the interaction of microorganisms with metal surfaces. By elucidating the definitions, examples, and factors influencing these corrosion mechanisms, this chapter provides a comprehensive understanding of the complex processes that drive corrosion and its significant economic, environmental, and safety implications.

2.   Corrosion Mechanisms

2.1   Uniform Corrosion

Uniform corrosion is a pervasive and insidious form of corrosion that uniformly attacks metal surfaces, leading to a gradual loss of material thickness and mechanical strength. The consequences of uniform corrosion can be severe, compromising the structural integrity and safety of affected components [14]. However, detection is often facilitated by visible signs, enabling timely intervention. Prevention and control of uniform corrosion necessitate a multifaceted approach [15]. Key strategies include selecting materials resilient to the corrosive environment, implementing proper design and engineering practices, applying surface treatments and coatings, and utilizing inhibitors to mitigate corrosion rates. Regular monitoring through inspections and non-destructive testing techniques is also crucial for identifying corrosion onset, measuring thickness loss, and informing corrective actions to extend component lifespan [16]. By adopting a proactive and comprehensive approach to uniform corrosion prevention and control, industries can minimize the economic, environmental, and safety implications associated with this pervasive form of corrosion [17].

Uniform corrosion, the most prevalent form of corrosion, results from a continuous attack on exposed metal surfaces, influenced by factors including oxygen, moisture, electrolytes, and metal type [18]. Oxygen facilitates oxidation reactions, releasing metal ions, while moisture leads to rust formation on metals like iron and steel [19]. The type and concentration of electrolytes, such as acidic or basic environments, saltwater, or saline solutions, also significantly impact the corrosion rate, with aggressive ions like chlorides accelerating corrosion. Furthermore, metal type plays a crucial role, with materials like aluminum, copper, and stainless-steel exhibiting higher resistance than carbon steel, iron, or galvanized steel, due to protective layers like chromium and natural oxide layers [20]. Understanding these factors is essential for selecting suitable materials and implementing measures to prevent or mitigate uniform corrosion, ultimately extending the service life of metal structures.

The corrosion mechanism in aqueous solutions is an example for uniform corrosion. A thin electrolyte layer facilitates corrosion, as demonstrated by placing a seawater drop on steel. On metal surfaces exposed to the atmosphere, limited water and dissolved ions are present, while oxygen access is unlimited. Corrosion products form near the metal surface, potentially preventing further corrosion by acting as a physical barrier, especially if insoluble, as seen with copper or lead. A simplified mechanism of aqueous corrosion of iron involves anodic reactions, where iron dissolves into ions, releasing electrons, while cathodic reactions involve oxygen reduction, consuming electrons and forming hydroxide ions. The corrosion mechanism of iron in aqueous solutions is an example of uniform corrosion, which involves a series of electrochemical reactions. The process can be divided into two main stages: anodic and cathodic reactions [21].

2.1.1 Anodic Reactions

At the anode, iron dissolves into ions, releasing electrons. This process is known as oxidation.

$$Fe \rightarrow Fe^{2+} + 2e^-$$

The resulting iron ions react with hydroxide ions to form ferrous hydroxide.

$$Fe^{2+} + 2OH^- \rightarrow Fe(OH)_2$$

2.1.2 Cathodic Reactions

At the cathode, oxygen reduction occurs, consuming electrons and forming hydroxide ions.

$$O_2 + 4H^+ + 4e^- \rightarrow 2H_2O$$

The resulting hydroxide ions react with iron ions to form ferric hydroxide.

$$Fe(OH)_2 + O_2 \rightarrow Fe(OH)_3$$

2.1.3 Rust Formation

Ferric hydroxide eventually forms hydrated ferric oxide, or rust.

$$Fe(OH)_3 + H_2O \rightarrow Fe_2O_3 \cdot 3H_2O$$

Rust formation occurs away from the corroding site, and the corrosion rate accelerates if ferrous ions are rapidly oxidized to ferric oxide.

2.2   Pitting Corrosion

Pitting corrosion is a localized, isolated form of corrosion that initiates with tiny pit formations and can be assessed using the pitting factor [22]. This type of corrosion commonly affects metals and alloys with a protective oxide layer, and is often triggered by oxygen or ion concentration differences. Chloride ions can exacerbate pitting corrosion by disrupting the protective oxide layer, creating a self-sustaining process. By evaluating the pitting potential and critical pitting temperature, the likelihood of pitting corrosion in a material can be predicted [23].

Pitting corrosion leads to the formation of pits on metal surfaces, ranging from microscopic to large enough to cause structural failure [24]. This type of corrosion is triggered by environmental factors that harm the metal, causing oxidation and localized acidity. The separation of anodic and cathodic reactions maintains the acidity, allowing corrosion to initiate at a single point and spread to other areas. As corrosion progresses, electromigration of anions occurs, leading to pit formation

Advances in Corrosion Science and Surface Engineering   Materials Research Forum LLC
Materials Research Foundations 188 (2026) 22-34   https://doi.org/10.21741/9781644903919-2

[25]. Over time, these pits can fill with corrosion byproducts, degrading the metal's quality and weakening its structure. Eventually, additional pits form, enlarge, and ultimately lead to metal failure.

2.2.1 Pitting Corrosion Process

Pitting corrosion is a localized corrosion phenomenon where small, isolated areas of the metal surface are attacked by a corrosive environment, resulting in the formation of tiny pits or holes that can compromise the metal's integrity [26].

Stage 1: Pitting Initiation

Pitting initiation occurs when a localized area of the metal surface is exposed to a corrosive environment, often due to factors such as inhomogeneities at the metal-corrosive interface, breakdown of the passive film, deposition of debris or solids on the metal surface, and formation of an active-passive cell with a significant potential difference, creating a vulnerable spot susceptible to pitting corrosion [27]. During pitting initiation, the metal surface undergoes anodic dissolution, releasing metal ions into the solution:

$$M \rightarrow Mn^+ + ne \tag{1}$$

This reaction is balanced by the cathodic reaction of oxygen on the adjacent surface:

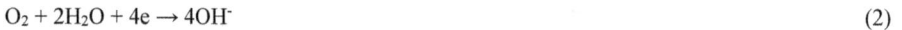

$$O_2 + 2H_2O + 4e \rightarrow 4OH^- \tag{2}$$

Stage 2: Pitting Propagation

Pitting propagation occurs when the pit grows in size due to the continued dissolution of the metal [28]. This stage is characterized by the accumulation of excess positive ions M^+ in the anodic area, leading to the migration of negative ions (anions) like chloride:

$$M+Cl^- + H_2O \rightarrow MOH + H^+ + Cl^- \tag{3}$$

The hydrolysis reaction generates free acid and lowers the pH value at the bottom of the pit:

$$M + Cl^- + H_2O \rightarrow M^+OH^- + H^+Cl^-$$

As the anodic dissolution rate accelerates, the migration of chloride ions also speeds up, rendering the reaction time-dependent.

Stage 3: Pitting Termination

Pitting termination occurs when the pit grows to a critical size, and the metal is perforated. At this stage, the reaction is terminated, and the pit is stabilized [29].

The mechanism of pitting corrosion on 316L stainless steel (SS) in chloride solutions involves the breakdown of the protective chromium oxide (Cr_2O_3) film often at manganese-rich sulfide (MnS) inclusions, which creates an area of high electrochemical activity (Fig.1) [30]. This initial stage is followed by the penetration of the chloride solution, initiating electrochemical reactions that oxidize iron and produce excess protons, further increasing reactivity. The corrosion process then propagates through the enlargement of the pit, driven by the influx of reaction residues and products, ultimately disrupting the oxide film and creating an autocatalytic process that maintains the aggressive conditions. Defects in the metal, such as crystal dislocations, secondary phases, and grain boundaries, can weaken the oxide film and create favorable sites for pit nucleation, while surface roughness can also contribute to pitting corrosion. The corrosion process is characterized by a positive feedback mechanism, where the reaction products continue to propagate the reaction under steady-state conditions, until the pit becomes wide enough, allowing fresh electrolyte solution to penetrate and transport away the enriched aggressive species, leading to the formation of a new oxide film and the cessation of corrosion activity.

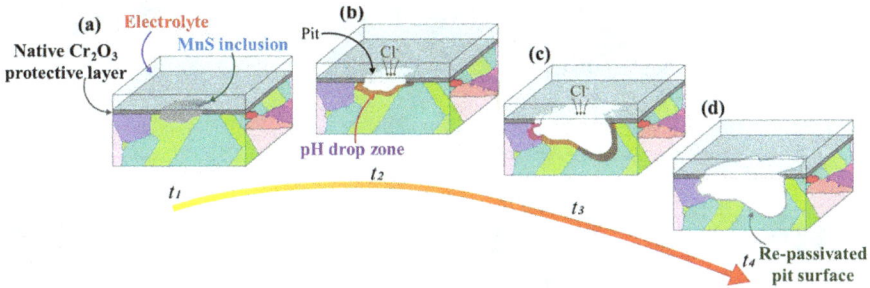

Figure 1. Initial Passive Film Formation: A protective oxide layer forms on the 316L SS surface . (Reproduced with permission from ref.30, Copyright© 2022 Springer)

2.3 Crevice Corrosion

Crevice corrosion is a localized electrochemical process occurring in stagnant solutions trapped in small spaces, such as pockets, corners, or under shields [31]. This type of corrosion is particularly hazardous, as its rate is 10-100 times higher than uniform corrosion. The presence of chloride, sulfate, or bromide ions in the electrolyte solution significantly accelerates crevice corrosion. Metals like stainless steel and aluminum alloys, which form passive oxide layers, are susceptible to crevice corrosion. The mechanism involves the dissolution of the passivating film and acidification of the electrolyte due to insufficient oxygen penetration [32]. In the presence of chloride ions, corrosion proceeds through an autocatalytic mechanism. The crevice corrosion process involves anodic reactions, where iron dissolves, releasing electrons:

$$Fe \rightarrow Fe^{2+} + 2e^-$$

The electrons flow to the cathode, where they are discharged, producing hydroxide ions:

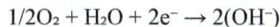

$$1/2O_2 + H_2O + 2e^- \rightarrow 2(OH^-)$$

As a result, the electrolyte within the crevice becomes positively charged, attracting negative chloride ions, which increases the acidity of the electrolyte:

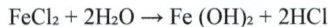

$$FeCl_2 + 2H_2O \rightarrow Fe(OH)_2 + 2HCl$$

This decrease in pH accelerates the corrosion process. The large ratio between the anode and cathode areas further increases the corrosion rate. Corrosion products, such as iron hydroxide, form within the crevice, leading to further separation of the electrolyte and exacerbating the corrosion process [33].

Advances in Corrosion Science and Surface Engineering   Materials Research Forum LLC
Materials Research Foundations 188 (2026) 22-34   https://doi.org/10.21741/9781644903919-2

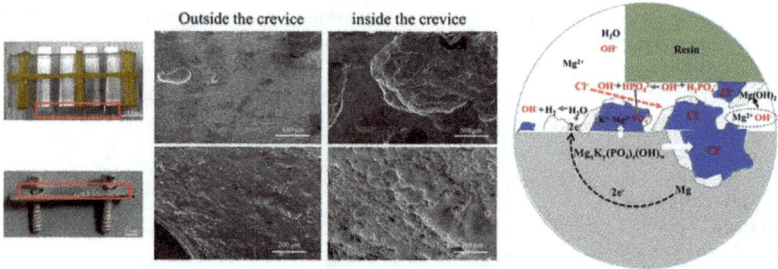

Figure 2. Crevice corrosion - the mechanism of degradation in biomedical magnesium.
(Reproduced with permission from ref.34, Copyright© 2019 Elsevier)

For example, the corrosion behavior of high-purity magnesium (HP-Mg) plates with different crevice thickness was investigated in vitro and in vivo (Figure 2). The results showed that the crevice accelerated the corrosion rate of HP-Mg through a novel mechanism [34]. When HP-Mg is immersed in a neutral or alkaline aqueous solution, two main half-reactions occur:

$$Mg \rightarrow Mg^{2+} + 2e^- \tag{4}$$

and

$$2H_2O + 2e^- \rightarrow H_2 + 2OH^- \tag{5}$$

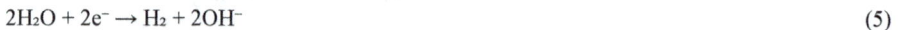

Reaction (3) increases the pH value of the solution, forming a stable corrosion product:

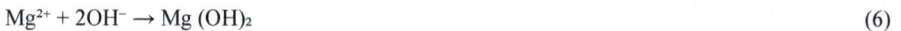

$$Mg^{2+} + 2OH^- \rightarrow Mg\,(OH)_2 \tag{6}$$

However, in the presence of a crevice, the corrosion product becomes more porous due to the penetration of chloride ions:

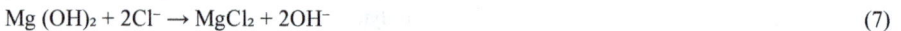

$$Mg\,(OH)_2 + 2Cl^- \rightarrow MgCl_2 + 2OH^- \tag{7}$$

The study on HP-Mg reveals a novel mechanism of crevice corrosion, where the porous corrosion layer accelerates degradation, generating more corrosion products and clogging the crevice, creating an increasingly alkaline environment that boosts the corrosion rate. The study observed deep holes with smooth or rough river-like patterns inside the crevice, while uniform corrosion with shallow pits occurred outside. Notably, the corrosion rate increased with growing crevice

Advances in Corrosion Science and Surface Engineering   Materials Research Forum LLC
Materials Research Foundations 188 (2026) 22-34   https://doi.org/10.21741/9781644903919-2

thickness, highlighting the significance of this new mechanism in understanding the crevice corrosion of magnesium in aqueous environments.

2.4   Galvanic Corrosion

Galvanic corrosion takes place when two dissimilar metals, having distinct electrochemical properties or corrosion susceptibilities, come into direct contact with each other in the presence of a corrosive electrolyte [35].

Galvanic corrosion occurs when two dissimilar metals, such as iron and copper, are in contact with each other in the presence of an electrolyte, like moisture [36]. The metal with the more positive potential, copper, acts as the cathode, while the metal with the more negative potential, iron, acts as the anode.

To understand the mechanism of galvanic corrosion, consider a galvanic cell consisting of:

1. A cathode (copper)

2. An anode (iron)

3. An electrolyte (moisture)

4. A metallic path for electron current

In this cell, iron corrodes, releasing Fe^{2+} ions into the electrolyte, while copper remains intact. The corrosion reaction at the anode is:

$$Fe \rightarrow Fe^{2+} + 2e^-$$

The Fe^{2+} ions combine with OH^- ions to form insoluble iron hydroxide:

$$Fe^{2+} + 2OH^- \rightarrow Fe(OH)_2$$

Meanwhile, hydrogen ions (H^+) are discharged at the copper cathode, releasing hydrogen gas:

$$2H^+ + 2e^- \rightarrow H_2$$

Alternatively, copper ions can be reduced at the cathode:

$$Cu^{2+} + 2e^- \rightarrow Cu$$

The flow of electric current in the galvanic cell occurs through:

1. Electron flow from anode to cathode in the external circuit

2. Positive ion (cation) flow from anode to cathode in the electrolyte

3. Negative ion (anion) flow from cathode to anode in the electrolyte

This process illustrates why iron corrodes and copper does not in a galvanic couple. When a steel pipe is joined to a copper pipe and exposed to moist soil, the steel pipe becomes the anode and corrodes, while the copper pipe remains intact [37].

2.5   Microbiologically Influenced Corrosion (MIC)

Microbiologically Influenced Corrosion (MIC) has been described in various ways, but the definitions share a common thread. Notably, the term "micro-organism" encompasses a broad

Advances in Corrosion Science and Surface Engineering   Materials Research Forum LLC
Materials Research Foundations 188 (2026) 22-34   https://doi.org/10.21741/9781644903919-2

range of microorganisms, including bacteria, cyanobacteria, algae, lichens, and fungi, all of which can contribute to MIC [38].

Microorganisms play a crucial role in microbial corrosion by creating surface conditions that sustain cathodic and/or anodic reactions, with various mechanisms proposed to reflect the diverse physiological activities of different microorganisms [39]. The corrosion of metals in the presence of microorganisms is driven by microbial alteration of the environment surrounding metal surfaces, influencing corrosion processes through direct modification of anodic and cathodic reactions, biofilm formation, production of corrosive metabolic by-products, alteration of resistance films, creation of galvanic cells, and modification of electrochemical properties [40]. Furthermore, microorganisms produce extracellular polymeric substances (EPS) that bind to metal ions, create corrosive environments, and influence corrosion rates through production of corrosive substances like hydrogen sulfide and ammonia, ultimately leading to a complex and multifaceted corrosion process [41].

Microbiologically Influenced Corrosion (MIC) is a complex process that involves the interaction of microorganisms, metallic surfaces, and the environment. The process begins with the formation of a biofilm on the metal surface, which is composed of microorganisms such as bacteria, archaea, and fungi [42]. These microorganisms produce extracellular polymeric substances (EPS) and enzymes, creating a complex array of microenvironments that can lead to the initiation of corrosion. The biofilm alters the electrochemical properties of the metal surface, leading to the formation of anodic and cathodic regions [43]. At the anode, the metal is oxidized, releasing metal ions into the solution

$$M \rightarrow M^{2+} + 2e^-,$$

while at the cathode, oxygen or other oxidizing species are reduced

$$O_2 + 2H_2O + 4e^- \rightarrow 4OH^-$$

The corrosion process is further propagated by the production of corrosive substances by microorganisms, such as hydrogen sulfide by sulfate-reducing bacteria

$$SO_4^{2-} + 4H_2 \rightarrow H_2S + 2H_2O + 2OH^-$$

and organic acids by acid-producing bacteria

$$C_6H_{12}O_6 \rightarrow 2CH_3COOH + 2H_2O$$

The formation of concentration cells, galvanic cells, and the production of auxiliary cathodic reactants also contribute to the corrosion process [44]. Additionally, microorganisms can influence the mass transport of reactants and products, further accelerating the corrosion process. The key factors that influence MIC include surface properties, environmental conditions, and the type and diversity of microorganisms present. For example, sulfate-reducing bacteria can cause corrosion of ferrous alloys in marine environments, while acid-producing bacteria can contribute to corrosion in industrial settings [45]. Overall, MIC is a complex process that requires a comprehensive understanding of the interactions between microorganisms, metallic surfaces, and the environment.

Advances in Corrosion Science and Surface Engineering   Materials Research Forum LLC
Materials Research Foundations 188 (2026) 22-34   https://doi.org/10.21741/9781644903919-2

Conclusion

In conclusion, corrosion is a complex and multifaceted phenomenon that affects various materials, leading to significant economic, environmental, and safety implications. This review has examined the fundamental principles and mechanisms underlying different types of corrosion, including uniform corrosion, pitting corrosion, crevice corrosion, galvanic corrosion, and microbiologically influenced corrosion. Understanding the interactions between materials, environments, and microorganisms is crucial for developing effective corrosion prevention and control strategies. By elucidating the underlying mechanisms of corrosion, this review aims to provide a comprehensive foundation for researchers, engineers, and practitioners to mitigate the adverse effects of corrosion and ensure the reliability and safety of infrastructure, equipment, and materials. Ultimately, a multidisciplinary approach that combines materials science, electrochemistry, and microbiology is necessary to tackle the complex challenges posed by corrosion.

References

[1] Harsimran, S., Santosh, K. and Rakesh, K., 2021. Overview of corrosion and its control: A critical review. Proc. Eng. Sci, 3(1), pp.13-24. https://doi.org/10.24874/PES03.01.002

[2] Wang, Q., Xie, J., Qin, Y., Kong, Y., Zhou, S., Li, Q., Sun, Q., Chen, B., Xie, P., Wei, Z. and Zhao, S., 2024. Recent Progress in High-Entropy Alloy Electrocatalysts for Hydrogen Evolution Reaction. Advanced Materials Interfaces, 11(14), p.2301020. https://doi.org/10.1002/admi.202301020

[3] Aggarwal, P., Awasthi, K., Sarkar, D. and Menezes, P.W., 2024. Introduction to single-atom catalysts. In Single Atom Catalysts (pp. 1-33). Elsevier. https://doi.org/10.1016/B978-0-323-95237-8.00010-0

[4] Tajuddin, A.A.B.H., 2024. Development of Corrosion-Resistant and Catalyst-Poisoning-Resistant Non-Noble-Metal Alloy-based Catalysts for PEM Electrolysers.

[5] Lele, N., 2023. Synthesis and Characterization of a Photocatalytic β-Cyclodextrin Mxene Nanocomposite Towards the Photodegradation of Emerging Pollutants in Water in a Wastewater Treatment Plant. University of Johannesburg (South Africa).

[6] Xia, Y., Zhou, D., Gao, Z. and Hu, W., 2022. Effect of $Cu2+$ on the corrosion behavior and mechanism of Al-2% Zn coatings in 3.5% NaCl solution. Journal of Electroanalytical Chemistry, 927, p.116976. https://doi.org/10.1016/j.jelechem.2022.116976

[7] Bhatt, M.D. and Lee, J.Y., 2020. Advancement of platinum (Pt)-free (non-Pt precious metals) and/or metal-free (non-precious-metals) electrocatalysts in energy applications: A review and perspectives. Energy & Fuels, 34(6), pp.6634-6695. https://doi.org/10.1021/acs.energyfuels.0c00953

[8] Sequeira, C.A.C., 2021. Austenitic Stainless Steels for Desalination Processes. In Marine Corrosion of Stainless Steels (pp. 96-114). CRC Press. https://doi.org/10.1201/9780138748104-11

[9] Malandruccolo, A., 2024. Properties of Superaustenitic Stainless Steels. In Superaustenitic Stainless Steels: A Comprehensive Overview (pp. 155-211). Cham: Springer Nature Switzerland. https://doi.org/10.1007/978-3-031-68744-0_3

[10] Kamachi Mudali, U., Jayaraj, J., Raman, R.S. and Raj, B., 2019. Corrosion: An Overview of Types, Mechanism, and Requisites of Evaluation. Non-Destructive Evaluation of Corrosion and Corrosion-assisted Cracking, pp.56-74. https://doi.org/10.1002/9781118987735.ch2

[11] Frankel, G.S., Vienna, J.D., Lian, J., Guo, X., Gin, S., Kim, S.H., Du, J., Ryan, J.V., Wang, J., Windl, W. and Taylor, C.D., 2021. Recent advances in corrosion science applicable to disposal of high-level nuclear waste. Chemical Reviews, 121(20), pp.12327-12383. https://doi.org/10.1021/acs.chemrev.0c00990

[12] Lobo, R.E., Guzmán, B., Orrillo, P.A., Domínguez, C.C., Jimenez, L.E. and Torino, M.I., 2024. Corrosion: Basics, Adverse Effects and Its Mitigation. In Sustainable Food Waste Management: Anti-corrosion Applications (pp. 3-22). Singapore: Springer Nature Singapore. https://doi.org/10.1007/978-981-97-1160-4_1

[13] Singh, A.P., Saxena, R., Saxena, S. and Maurya, N.K., 2024. The Future of Protection: Unleashing the Power of Nanotech against Corrosion. Asian Journal of Current Research, 9(3), pp.23-44. https://doi.org/10.56557/ajocr/2024/v9i38727

[14] Aljibori, H., Al-Amiery, A. and Isahak, W.N.R., 2024. Advancements in corrosion prevention techniques. Journal of Bio-and Tribo-Corrosion, 10(4), p.78. https://doi.org/10.1007/s40735-024-00882-w

[15] Odeyemi, O.O. and Alaba, P.A., 2024. Efficient and reliable corrosion control for subsea assets: challenges in the design and testing of corrosion probes in aggressive marine environments. Corrosion Reviews, (0). https://doi.org/10.1515/corrrev-2024-0046

[16] Thakur, A., Kaya, S. and Kumar, A., 2023. Recent trends in the characterization and application progress of nano-modified coatings in corrosion mitigation of metals and alloys. Applied Sciences, 13(2), p.730. https://doi.org/10.3390/app13020730

[17] Onuoha, D.O., Mgbemena, C.E., Godwin, H.C. and Okeagu, F.N., 2022. Application Of Industry 4.0 Technologies For Effective Remote Monitoring Of Cathodic Protection System Of Oil And Gas Pipelines-A Systematic Review. International journal of industrial and production engineering, 1(2).

[18] Popova, K. and Prošek, T., 2022. Corrosion monitoring in atmospheric conditions: a review. Metals, 12(2), p.171. https://doi.org/10.3390/met12020171

[19] Furlan, P.Y., Jaravata, E.J., Furlan, A.Y. and Kahl, P., 2023. Will it rust? A set of simple demonstrations illustrating iron corrosion prevention strategies at sea. Journal of Chemical Education, 100(2), pp.1081-1088. https://doi.org/10.1021/acs.jchemed.2c00802

[20] Prabhakar, J.M., Varanasi, R.S., da Silva, C.C., Saood, S., de Vooys, A., Erbe, A. and Rohwerder, M., 2021. Chromium coatings from trivalent chromium plating baths: Characterization and cathodic delamination behaviour. Corrosion Science, 187, p.109525. https://doi.org/10.1016/j.corsci.2021.109525

[21] Perez, N., 2024. Electrochemical corrosion. In Materials Science: Theory and Engineering (pp. 835-898). Cham: Springer Nature Switzerland. https://doi.org/10.1007/978-3-031-57152-7_16

Advances in Corrosion Science and Surface Engineering
Materials Research Forum LLC
Materials Research Foundations 188 (2026) 22-34
https://doi.org/10.21741/9781644903919-2

[22] Akpanyung, K.V. and Loto, R.T., 2019, December. Pitting corrosion evaluation: a review. In Journal of Physics: Conference Series (Vol. 1378, No. 2, p. 022088). IOP Publishing. https://doi.org/10.1088/1742-6596/1378/2/022088

[23] Jafarzadeh, S., Chen, Z. and Bobaru, F., 2019. Computational modeling of pitting corrosion. Corrosion reviews, 37(5), pp.419-439. https://doi.org/10.1515/corrrev-2019-0049

[24] Songbo, R., Ying, G., Chao, K., Song, G., Shanhua, X. and Liqiong, Y., 2021. Effects of the corrosion pitting parameters on the mechanical properties of corroded steel. Construction and Building Materials, 272, p.121941. https://doi.org/10.1016/j.conbuildmat.2020.121941

[25] Li, K., Sun, L., Cao, W., Chen, S., Chen, Z., Wang, Y. and Li, W., 2022. Pitting corrosion of 304 stainless steel in secondary water supply system. Corrosion Communications, 7, pp.43-50. https://doi.org/10.1016/j.corcom.2021.11.010

[26] Akhlaghi, B., Mesghali, H., Ehteshami, M., Mohammadpour, J., Salehi, F. and Abbassi, R., 2023. Predictive deep learning for pitting corrosion modeling in buried transmission pipelines. Process Safety and Environmental Protection, 174, pp.320-327. https://doi.org/10.1016/j.psep.2023.04.010

[27] Kovalov, D., Taylor, C.D., Heinrich, H. and Kelly, R.G., 2022. Operando electrochemical TEM, ex-situ SEM and atomistic modeling studies of MnS dissolution and its role in triggering pitting corrosion in 304L stainless steel. Corrosion Science, 199, p.110184. https://doi.org/10.1016/j.corsci.2022.110184

[29] Talebian, M., Raeissi, K., Atapour, M., Fernández-Pérez, B.M., Betancor-Abreu, A., Llorente, I., Fajardo, S., Salarvand, Z., Meghdadi, S., Amirnasr, M. and Souto, R.M., 2019. Pitting corrosion inhibition of 304 stainless steel in NaCl solution by three newly synthesized carboxylic Schiff bases. Corrosion Science, 160, p.108130. https://doi.org/10.1016/j.corsci.2019.108130

[30] Voisin, T., Shi, R., Zhu, Y., Qi, Z., Wu, M., Sen-Britain, S., Zhang, Y., Qiu, S.R., Wang, Y.M., Thomas, S. and Wood, B.C., 2022. Pitting corrosion in 316L stainless steel fabricated by laser powder bed fusion additive manufacturing: a review and perspective. Jom, 74(4), pp.1668-1689. https://doi.org/10.1007/s11837-022-05206-2

[31] Datta, M., 2020. Electrodissolution processes: fundamentals and applications. CRC Press. https://doi.org/10.1201/9780367808594

[32] Xu, W., Zhang, B., Addison, O., Wang, X., Hou, B. and Yu, F., 2023. Mechanically-assisted crevice corrosion and its effect on materials degradation. Corrosion Communications. https://doi.org/10.1016/j.corcom.2023.01.002

[33] Wu, X., Liu, Y., Sun, Y., Dai, N., Li, J. and Jiang, Y., 2021. A discussion on evaluation criteria for crevice corrosion of various stainless steels. Journal of Materials Science & Technology, 64, pp.29-37. https://doi.org/10.1016/j.jmst.2020.04.017

[34] Wu, H., Zhang, C., Lou, T., Chen, B., Yi, R., Wang, W., Zhang, R., Zuo, M., Xu, H., Han, P. and Zhang, S., 2019. Crevice corrosion-a newly observed mechanism of degradation in biomedical magnesium. Acta biomaterialia, 98, pp.152-159. https://doi.org/10.1016/j.actbio.2019.06.013

[35] Chen, H., Lv, Z., Lu, L., Huang, Y. and Li, X., 2021. Correlation of micro-galvanic corrosion behavior with corrosion rate in the initial corrosion process of dual phase steel. Journal of Materials Research and Technology, 15, pp.3310-3320. https://doi.org/10.1016/j.jmrt.2021.09.123

[36] Snihirova, D., Höche, D., Lamaka, S., Mir, Z., Hack, T. and Zheludkevich, M.L., 2019. Galvanic corrosion of Ti6Al4V-AA2024 joints in aircraft environment: Modelling and experimental validation. Corrosion Science, 157, pp.70-78. https://doi.org/10.1016/j.corsci.2019.04.036

[37] Wu, X., Sun, J., Wang, J., Jiang, Y. and Li, J., 2019. Investigation on galvanic corrosion behaviors of CFRPs and aluminum alloys systems for automotive applications. Materials and Corrosion, 70(6), pp.1036-1043. https://doi.org/10.1002/maco.201810635

[38] Jia, R., Unsal, T., Xu, D., Lekbach, Y. and Gu, T., 2019. Microbiologically influenced corrosion and current mitigation strategies: a state of the art review. International biodeterioration & biodegradation, 137, pp.42-58. https://doi.org/10.1016/j.ibiod.2018.11.007

[39] Little, B.J., Blackwood, D.J., Hinks, J., Lauro, F.M., Marsili, E., Okamoto, A., Rice, S.A., Wade, S.A. and Flemming, H.C., 2020. Microbially influenced corrosion-Any progress?. Corrosion Science, 170, p.108641. https://doi.org/10.1016/j.corsci.2020.108641

[40] Lou, Y., Chang, W., Cui, T., Wang, J., Qian, H., Ma, L., Hao, X. and Zhang, D., 2021. Microbiologically influenced corrosion inhibition mechanisms in corrosion protection: A review. Bioelectrochemistry, 141, p.107883. https://doi.org/10.1016/j.bioelechem.2021.107883

[41] Liu, L., Wu, X., Wang, Q., Yan, Z., Wen, X., Tang, J. and Li, X., 2022. An overview of microbiologically influenced corrosion: Mechanisms and its control by microbes. Corrosion Reviews, 40(2), pp.103-117. https://doi.org/10.1515/corrrev-2021-0039

[42] Khan, M.A.A., Hussain, M. and Djavanroodi, F., 2021. Microbiologically influenced corrosion in oil and gas industries: A review. International Journal of Corrosion and Scale Inhibition, 10(1), pp.80-106. https://doi.org/10.17675/2305-6894-2021-10-1-5

[43] Spark, A., Wang, K., Cole, I., Law, D. and Ward, L., 2020. Microbiologically influenced corrosion: a review of the studies conducted on buried pipelines. Corrosion Reviews, 38(3), pp.231-262. https://doi.org/10.1515/corrrev-2019-0108

[44] Yazdi, M., Khan, F. and Abbassi, R., 2021. Microbiologically influenced corrosion (MIC) management using Bayesian inference. Ocean Engineering, 226, p.108852. https://doi.org/10.1016/j.oceaneng.2021.108852

[45] Chang, W., Qian, H., Li, Z., Mol, A. and Zhang, D., 2024. Application and prospect of localized electrochemical techniques for microbiologically influenced corrosion: A review. Corrosion Science, p.112246. https://doi.org/10.1016/j.corsci.2024.112246

Advances in Corrosion Science and Surface Engineering
Materials Research Foundations 188 (2026) 35-54

Materials Research Forum LLC
https://doi.org/10.21741/9781644903919-3

Chapter 3

Factors Affecting Corrosion

P.S. Srikanth[1]*, Kingshuk Modak[2], Rajender Boddula[3], Ramyakrishna Pothu[4,5]

[1]Ecology & Climate Change Division, ICFRE—Institute of Forest Biodiversity, Hyderabad 500100, Telangana, India

[2]Division of Forest Ecology and Climate Change, Rain Forest Research Institute, Jorhat, Assam 785001, India

[3]School of Sciences, Woxsen University, Hyderabad - 502345, Telangana, India

[4]Center for Innovation and Inclusive Research, Sharda University, Greater Noida – 201310, India

[5]School of Physics and Electronics, College of Chemistry and Chemical Engineering, Hunan University, Changsha 410082, P.R. China

srikanthpandala@gmail.com

Abstract

Corrosion is a complex process that causes materials to deteriorate over time, resulting in significant economic and safety consequences. It occurs when a material reacts with its environment, leading to a loss of its properties and structure. Environmental factors such as temperature, humidity, and air pollution play a crucial role, as do material-specific characteristics like chemical composition, surface finish, and porosity. Electrochemical reactions, mechanical stress, and crevices also contribute to corrosion. Furthermore, microorganisms can accelerate corrosion through biofilm formation. Understanding the interplay between these factors is essential for developing effective corrosion prevention strategies, including material selection, surface treatment, and environmental control. By mitigating corrosion, we can extend the lifespan of materials, reduce maintenance costs, and ensure public safety.

Keywords

Environmental Factors, Material Properties, Biofilms, Air Pollution

Contents

1.   Introduction

Corrosion is a multifaceted phenomenon that affects materials, leading to significant economic and safety consequences [1]. It is estimated that corrosion costs the global economy over $2.5 trillion annually [2]. Understanding the factors that influence corrosion is crucial for developing effective prevention strategies. Corrosion is an electrochemical process that occurs when a material reacts with its environment, leading to a loss of its properties and structure [3]. It is a natural process that can be accelerated by various factors, including environmental conditions, material properties, and mechanical stress [4]. Environmental factors such as temperature, humidity, and air pollution play a significant role in corrosion. Temperature can influence corrosion rates, with higher temperatures generally leading to faster corrosion [5]. Humidity and air pollution can also contribute to corrosion, particularly in coastal or industrial environments (Fig. 1). Material properties, such as chemical composition and surface finish, also affect corrosion. Surface finish can also impact corrosion, with smooth surfaces being more resistant than rough or porous surfaces [6]. Certain materials, like stainless steel and titanium, are more resistant to corrosion due to their chemical composition [7]. Mechanical stress, including stress, strain rate, and crevices, can also contribute to corrosion. Stress can lead to corrosion fatigue, while crevices can provide sites for corrosion to initiate [8]. Microbiologically influenced corrosion (MIC) is another significant factor, particularly in environments where microorganisms are present. Microorganisms can form biofilms that accelerate corrosion through various mechanisms [9].

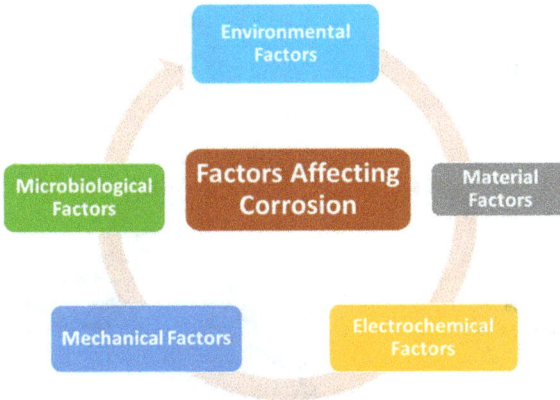

Figure 1. Major factors for corrosion.

2. Factors affecting corrosion

Environmental factors play a crucial role in corrosion, and understanding these factors is essential for developing effective corrosion prevention strategies. Temperature, humidity, air pollution, and water exposure are all environmental factors that can contribute to corrosion. Temperature is a critical factor in corrosion, as it can influence the rate of corrosion reactions (**Fig.2**). Higher temperatures generally lead to faster corrosion rates, while lower temperatures slow down corrosion [10]. The temperature of the environment can also influence the solubility of corrosive species, such as oxygen and carbon dioxide, which can accelerate corrosion [11]. For example, in marine environments, the temperature of the seawater can affect the corrosion rate of steel structures, with higher temperatures leading to faster corrosion rates. Humidity is another important environmental factor that can contribute to corrosion. High humidity environments can lead to faster corrosion rates, particularly in coastal or industrial areas [12]. This is because humidity provides a medium for corrosion reactions to occur, allowing corrosive species to come into contact with the metal surface. The humidity of the environment can also influence the formation of corrosive films on the metal surface, which can accelerate corrosion [13]. Air pollution is also a significant environmental factor that can accelerate corrosion. Particulates, sulphur dioxide, and nitrogen oxides are common air pollutants that can contribute to corrosion. These pollutants can react with moisture in the air to form corrosive species, which can then come into contact with the metal surface [14].

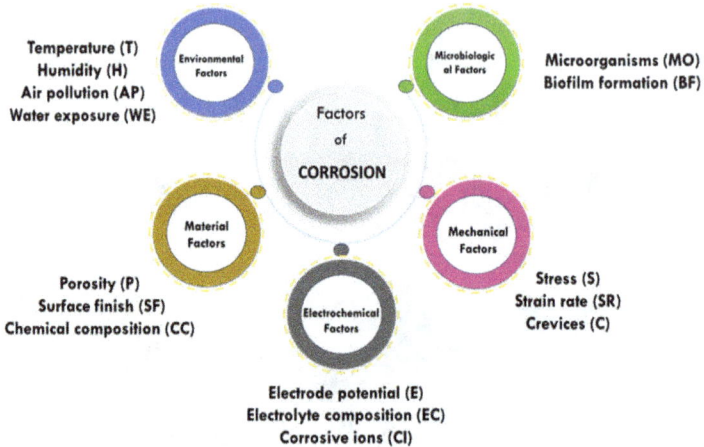

Figure 2. Factors affecting corrosion in various forms.

For example, sulfur dioxide can react with moisture to form sulfuric acid, which can accelerate corrosion. Water exposure is another critical environmental factor that can lead to corrosion. Seawater, in particular, contains corrosive species like chloride ions, which can accelerate corrosion [15]. Freshwater can also contribute to corrosion, particularly if it contains corrosive species like oxygen or carbon dioxide [16]. The pH of the water can also influence corrosion, with acidic or basic environments leading to faster corrosion rates [17].

3. Environmental factors

3.1 Temperature

Elevated temperatures accelerate the corrosion process due to increased reactivity between metals and atmospheric agents (such as oxygen and moisture). The higher kinetic energy at elevated temperatures promotes faster oxidation and dissolution of metal ions, leading to increased corrosion rates. Arrhenius equation to quantify the temperature effect on corrosion kinetics (Eq.1). It states that the rate of corrosion approximately doubles for every 10-degree Celsius increase in temperature [18]. Increase in temperature generally increases the rate of corrosion, because it provides the energy needed to overcome the activation energy barrier more frequently. This is why metals tend to corrode faster in at high temperature or environments. It suggests that any factor that lowers the activation energy, such as the presence of a catalyst or corrosive agent, will increase the rate of corrosion.

$$k = A \cdot e^{-\frac{E_a}{RT}} \tag{1}$$

Advances in Corrosion Science and Surface Engineering   Materials Research Forum LLC
Materials Research Foundations 188 (2026) 35-54   https://doi.org/10.21741/9781644903919-3

This phenomenon is particularly evident in environments where metals are exposed to varying temperatures, such as in industrial processes or in household appliances like water heaters. In the case of galvanized steel water tanks, which historically operated below 60°C, an increase in temperature to around 80°C due to modern appliances led to a reversal of the anodic and cathodic roles of iron and zinc. This resulted in the zinc coating no longer providing cathodic protection, leading to increased corrosion and tank failures

The role of temperature in the corrosion of metallic components is dominant, particularly when operating at elevated temperatures up to about 1100°C. At such high temperatures, the formation of a protective oxide scale is crucial to shield materials from corrosive agents like O_2, H_2O, N_2, and SO_2 [19]. Alloys used in these high-temperature environments vary, with carbon and low alloy steels being suitable up to 650°C, while NiCr(Al)-based alloys are preferred above 800°C due to their superior microstructural stability and resistance to thermal degradation. The efficacy of the protective oxide scale, whether it's a Cr-rich scale or an external chromia (Cr_2O_3) or alumina (Al_2O_3) layer, is highly dependent on the temperature of exposure [20]. Alumina scales are often favourable for their thermodynamic stability and slower growth rate compared to chromia scales. However, the alloy's content of Al and Cr, crucial for forming these scales, is limited by their impact on mechanical properties such as ductility. Temperature not only influences the formation of these scales, but also affects the microstructural integrity of the alloys. High temperatures can lead to external oxidation, internal corrosion processes like carburization and nitridation, and phase transformations that weaken the material's structure [21]. The depletion of oxide-forming elements due to scale formation or failure through spallation or evaporation exacerbates this degradation. Moreover, the presence of other elements can form internal precipitates, which can cause volume changes and further impair the material's properties. Temperature influence formation and stability of protective oxide scales, the microstructural degradation of alloys, and ultimately, the longevity of components used in high-temperature applications [22]. The intricate balance between maintaining the protective qualities of these scales and preserving the mechanical properties of the alloys underscores the complexity of material selection and design for high-temperature service environments.

3.2 Humidity

Humidity can create a corrosive environment, particularly when it condenses on metal surfaces, forming a thin film of moisture that facilitates the electrochemical reactions responsible for corrosion [23]. This film can become highly concentrated with corrosive contaminants, especially in conditions where there's alternating wetting and drying. The severity of corrosion is influenced by the relative humidity (RH), the nature of the material, and the presence of pollutants [24]. For instance, steel starts to corrode at a critical humidity level of 60% in clean air. The dew point, which is the temperature at which condensation begins, is also a key consideration. Keeping the temperature of a metal surface above the dew point can prevent condensation and thus corrosion [25]. Indoor spaces, if not properly ventilated, can accumulate fumes and moisture, further exacerbating corrosion, especially when condensation forms on colder surfaces like metal. The dew point—the temperature at which air moisture condenses—becomes a critical factor. In environments with high humidity, such as arid or tropical climates, the air's moisture content can lead to frequent condensation, creating conditions that are highly conducive to corrosion. This is particularly true in areas where the daily temperature cycle causes long periods of condensation, which can persist and create a continuous corrosive effect, similar to that of constant immersion.

Advances in Corrosion Science and Surface Engineering   Materials Research Forum LLC
Materials Research Foundations 188 (2026) 35-54   https://doi.org/10.21741/9781644903919-3

The relative humidity (RH) is the measure of water vapor in the air relative to the temperature, and it significantly affects the corrosion process. A thin film of electrolyte, which forms on metal surfaces at a certain RH level, carries corrosive contaminants that can concentrate and accelerate corrosion. For steel, a critical RH level is around 60% in pollutant-free environments. Below this level, the risk of corrosion decreases.

3.3   Air pollution

Air pollutants play a significant role in the corrosion of materials, particularly metals. Most common air pollutants SO_2, CO_2, NO_X and Particulate Matter (PM) form their corresponding acids and corrosive substances when these pollutants contact with moisture. These pollutants can cause substantial deterioration of outdoor materials, including construction materials, stone facades, and metal structures. The presence of these contaminants, along with climatic conditions, can significantly deteriorate materials exposed to the environment. The effects of air pollutants on corrosion are not limited to metals. All natural and man-made materials are subject to decay, and the level of pollutants in the air can speed up this process due to chemical reactions between liquids and solids [26]. The atmospheric environmental corrosion grade of substations found that the corrosion of metal equipment in transmission and transformation power stations is significantly influenced by SO_2, Cl^-, and dust. The research highlighted the need for tailored anti-corrosion measures that consider the specific atmospheric conditions of each region to enhance protection effectiveness [27].

3.4   Water exposure

The presence of water, whether as humidity, condensation, or immersion, provides the necessary medium for electrochemical reactions that lead to corrosion. Water, a ubiquitous substance, interacts with metals in various ways, influencing their corrosion behaviour, water acts as an electrolyte, facilitating electrochemical reactions that lead to corrosion. Dissolved ions in water such as chloride, sulphate, and bicarbonate enhance the electrochemical activity at metal surfaces, promoting corrosion [28]. The flow rate of water over a metal surface can influence corrosion. High flow rates can remove protective films from the metal surface, exposing fresh metal to the corrosive environment. Turbulence can increase the rate of corrosion by enhancing mass transfer and oxygen availability [29]. Higher temperatures generally increase the rate of corrosion due to enhanced ion mobility and reaction kinetics. Freshwater is less corrosive than seawater, but still contains dissolved oxygen and ions that can initiate corrosion. Seawater is highly corrosive, its high salt content due to mainly sodium chloride. Further presence of chlorides accelerates the corrosion process, leading to rapid material degradation [30]. Seawater can cause pitting corrosion, where small, localized pits form on the metal surface. These pits can penetrate deeply, compromising structural integrity. Seawater exposure in crevices (such as gaps, joints, or under deposits) exacerbates corrosion due to restricted oxygen availability and concentration of corrosive ions [31]. When dissimilar metals are in contact in seawater, galvanic corrosion occurs. The more noble (less reactive) metal acts as the cathode, while the less noble (more reactive) metal becomes the anode and corrodes [29]. Salt spray from ocean waves or coastal environments accelerates atmospheric corrosion. The combination of moisture, salt, and oxygen leads to the formation of corrosion products (such as rust) on metal surfaces. Submerged structures such as ship hulls, offshore platforms, and underwater pipelines experience continuous water exposure [32]. This water exposure significantly impacts corrosion, especially in marine environments. Engineers

Advances in Corrosion Science and Surface Engineering   Materials Research Forum LLC
Materials Research Foundations 188 (2026) 35-54   https://doi.org/10.21741/9781644903919-3

must consider the corrosive effects of water when designing structures, selecting materials, and implementing protective measures.

4.   Electrochemical factors

Corrosion is depending on the position of the metal in the galvanic series, its purity, the nature of the surface film, and the characteristics of the corrosive product. According to the galvanic series, metals can be arranged in order of their electrochemical reactivity, with metals like zinc and magnesium being more reactive and noble metals like gold and silver being less reactive [33]. When two metals with different reactivity are in contact, the more reactive metal will corrode faster, as seen in the case of cast iron and copper. The purity of the metal also plays a significant role, as impurities can increase the corrosion rate by forming electrochemical cells [34]. For example, zinc containing impurities like iron or lead will corrode faster. The nature of the surface film, specifically the specific volume ratio of metal oxide to metal, affects the oxidation rate, with higher ratios resulting in slower oxidation [35]. For instance, tungsten has a high specific volume ratio, which makes it more resistant to corrosion even at high temperatures. Lastly, the characteristics of the corrosive product, such as its solubility and volatility, can also impact the corrosion rate. If the product is soluble or volatile, it can accelerate the corrosion process [36]. Galvanic Corrosion Occurs when two dissimilar metals with different electric potentials are in contact within an electrolyte solution. The metal with the higher potential acts as the cathode, and the one with the lower potential becomes the anode, leading to accelerated corrosion at the anodic metal [37].

The understanding of the microstructural factors that influence localized corrosion, developing new alloy compositions with improved corrosion resistance, and enhancing corrosion monitoring and prediction techniques [38]. Advanced simulation methods such as finite element modelling and phase field models are being employed to predict the behaviour of corrosion processes and devise more effective mitigation strategies [39].

4.1   Electrolyte composition

Electrolytes are conductive solutions that facilitate the flow of ions, and their composition can significantly influence the rate and form of corrosion. The ionic concentration within the electrolyte affects its electrical conductivity, which in turn impacts the rate of electrochemical reactions. Higher ionic concentrations typically increase corrosion rates due to enhanced conductivity, allowing for more rapid electron transfer between anodic and cathodic sites [40]. The pH level of the electrolyte is another influential factor; acidic conditions (low pH) generally accelerate corrosion by increasing the hydrogen ion concentration, which can lead to increased hydrogen evolution reactions at the cathode. Conversely, alkaline conditions (high pH) can induce passivation in certain metals, slowing down corrosion rates or even halting them altogether [41]. Specific ions present in the electrolyte, such as chloride ions, are particularly aggressive and can penetrate protective oxide layers on metals like stainless steel, leading to localized forms of corrosion such as pitting or crevice corrosion. Sulphate and nitrate ions can also contribute to corrosion under specific conditions, altering the local electrochemical environment and potentially exacerbating corrosive attacks [42]. Dissolved oxygen in the electrolyte can lead to the formation of rust in ferrous materials through oxidation reactions. The availability of oxygen is a key factor in determining the type and rate of corrosion that occurs. Water hardness, influenced by the presence of calcium and magnesium ions, can affect the formation of scale on metal surfaces [43]. This scale can either protect the metal from further corrosion or create differential aeration cells that lead to localized corrosion. Temperature, while not a component of the electrolyte itself, affects the solubility and mobility of ions, thereby influencing corrosion rates. Higher temperatures generally increase corrosion due to faster reaction kinetics [44].

Electrochemical corrosion in oil and gas wells is a complex interplay of chemical reactions and material properties. Understanding these processes is crucial for developing strategies to mitigate corrosion and extend the lifespan of critical infrastructure in the industry. Electrochemical corrosion is a pervasive issue in the oil and gas industry, affecting the integrity of steel structures such as tubing, casing, and equipment. This type of corrosion occurs when steel comes into contact with electrolytic solutions containing dissolved salts, CO_2, H_2S, and other corrosive substances. The protective oxide film that forms on steel in the presence of air can be compromised, leading to the formation of a galvanic cell where metal ions dissolve into the electrolyte, creating ferrous salts and leaving behind electrons at the metal surface, thus establishing an anodic area [45]. At the cathodic area, these electrons facilitate reduction reactions, typically binding with hydrogen ions to form hydrogen gas or, in oxygen-rich environments, hydroxide radicals. The corrosion products, which may include various hydrated forms of iron oxide such as $Fe_2O_3 \cdot (H_2O)_x$ and FeS_x, can deposit on the metal surface, sometimes forming a protective layer that may slow down or alter the course of further corrosion [46]. Uniform electrochemical corrosion affects the entire metal surface evenly and is generally more predictable. Modern corrosion prediction software focuses on this type of corrosion due to its uniform nature, allowing for preventive measures such as increasing the material's wall thickness or applying cathodic protection. Cathodic protection involves imposing an electric field to counteract the electrochemical potential that drives the corrosion process [47].

4.2   Electrode potential

Electrode potential, also known as electrochemical potential, is a fundamental concept in the study of corrosion. It refers to the voltage difference between a metal immersed in a given environment and a standard reference electrode (SRE), which has a stable and well-known electrode potential. This potential difference is crucial because it determines the tendency of a metal to lose electrons and corrode [48]. In an electrolytic solution, different metals or alloys will have different corrosion rates based on their electrode potentials. The rate of corrosion depends on this potential, which can be measured by connecting the metal to an SRE and using a voltmeter to measure the potential difference. The magnitude and sign of the voltage are essential for measuring and reporting corrosion potentials [49]. The electrode potential is influenced by several factors, including the chemical composition of the metal, the temperature, the pH of the solution, and the concentration of ions in the electrolyte. Metals with a more negative standard electrode potential are more likely to undergo oxidation (corrode) when in contact with metals with a more positive potential. This is why, for example, zinc is often used as a sacrificial anode to protect iron from corrosion, as zinc has a more negative electrode potential than iron [50].

The standard electrode potential is set to zero for the standard hydrogen electrode (SHE), and all other potential measurements are made relative to this standard. The electrochemical series is a list of metals arranged in order of their standard potentials relative to the hydrogen electrode. Metals higher up in the series displace metals lower in the series, which means that when connecting two metals with different potentials, the metal with the lowest potential corrodes [51]. Understanding electrode potentials is vital for predicting and preventing corrosion. For instance, in galvanic corrosion, when two metals with a significant difference in electrode potentials are connected in an electrolyte, the one with the lower potential acts as the anode and corrodes, while the one with the higher potential acts as the cathode and is protected. This principle is widely used in cathodic protection systems to prevent corrosion in underground pipelines and other metallic structures.

4.3   Corrosive Ions

Corrosive ions are a pivotal factor in the corrosion process, significantly influencing the rate and mechanism by which metals degrade. Among these ions, chlorides are particularly notorious for their role in pitting corrosion, a localized form of corrosion that leads to the formation of small, often hard-to-detect pits on the metal surface. Chloride ions are known to penetrate protective oxide layers on metals, destabilizing these layers and initiating the corrosion process. Once the passive film is compromised, the underlying metal is exposed to further attack, leading to the growth of pits [30]. Sulphate and nitrate ions also contribute to corrosion, particularly in environments where they can form aggressive acidic or basic conditions that facilitate metal dissolution. These ions can interact with the metal surface and influence the electrochemical reactions that lead to corrosion. For example, sulphate-reducing bacteria can produce hydrogen sulfide, which can lead to sulfide stress cracking, a form of hydrogen embrittlement [52]. The presence of oxygen in the electrolyte can exacerbate the effects of corrosive ions by providing a reactant for the cathodic process, which is part of the corrosion reaction. Oxygen reduction at the cathode can lead to the formation of hydroxide ions, which can increase the local pH and promote the formation of metal hydroxides. These reactions can further influence the local electrochemical environment and accelerate the corrosion process [53].

5.   Material Factors

The environmental compatibility of a material, its ability to withstand the specific chemical, thermal, and mechanical conditions of its environment, is dominant parameter in determining its longevity and performance. Metallic materials generally exposed to atmosphere, water and soil leads to corrosion. Soil itself acts as corrosive medium in in the ambient temperature environment [54]. Soil is a complex medium composed of gaseous, liquid, and solid phases. Its nonuniform and highly variable composition makes the corrosion process intricate. Properties such as porosity, water content, oxygen levels, salt content, electrical conductivity, acidity/alkalinity, temperature, and microorganisms significantly influence soil corrosion rates [55].

5.1   Chemical composition

Corrosion is an inevitable process for metals and alloys, surpassing other failure modes like fatigue and creep. Metals, excluding noble metals, predominantly exist as compounds (such as oxides and sulfides) due to their thermodynamic stability [56]. When exposed to their environment, metals tend to revert to their more energetically favourable compound states through corrosion [57]. The consequences of corrosion extend beyond replacement costs and include production losses, elevated maintenance expenses, compliance challenges, compromised product quality, increased fuel and energy costs from corroded pipes, and the need for additional working capital [58]. Corrosion involves electrochemical reactions, leading to material degradation. Understanding material-environment interactions, including metallurgy, stress, and environmental chemistry, is crucial for effective corrosion management. All engineering materials are chemically reactive, and their strength depends on environmental influences over time. Therefore, defining material strength for corrosion-resistant design requires considering various factors that contribute to the corrosion process [59].

Iron forms a protective oxide film on its surface when exposed to strongly oxidizing environments, termed as passivation. This film isolates the metal surface from corrosive agents, preventing further corrosion [60]. The oxide film is not static, it constantly repairs and renews itself, maintaining its protective function. Passivated iron is covered by a thin film of γ-Fe_2O_3/Fe_3O_4, existing in a cubic system with a thickness of 1-3 nanometers. Spectroscopic and microscopic studies generally support these characteristics, although variations exist based on iron treatment and solution type [61]. This dynamic barrier plays a crucial role in materials science and corrosion protection. In atmospheric corrosion, ordinary structural steels (not necessarily in strongly oxidizing environments) corrode to form rust. These rusts lack the properties necessary to shield the steel from its surroundings and do not possess the self-maintaining characteristics of passive oxide films. Weathering steels, however, exhibit corrosion resistance due to the formation of special rust types. These rusts include α-, β-, and γ-FeOOH, γ-Fe_2O_3, Fe_3O_4, and amorphous oxyhydroxides. The densely packed, nano-sized rust particles create a tightly adherent film that acts as a barrier against further corrosion. Despite this, rust components remain somewhat soluble in water [62]. If a protective rust film forms and corrosion stops, the absence of iron ions and hydroxide ions leads to film deterioration over time, resulting in defects. Consequently, rust films alone, even if mechanically dense, are insufficient for long-lasting protection of iron and steel against water and oxygen [63].

5.2   Porosity

The presence of voids or pores within a material able to influence corrosion, particularly in coatings and metallic structures. Porosity can create pathways for corrosive agents to penetrate deeper into the material, leading to accelerated degradation [64]. In coatings, porosity is often associated with a higher number of solidified particles that become trapped, resulting in poor cohesion and increased corrosion rate and wear. The size, distribution, and connectivity of pores can affect the material's ability to resist corrosive environments [65]. For example, high porosity levels in metallic components can lead to a larger surface area exposed to corrosive elements, which can increase the overall corrosion rate [66]. Moreover, in porous metals used for biomedical implants, such as titanium alloys, the porosity must be carefully controlled to balance biomechanical compatibility with corrosion resistance, as excessive porosity may reduce the material's ability to withstand corrosive bodily fluids [65]. Porosity can be beneficial for certain applications by reducing weight or increasing surface area, it generally has a detrimental effect on corrosion resistance, making it a critical factor to consider in material design and engineering.

5.3   Surface finish

Surface finishes, encompassing the topographical characteristics and texture of a material's surface. The degree of smoothness or roughness, presence of grooves, pits, and other surface irregularities can alter the electrochemical behaviour of the material, when exposed to a corrosive environment [67]. A smoother surface finish generally offers better corrosion resistance as it reduces the surface area where corrosive agents can attack and minimizes the likelihood of initiating localized corrosion forms such as pitting or crevice corrosion [68]. Rougher surfaces, on the other hand, can corrosive particles, retain moisture, and facilitate the formation of micro-

Advances in Corrosion Science and Surface Engineering   Materials Research Forum LLC
Materials Research Foundations 188 (2026) 35-54   https://doi.org/10.21741/9781644903919-3

environments that are more aggressive than the bulk solution, leading to accelerated corrosion rates [69]. Moreover, surface imperfections can act as stress concentrators, potentially leading to stress corrosion cracking under tensile loading conditions [70]. In manufacturing and material processing, achieving the desired surface finish is crucial for extending the service life of components, especially in aggressive environments. Processes such as polishing, grinding, and coating are employed to improve surface finish and enhance corrosion resistance [69]. The choice of surface finishing technique and the resulting surface characteristics must be carefully considered in the context of the expected service environment to ensure optimal performance and durability of the material

6. Mechanical factors

Environment-assisted fractures, including stress corrosion and hydrogen embrittlement, is a critical concern in the integrity of metallic structures such as tubing, casing, and surface assemblies in various industries. These phenomena lead to the degradation of the material's physical and mechanical properties, particularly toughness, due to the presence of certain chemical elements or compounds.

6.1 Stress

Stress corrosion is sudden and brittle fracture of materials under the simultaneous influence of tensile stress and a corrosive environment. The occurrence of stress corrosion is highly dependent on factors such as the type of material, the presence of specific corrosive media, temperature, and pH levels [71]. For example, chloride-containing environments are notorious for inducing SCC in certain stainless steels and high-strength alloys. Conversely, low-alloy steels like J55, N80, and P110 are less susceptible to this form of cracking.

Hydrogen Embrittlement is another form of environment-assisted fracture where hydrogen atoms diffuse into the metal lattice, leading to embrittlement and potential failure [72]. This process can occur simultaneously with SCC or act as a contributing factor to it. Recent research has shown that adding elements like molybdenum to steel can enhance its ability to trap hydrogen, potentially mitigating hydrogen embrittlement [73].

Fatigue Corrosion is a phenomenon where cyclic loading and corrosive environments interact, leading to accelerated material fatigue. Advances in understanding the fatigue life and corrosion properties of materials have been made by adding different elements to alloys to improve their resistance to such conditions [74].

Impingement Corrosion, also known as erosion-corrosion, occurs when high-velocity fluids carrying particles impact a material's surface, leading to increased corrosion rates. Studies have found that the impingement angle and velocity significantly affect the erosion-corrosion performance of materials like carbon stee [75].

6.2 Strain Rate

The strain rate is a significant factor in the corrosion process, particularly in stress corrosion cracking (SCC). It refers to the speed at which a material is deformed under stress. The strain rate can influence the initiation and propagation of cracks in materials exposed to corrosive environments.

Advances in Corrosion Science and Surface Engineering   Materials Research Forum LLC
Materials Research Foundations 188 (2026) 35-54   https://doi.org/10.21741/9781644903919-3

High Strain Rate: A study on additively manufactured maraging steels showed that samples with higher titanium content exhibited better dynamic behaviour at high strain rates of 5000 s^{-1}. These samples did not fracture until the strain rate reached this high level. Additionally, the corrosion resistance of these materials improved after dynamic loading due to grain recrystallization during the deformation process [76].

Slow Strain Rate: Research on austenitic stainless steel produced by wire laser additive manufacturing (WLAM) revealed that slow strain rates adversely affect the stress corrosion resistance of the material. The WLAM samples demonstrated inferior mechanical properties, general corrosion resistance, and stress corrosion performance at slow strain rates compared to regular AISI 316L alloy. The differences were attributed to phase compositions, structural morphology, and inherent defects [77].

Strain Rate Sensitivity: It has been established that many metal-environment combinations sensitive to SCC are also sensitive to the strain rate. This is particularly evident in dissolution-controlled SCC, where the strain rate associated with applied stress can significantly impact the crack growth kinetics [78].

Correlation with Crack Growth: Laboratory work involving realistic stressing conditions, including low-frequency cyclic loading, has shown that strain or creep rates provide good correlation with thresholds for cracking and crack growth kinetics [79].

Influence of Stress and Bacteria: A study on the corrosion behaviour of X70 pipeline steel indicated that stress is the most relevant factor, followed by the presence of sulfate-reducing bacteria (SRB) and strain rate. At high stresses, stress dominated the corrosion behaviour, but at low stress, SRB played a more significant role [80].

6.3   Crevices

Develops when there are defects or crevices in the metal that are exposed to the electrolyte. These confined spaces can harbour aggressive conditions that lead to localized attack and potentially severe corrosion [81]. Crevice corrosion, a type of localized corrosion, results in the formation of pits or holes at metal junctions or crevices, exacerbated by the differential aeration and concentration of ions within these confined spaces. Crevice corrosion is a localized form of corrosion that occurs in confined spaces, or crevices, where a stagnant microenvironment is created. This type of corrosion is particularly insidious because it can lead to the rapid deterioration of metals, often going undetected until significant damage has occurred. The crevices can be inherent in the design, such as those found at flanges, gaskets, and under washers, or they can be formed by the deposition of particles, corrosion products, or biological materials.

The process of crevice corrosion typically begins when the crevice environment becomes more acidic and depleted in oxygen compared to the bulk environment. This is due to the hydrolysis of metal cations that migrate out of the crevice, which leads to a decrease in pH [82]. Hazardous anions, such as chloride ions, play a crucial role in the initiation and propagation of crevice corrosion. They can penetrate the passive film on metals like stainless steel, leading to localized breakdown of passivity and initiation of corrosion. Recent studies have provided deeper insights into the mechanisms of crevice corrosion. The geometry of the crevice can restrict the diffusion of metal ions and create an autocatalytic effect that promotes the development of crevice corrosion. Elements such as Chromium (Cr), Molybdenum (Mo), and Nitrogen (N) have been found to have

Advances in Corrosion Science and Surface Engineering   Materials Research Forum LLC
Materials Research Foundations 188 (2026) 35-54   https://doi.org/10.21741/9781644903919-3

a synergistic effect that improves resistance to crevice corrosion [83]. Moreover, the corrosion behaviour is most severe at the crevice mouth position, where the concentration of aggressive ions is highest and the availability of oxygen is lowest. In the context of oil and gas wells, crevice corrosion is a significant concern due to the harsh environments these materials are exposed to, which often include high temperatures, pressures, and the presence of corrosive substances like hydrogen sulfide and carbon dioxide. The threaded connections of tubing and casing, as well as the drill pipe subs, are particularly susceptible to crevice corrosion due to the presence of narrow gaps and the accumulation of deposits and adhesions [84].

7. Microbiological Factors

Microorganisms, bloom across diverse environments, can infiltrate industrial systems, leading to Microbiologically Influenced Corrosion (MIC) through their ability to colonize nutrient-rich, water-exposed surfaces [85].

7.1 Microorganisms

Microbes, including bacteria, archaea, fungi, and some algae, inhabit at micro size range. They exhibit fast growth, adaptability, and the ability to survive in diverse habitats. These organisms take up nutrients from their environment and produce metabolites during metabolism. Microbes communicate via chemical signalling molecules and can evolve through mechanisms like horizontal gene transfer. In the situation of corrosion studies, aerobic microbial communities are significant in marine environments due to the high dissolved oxygen levels. Each environment maintains a specific microbial community, with different species contributing to the production of various metabolites.

7.2 Biofilm formation

Microbial cells can attach to metal surfaces, forming biofilms that secrete extracellular polymeric substances (EPS). These biofilms are thought to enhance the corrosion process. Bacteria, especially sulfate-reducing bacteria (SRB), sulfur-oxidizing bacteria (SOB), iron-oxidizing/reducing bacteria, manganese-oxidizing bacteria (MOB), and acid-producing bacteria (APB), have been studied for their involvement in microbiologically influenced corrosion (MIC) and atmospheric localised corrosion (ALWC). Iron-oxidizing genera like Clonothrix, Gallionella, Leptothrix, and Sphaerotilus convert ferrous (Fe^{+2}) to ferric (Fe^{+3}) iron, contributing to tubercle formation on stainless steel weld seams. The mechanisms consist of chemical attacks from microbial by-products like acids and gases, degradation of organic coatings and inhibitors, depassivation and depolarization of metal surfaces, and synergistic microbial-metal interactions [86]. Signs of MIC include accelerated steel corrosion rates, premature pipe failures, tubercular deposits, high microbial counts, and pinhole leaks. These microorganisms' capacity to concentrate essential nutrients facilitates their persistence and impact on large-scale engineering structures [87].

References

[1] P.R. Roberge, P. Eng. Corrosion engineering: McGraw-Hill New York, NY, USA.; 2008.

Materials Research Forum LLC
https://doi.org/10.21741/9781644903919-3

[2] G. Koch, M. Brongers, N. Thompson, Y. Virmani, J. Payer, Corrosion costs and preventive strategies in the United States, Publication No, FHWARD-01 (Washington, DC: FHWA. 156 (2002).

[3] D.A. Jones, Principles and prevention, Corrosion. 2 (1996) 168.

[4] L. Li, C.-Q. Li, M. Mahmoodian, Effect of applied stress on corrosion and mechanical properties of mild steel, Journal of Materials in Civil Engineering. 31 (2019) 04018375. https://doi.org/10.1061/(ASCE)MT.1943-5533.0002594

[5] T.W. Quadri, E.D. Akpan, L.O. Olasunkanmi, O.E. Fayemi, E.E. Ebenso. Fundamentals of corrosion chemistry. Environmentally Sustainable Corrosion Inhibitors: Elsevier; 2022. p. 25-45. https://doi.org/10.1016/B978-0-323-85405-4.00019-7

[6] E. Ghali, V.S. Sastri, M. Elboujdaini. Corrosion prevention and protection: practical solutions: John Wiley & Sons; 2007.

[7] M.F. McGuire. Stainless steels for design engineers: Asm International; 2008. https://doi.org/10.31399/asm.tb.ssde.9781627082860

[8] D.E. Talbot, J.D. Talbot. Corrosion science and technology: CRC press; 2018.

[9] A. Gruca, A brief review of microbial induced corrosion research, Annales Universitatis Paedagogicae Cracoviensis Studia Naturae. (2018). https://doi.org/10.24917/25438832.3.12

[10] P. Kritzer, N. Boukis, E. Dinjus, Factors controlling corrosion in high-temperature aqueous solutions: a contribution to the dissociation and solubility data influencing corrosion processes, The Journal of supercritical fluids. 15 (1999) 205-227. https://doi.org/10.1016/S0896-8446(99)00009-1

[11] Z. Han, C. He, J. Lian, Y. Zhao, X. Chen, Effects of temperature on corrosion behaviour of 2205 duplex stainless steel in carbon dioxide-containing environments, International Journal of Electrochemical Science. 15 (2020) 3627-3645. https://doi.org/10.20964/2020.05.73

[12] A.R. Mendoza, F. Corvo, Outdoor and indoor atmospheric corrosion of non-ferrous metals, Corrosion science. 42 (2000) 1123-1147. https://doi.org/10.1016/S0010-938X(99)00135-3

[13] J.R. Davis. Corrosion: Understanding the basics: Asm International; 2000. https://doi.org/10.31399/asm.tb.cub.9781627082501

[14] Y. Cai, Y. Xu, Y. Zhao, X. Ma, Atmospheric corrosion prediction: a review, Corrosion Reviews. 38 (2020) 299-321. https://doi.org/10.1515/corrrev-2019-0100

[15] F. Qu, W. Li, W. Dong, V.W. Tam, T. Yu, Durability deterioration of concrete under marine environment from material to structure: A critical review, Journal of Building Engineering. 35 (2021) 102074. https://doi.org/10.1016/j.jobe.2020.102074

[16] J. Rossum, Fundamentals of metallic corrosion in fresh water, CiteSeerX. 1983 (1983) 1-18.

[17] M. Wasim, S. Shoaib, N. Mubarak, Inamuddin, A.M. Asiri, Factors influencing corrosion of metal pipes in soils, Environmental Chemistry Letters. 16 (2018) 861-879. https://doi.org/10.1007/s10311-018-0731-x

[18] C.D. Taylor, B.M. Tossey, High temperature oxidation of corrosion resistant alloys from machine learning, npj Materials Degradation. 5 (2021) 38. https://doi.org/10.1038/s41529-021-00184-3

[19] B. Gleeson, High-temperature corrosion of metallic alloys and coatings, Materials Science and Technology: A Comprehensive Treatment: Corrosion and Environmental Degradation, Volumes I+ II. 1 (2000) 173-228. https://doi.org/10.1002/9783527619306.ch14

[20] P. Huczkowski, N. Christiansen, V. Shemet, J. Piron-Abellan, L. Singheiser, W. Quadakkers, Oxidation induced lifetime limits of chromia forming ferritic interconnector steels. (2004). https://doi.org/10.1115/1.1782925

[21] J. Barnes, G. Lai, Factors affecting the nitridation behavior of Fe-base, Ni-base, and Co-base alloys in pure nitrogen, Le Journal de Physique IV. 3 (1993) C9-167-C169-174. https://doi.org/10.1051/jp4:1993915

[22] R. Pillai, A. Chyrkin, W.J. Quadakkers, Modeling in high temperature corrosion: a review and outlook, Oxidation of Metals. 96 (2021) 385-436. https://doi.org/10.1007/s11085-021-10033-y

[23] K. Popova, T. Prošek, Corrosion monitoring in atmospheric conditions: a review, Metals. 12 (2022) 171. https://doi.org/10.3390/met12020171

[24] D.E. Klenam, M.O. Bodunrin, S. Akromah, E. Gikunoo, A. Andrews, F. McBagonluri, Ferrous materials degradation: characterisation of rust by colour-an overview, Corrosion Reviews. 39 (2021) 297-311. https://doi.org/10.1515/corrrev-2021-0005

[25] W. Zuo, X. Zhang, Y. Li, Review of flue gas acid dew-point and related low temperature corrosion, Journal of the Energy Institute. 93 (2020) 1666-1677. https://doi.org/10.1016/j.joei.2020.02.004

[26] V. Kucera, E. Mattsson. Atmospheric corrosion. Corrosion mechanisms: CRC Press; 2020. p. 211-284.

[27] X. Chen, Z. Zhang, H. Zhang, H. Yan, F. Liu, S. Tu, Influence of air pollution factors on corrosion of metal equipment in transmission and transformation power stations, Atmosphere. 13 (2022) 1041. https://doi.org/10.3390/atmos13071041

[28] C.J. Lytle, M.A. Edwards, Phosphate Chemical Use for Sequestration, Scale Inhibition, and Corrosion Control, ACS ES&T Water. 3 (2023) 893-907. https://doi.org/10.1021/acsestwater.2c00570

[29] R. Bender, D. Féron, D. Mills, S. Ritter, R. Bäßler, D. Bettge, et al., Corrosion challenges towards a sustainable society, Materials and corrosion. 73 (2022) 1730-1751. https://doi.org/10.1002/maco.202213140

[30] R.E. Lobo, B. Guzmán, P.A. Orrillo, C.C. Domínguez, L.E. Jimenez, M.I. Torino. Corrosion: Basics, Adverse Effects and Its Mitigation. Sustainable Food Waste Management: Anti-corrosion Applications: Springer; 2024. p. 3-22. https://doi.org/10.1007/978-981-97-1160-4_1

[31] H.M. Hussein Farh, M.E.A. Ben Seghier, R. Taiwo, T. Zayed, Analysis and ranking of corrosion causes for water pipelines: a critical review, npj Clean Water. 6 (2023) 65. https://doi.org/10.1038/s41545-023-00275-5

[32] D.G. Enos, Atmospheric Corrosion in Marine Environments, LaQue's Handbook of Marine Corrosion. (2022) 49-61. https://doi.org/10.1002/9781119788867.ch3

[33] A. Zupanc, J. Install, M. Jereb, T. Repo, Sustainable and selective modern methods of noble metal recycling, Angewandte Chemie. 135 (2023) e202214453. https://doi.org/10.1002/ange.202214453

[34] G.-S. Peng, J. Huang, Y.-C. Gu, G.-S. Song, Self-corrosion, electrochemical and discharge behavior of commercial purity Al anode via Mn modification in Al-air battery, Rare Metals. 40 (2021) 3501-3511. https://doi.org/10.1007/s12598-020-01687-9

[35] J.R. Xavier, Influence of surface modified mixed metal oxide nanoparticles on the electrochemical and mechanical properties of polyurethane matrix, Frontiers of Chemical Science and Engineering. 17 (2023) 1-14. https://doi.org/10.1007/s11705-022-2176-9

[36] S. Zehra, M. Mobin, R. Aslam. Corrosion prevention and protection methods. Eco-Friendly Corrosion Inhibitors: Elsevier; 2022. p. 13-26. https://doi.org/10.1016/B978-0-323-91176-4.00023-4

[37] Y. Huang, J. Zhang, F.-Z. Xuan, Y. Ma, Modeling and Prediction of Galvanic Corrosion for an Overlaying Welded Structure, Journal of Materials Engineering and Performance. (2023) 1-13.

[38] C. Dong, Y. Ji, X. Wei, A. Xu, D. Chen, N. Li, et al., Integrated computation of corrosion: Modelling, simulation and applications, Corrosion Communications. 2 (2021) 8-23. https://doi.org/10.1016/j.corcom.2021.07.001

[39] A. Kosari, H. Zandbergen, F. Tichelaar, P. Visser, P. Taheri, H. Terryn, et al., In-situ nanoscopic observations of dealloying-driven local corrosion from surface initiation to in-depth propagation, Corrosion Science. 177 (2020) 108912. https://doi.org/10.1016/j.corsci.2020.108912

[40] Z. Chen, W. Lai, Y. Xu, G. Xie, W. Hou, P. Zhanchang, et al., Anodic oxidation of ciprofloxacin using different graphite felt anodes: kinetics and degradation pathways, Journal of Hazardous Materials. 405 (2021) 124262. https://doi.org/10.1016/j.jhazmat.2020.124262

[41] B.E. Ibrahimi, J.V. Nardeli, L. Guo, An Overview of Corrosion, Sustainable Corrosion Inhibitors I: Fundamentals, Methodologies, and Industrial Applications. (2021) 1-19. https://doi.org/10.1021/bk-2021-1403.ch001

[42] H. Veselivska, V. Hvozdetskyi, M. Student, K.R. Zadorozhna, Y.V. Dzioba, The Influence of the Electrolyte Composition for Hard Anodizing of Aluminum on Corrosion Resistance of Synthesized Coatings, Materials Science. 59 (2023) 228-233. https://doi.org/10.1007/s11003-024-00767-w

[43] D.-H. Xia, C.-M. Deng, D. Macdonald, S. Jamali, D. Mills, J.-L. Luo, et al., Electrochemical measurements used for assessment of corrosion and protection of metallic materials in the field: A critical review, Journal of Materials Science & Technology. 112 (2022) 151-183. https://doi.org/10.1016/j.jmst.2021.11.004

Advances in Corrosion Science and Surface Engineering   Materials Research Forum LLC
Materials Research Foundations 188 (2026) 35-54   https://doi.org/10.21741/9781644903919-3

[44] J.R. Scully, C.F. Glover, R.J. Santucci, | Electrochemical Tests. (2022). https://doi.org/10.1520/MNL202NDSUP20190036

[45] L. Yang. Techniques for corrosion monitoring: Woodhead Publishing; 2020.

[46] L. Fan, X. Shi, Techniques of corrosion monitoring of steel rebar in reinforced concrete structures: A review, Structural health monitoring. 21 (2022) 1879-1905. https://doi.org/10.1177/14759217211030911

[47] J. Aaziz, E.-A. Abdallah, Corrosion Processes and Strategies for Protection, Anti-Corrosive Nanomaterials: Design, Characterization, Mechanisms and Applications. (2023) 25. https://doi.org/10.1201/9781003331124-3

[48] M. Sophocleous, J.K. Atkinson, A review of screen-printed silver/silver chloride (Ag/AgCl) reference electrodes potentially suitable for environmental potentiometric sensors, Sensors and Actuators A: Physical. 267 (2017) 106-120. https://doi.org/10.1016/j.sna.2017.10.013

[49] R. Buchanan, E. Stansbury, Electrochemical corrosion, Handbook of environmental degradation of materials. 4 (2012) 87-125. https://doi.org/10.1016/B978-1-4377-3455-3.00004-3

[50] F.J. Ansuini, J.R. Dimond, Factors affecting the accuracy of reference electrodes, Materials performance. 33 (1994) 14-17. https://doi.org/10.5006/C1994-94323

[51] N. Perez, N. Perez, Electrochemical corrosion, Electrochemistry and corrosion science. (2016) 1-23. https://doi.org/10.1007/978-3-319-24847-9_1

[52] F.-Y. Ma, Corrosive effects of chlorides on metals, Pitting corrosion. 294 (2012) 139-178.

[53] C.-Q. Li, W. Yang. Steel corrosion and degradation of its mechanical properties: CRC Press; 2021.

[54] H.J. Jun, J.K. Park, C.H. Bae, Factors affecting steel water-transmission pipe failure and pipe-failure mechanisms, Journal of Environmental Engineering. 146 (2020) 04020034. https://doi.org/10.1061/(ASCE)EE.1943-7870.0001692

[55] X. Zhou, Z. Zhou, T. Wu, C. Li, Z. Li, Effects of non-viable microbial film on corrosion of pipeline steel in soil environment, Corrosion Communications. 3 (2021) 23-33. https://doi.org/10.1016/j.corcom.2021.11.003

[56] M. Zhang, X. Li, J. Zhao, X. Han, C. Zhong, W. Hu, et al., Surface/interface engineering of noble-metals and transition metal-based compounds for electrocatalytic applications, Journal of Materials Science & Technology. 38 (2020) 221-236. https://doi.org/10.1016/j.jmst.2019.07.040

[57] S.J. Price, R.B. Figueira, Corrosion protection systems and fatigue corrosion in offshore wind structures: current status and future perspectives, Coatings. 7 (2017) 25. https://doi.org/10.3390/coatings7020025

[58] M. Sofian, M.B. Haq, D. Al Shehri, M.M. Rahman, N.S. Muhammed, A review on hydrogen blending in gas network: Insight into safety, corrosion, embrittlement, coatings and liners, and bibliometric analysis, International Journal of Hydrogen Energy. 60 (2024) 867-889. https://doi.org/10.1016/j.ijhydene.2024.02.166

[59] M. Elboujdaini, Challenges of Materials and Corrosion Management: Stress Corrosion Cracking (SCC) and Gaping in our Understanding of the Subject, Procedia Structural Integrity. 42 (2022) 1033-1039. https://doi.org/10.1016/j.prostr.2022.12.130

[60] S.K. Dhawan, H. Bhandari, G. Ruhi, B.M.S. Bisht, P. Sambyal. Corrosion Preventive Materials and Corrosion Testing: CRC Press; 2020. https://doi.org/10.1201/9781315101217

[61] M. Rozana, A. Matsuda, G. Kawamura, W.K. Tan, Z. Lockman. Formation of Nanoporous α-Fe2O3 Thin Film as Photoanode by Anodic Oxidation on Iron. Two-Dimensional Nanostructures for Energy-Related Applications: CRC Press; 2017. p. 240-268. https://doi.org/10.1201/9781315369877-9

[62] V.S. Saji, Temporary rust preventives-A retrospective, Progress in Organic Coatings. 140 (2020) 105511. https://doi.org/10.1016/j.porgcoat.2019.105511

[63] H. Tamura, The role of rusts in corrosion and corrosion protection of iron and steel, Corrosion Science. 50 (2008) 1872-1883. https://doi.org/10.1016/j.corsci.2008.03.008

[64] O. Oladijo, B. Obadele, A. Venter, L. Cornish, Investigation the effect of porosity on corrosion resistance and hardness of wc-co coatings on metal substrates. (2016).

[65] W. Xu, X. Lu, B. Zhang, C. Liu, S. Lv, S. Yang, et al., Effects of porosity on mechanical properties and corrosion resistances of PM-fabricated porous Ti-10Mo alloy, Metals. 8 (2018) 188. https://doi.org/10.3390/met8030188

[66] D. Yang, X. Kan, P. Gao, Y. Zhao, Y. Yin, Z. Zhao, et al., Influence of porosity on mechanical and corrosion properties of SLM 316L stainless steel, Applied Physics A. 128 (2022) 1-9. https://doi.org/10.1007/s00339-021-05118-z

[67] Z. Liu, T. Jackson, editors. Impact of Surface Finish on Carbon Steel Corrosion and Corrosion Inhibition by Chemicals in Carbon Dioxide Environment. NACE CORROSION; 2017: NACE. https://doi.org/10.5006/C2017-09175

[68] D. Dwivedi, K. Lepková, T. Becker, Carbon steel corrosion: a review of key surface properties and characterization methods, RSC advances. 7 (2017) 4580-4610. https://doi.org/10.1039/C6RA25094G

[69] E. Messinese, L. Casanova, L. Paterlini, F. Capelli, F. Bolzoni, M. Ormellese, et al., A comprehensive investigation on the effects of surface finishing on the resistance of stainless steel to localized corrosion, Metals. 12 (2022) 1751. https://doi.org/10.3390/met12101751

[70] R.N. Asma, P. Yuli, C. Mokhtar, Study on the effect of surface finish on corrosion of carbon steel in CO2 environment, Journal of Applied Sciences. 11 (2011) 2053-2057. https://doi.org/10.3923/jas.2011.2053.2057

[71] M. Hussain, T. Zhang, M. Chaudhry, I. Jamil, S. Kausar, I. Hussain, Review of prediction of stress corrosion cracking in gas pipelines using machine learning, Machines. 12 (2024) 42. https://doi.org/10.3390/machines12010042

[72] B. Sun, D. Wang, X. Lu, D. Wan, D. Ponge, X. Zhang, Current challenges and opportunities toward understanding hydrogen embrittlement mechanisms in advanced high-strength steels: a review, Acta Metallurgica Sinica (English Letters). 34 (2021) 741-754. https://doi.org/10.1007/s40195-021-01233-1

[73] W. Liu, L. Sun, Z. Li, M. Fujii, Y. Geng, L. Dong, et al., Trends and future challenges in hydrogen production and storage research, Environmental science and pollution research. 27 (2020) 31092-31104. https://doi.org/10.1007/s11356-020-09470-0

[74] K. Reza Kashyzadeh, N. Amiri, E. Maleki, O. Unal, A Critical Review on Improving the Fatigue Life and Corrosion Properties of Magnesium Alloys via the Technique of Adding Different Elements, Journal of Marine Science and Engineering. 11 (2023) 527. https://doi.org/10.3390/jmse11030527

[75] H.M. Irshad, I.U. Toor, H.M. Badr, M.A. Samad, Evaluating the Flow Accelerated Corrosion and Erosion-Corrosion Behavior of a Pipeline Grade Carbon Steel (AISI 1030) for Sustainable Operations, Sustainability. 14 (2022) 4819. https://doi.org/10.3390/su14084819

[76] S. Dehgahi, A. Shahriari, A. Odeshi, M. Mohammadi, Influence of Ti Content on High Strain Rate Mechanical and Corrosion Behavior of Additively Manufactured Maraging Steels, Journal of Materials Engineering and Performance. 32 (2023) 1169-1184. https://doi.org/10.1007/s11665-022-07166-9

[77] M. Bassis, A. Kotliar, R. Koltiar, T. Ron, A. Leon, A. Shirizly, et al., The effect of a slow strain rate on the stress corrosion resistance of austenitic stainless steel produced by the wire laser additive manufacturing process, Metals. 11 (2021) 1930. https://doi.org/10.3390/met11121930

[78] C. Zhang, M. Ran, Y. Wang, W. Zheng, Microstructural effects in the development of near-neutral pH stress corrosion cracks in pipelines, Materials. 15 (2022) 4372. https://doi.org/10.3390/ma15134372

[79] U. Martin, J. Ress, J. Bosch, D. Bastidas, Stress corrosion cracking mechanism of AISI 316LN stainless steel rebars in chloride contaminated concrete pore solution using the slow strain rate technique, Electrochimica Acta. 335 (2020) 135565. https://doi.org/10.1016/j.electacta.2019.135565

[80] D. Sun, D. Wang, L. Li, K. Gong, S. Ren, F. Xie, et al., Study on stress corrosion behavior and mechanism of X70 pipeline steel with the combined action of sulfate-reducing bacteria and constant load, Corrosion Science. 213 (2023) 110968. https://doi.org/10.1016/j.corsci.2023.110968

[81] P. Liu, Q.-H. Zhang, Y. Watanabe, T. Shoji, F.-H. Cao, A critical review of the recent advances in inclusion-triggered localized corrosion in steel, npj Materials Degradation. 6 (2022) 81. https://doi.org/10.1038/s41529-022-00294-6

[82] E.M. Costa, B.A. Dedavid, C.A. Santos, N.F. Lopes, C. Fraccaro, T. Pagartanidis, et al., Crevice corrosion on stainless steels in oil and gas industry: A review of techniques for evaluation, critical environmental factors and dissolved oxygen, Engineering Failure Analysis. 144 (2023) 106955. https://doi.org/10.1016/j.engfailanal.2022.106955

[83] Y. Hu, H. Zhang, H. Zhang, M. Ouyang, Y. Hu, L. Wang, et al., Crevice Corrosion Behavior of Stainless Steels in a Flue Gas Desulfurization Environment, Journal of Materials Engineering and Performance. 32 (2023) 10567-10581. https://doi.org/10.1007/s11665-023-07888-4

[84] N. Larché, D. Thierry, C. Leballeur, E. Diler, editors. Crevice Corrosion of High-Grade Stainless Steels in Seawater: A Comparison Between Temperate and Tropical Locations. AMPP CORROSION; 2022: AMPP. https://doi.org/10.5006/C2022-17828

[85] B.J. Little, J.S. Lee, Microbiologically influenced corrosion: an update, International Materials Reviews. 59 (2014) 384-393. https://doi.org/10.1179/1743280414Y.0000000035

[86] H.-C. Flemming, J. Wingender, The biofilm matrix, Nature reviews microbiology. 8 (2010) 623-633. https://doi.org/10.1038/nrmicro2415

[87] Y. Ma, Y. Zhang, R. Zhang, F. Guan, B. Hou, J. Duan, Microbiologically influenced corrosion of marine steels within the interaction between steel and biofilms: a brief view, Applied microbiology and biotechnology. 104 (2020) 515-525. https://doi.org/10.1007/s00253-019-10184-8

Advances in Corrosion Science and Surface Engineering   Materials Research Forum LLC
Materials Research Foundations 188 (2026) 55-67   https://doi.org/10.21741/9781644903919-4

Chapter 4

Techniques in Corrosion Assessment and Control

Utchimahali Muthu Raja P[1], G. Kausalya Sasikumar[1], R.R. Shenthilkumar[1]*, C. Senthilkumar [2], Rajender Boddula[3]

[1]Centre for Research and Development, KPR Institute of Engineering and Technology, Coimbatore, India

[2]Department of Mechanical Engineering, RVS College of Engineering and Technology, Coimbatore-641402, India

[3]School of Sciences, Woxsen University, Hyderabad - 502345, Telangana, India

senthilkumar.r@kpriet.ac.in

Abstract

Corrosion is a natural process that deteriorates materials, which presents some serious problems to many businesses environmental, economical and safety. The following study considers in-depth corrosion testing and evaluation techniques, such as electrochemical methods, coating application, field monitoring, and salt spray testing. It covers economic costs, environmental effects, and environmental factors that contribute to corrosion along with advanced technologies, including AI-driven models and smart coatings. The document focuses on best practices and solutions for proactive management that prevent corrosion risks and uses case studies and innovative approaches. This analysis provides an all-inclusive framework for augmenting environmental stewardship, cost-effectiveness, and the durability of materials.

Keywords

Corrosion Testing, Electrochemical Methods, Coating Application, Smart Coatings, Sustainability

Contents

1.   Introduction

This process is referred to as corrosion and is defined as the slow degradation of materials, mostly metals, due to interaction with ambient elements including oxygen, moisture, contaminants, or biological agents. This is the naturally occurring cause of the thermodynamic instability of the majority of metals in their refined, high-energy forms, which causes these metals to return to their stable, natural forms that are mostly oxides, hydroxides, or other compounds. It results in this tendency to oxidize and degrade, leading to material deterioration, reduced mechanical strength, and very often total breakdown of structures or systems [1].

In simple words, corrosion is a kind of destructive process that occurs every time metals and alloys are exposed to their surroundings and gradually deteriorate, taking on myriad different forms depending upon the characteristics of the material, its surroundings, and the mechanisms at play. Although most people think of the obvious rusting of steel, the broad parameters of the corrosion process actually encompass so much more, such as pitting, cracking, and degradation of non-metallic materials, including ceramics and polymers [2].

Corrosion has many safety and economic implications. Injuries related to corrosion result in major losses in most sectors such as construction, transport, energy, oil and gas, and the marine sector. Research estimates that the annual worldwide cost of corrosion is between 3 and 4 percent of the world's gross domestic product (GDP). Besides the financial implications, corrosion puts the integrity and safety of infrastructure at risk as well as enhancing the potential for catastrophic failure and environmental damage [3].

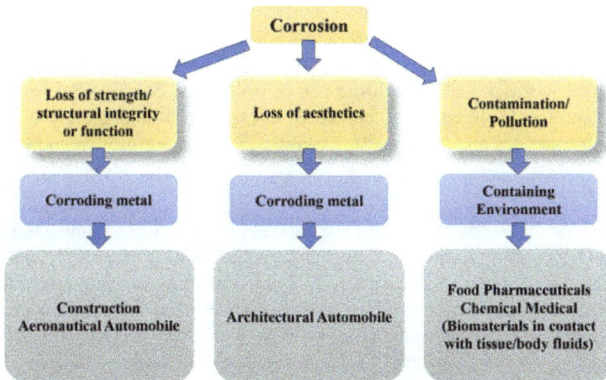

Figure 1. An illustration of corrosion's three main impacts and their applicability towards various industries.

Advances in Corrosion Science and Surface Engineering   Materials Research Forum LLC
Materials Research Foundations 188 (2026) 55-67   https://doi.org/10.21741/9781644903919-4

The three basic consequences of material corrosion and degradation are the loss of structural integrity or function, a decrease in or loss of aesthetic quality, and contamination or pollution from materials leaking out of the corroding material or from the contents that the corroding material has encased escaping into the environment (Figure 1). Though such a corrosion event may cause all of these effects simultaneously, some impacts will be greater than others depending on the application or industry. Structural integrity is the most important factor for both the building and aviation industries, whereas contamination is crucial for food and pharmaceutical sectors, as shown in Figure 1 [4].

1.1   Principles of Corrosion

The basics underlying corrosion are its electrochemical origin since an electrolyte enables reactions that occur within a metal's surface anodic and cathodic regions, further destroying it. At the anodic region, the metal undergoes oxidation, losing electrons: $M \rightarrow M^{n+} + ne^-$. Such electrons consumed at the cathodic region are used in reduction, such as reduction of oxygen: $O_2 + 2H_2O + 4e^- \rightarrow 4OH^-$ or hydrogen ion reduction: $2H^+ + 2e^- \rightarrow H_2$ [5]. The substance that will enable the flow of ions is the electrolyte; it could be water, acids, or industrial fluids. This completes the circuit. Unchecked corrosion might cause the material to deteriorate and fail structurally. Consequently, financial losses will be incurred. It is caused by the interplay between anodic dissolution and cathodic reduction [6].

1.2   Types of Corrosion

An understanding of the various types of corrosion is needed for the successful prevention and mitigation of techniques. Every type of corrosion affects materials in a different manner and forms under specific conditions. The following explains the main types of corrosion:

1.2.1   Uniform Corrosion

Uniform corrosion is usually the most predicted and expected corrosion form, showing uniform material loss over the surface. This generally occurs because every part of the material will react uniformly to the environment in which it operates. Such type of corrosion arises because of exposure of a material to air, moisture, or some other corrosive chemicals [4]. One common example of uniform corrosion is the rusting of steel in air environments. Even though it progressively degrades the material, with the use of protective coatings, corrosion inhibitors, or appropriate maintenance, uniform corrosion can be controlled. It is relatively easy to predict and prepare for its consequences and implement necessary countermeasures by industries.

1.2.2   Galvanic Corrosion

Galvanic corrosion occurs when two dissimilar metals are electrically connected in the presence of an electrolyte[7]. The more noble metal acts as a cathode and is protected, while the less noble metal acts as an anode and corrodes preferentially. For instance, steel can corrode faster when steel fasteners contact aluminium structures in a damp climate [8]. The primary type of corrosion associated with mixed-metal assemblies is galvanic, which often takes place in damp or marine environments. Such corrosion can be significantly reduced through protective coatings, proper selection of suitable metals that possess electrochemical potential close to that of each other, or separation of the two metals with suitable insulating material [9].

Advances in Corrosion Science and Surface Engineering   Materials Research Forum LLC
Materials Research Foundations 188 (2026) 55-67   https://doi.org/10.21741/9781644903919-4

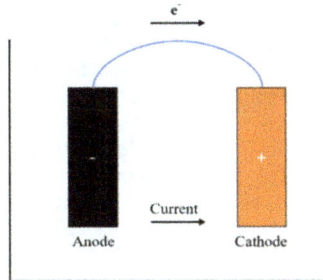

Figure 2. Simple cell for galvanic corrosion.

1.2.3 Pitting Corrosion

Pitting corrosion is the sort of localized corrosion that creates small holes or pits on the surface of material. In many cases, this type of corrosion commences with the infiltration of aggressive ions, such as chlorides, into the protective oxide layer of a material, like stainless steel. Overall material loss is marginal, but the pits can grow deeper and larger with time, causing such a material to fracture rapidly. Stainless steel, for instance, is notoriously susceptible to the pitting corrosion by environments filled with chlorides, such as seawater. Pitting is hard to trace inasmuch as it is encapsulated and many times hidden; prevention involves managing the environment with protective coatings serviced and the materials chosen resistant [10].

1.2.4 Crevice Corrosion

Crevice corrosion occurs in low oxygen environments that are enclosed or shielded, such as beneath gaskets, washers, or fasteners. Here, degradation is accelerated by a localized corrosive environment created by a change in oxygen content. An example of this is the rust that forms under a washer or bolt head in a marine environment [11]. Crevice corrosion often goes unnoticed until serious harm has been done. Mitigation techniques include the use of sealants, proper design to reduce crevices, and the use of materials such as stainless steel that are resistant to corrosion in crevices [12].

1.2.5 Intergranular Corrosion

Intergranular corrosion attacks the grain boundary of an alloy; this type of attack is often induced by improper heat treatment or the presence of contaminants. For example, in stainless steel that has been underheated heat treatment, it may precipitate chromium carbide at grain boundaries, lowering its corrosion resistance. This phenomenon, known as sensitization, makes the material vulnerable to intergranular corrosion when exposed to corrosive environments [13]. Without obviously changing the surface of a component, intergranular corrosion can very seriously compromise the integrity of its structure. It is prevented using stable or low-carbon grades of stainless steel and proper heat treatments and production procedures [14].

Advances in Corrosion Science and Surface Engineering   Materials Research Forum LLC
Materials Research Foundations 188 (2026) 55-67   https://doi.org/10.21741/9781644903919-4

1.2.6   Stress Corrosion Cracking (SCC)

The combination of tensile stress and corrosive environment is responsible for stress corrosion cracking (SCC), a failure mechanism. The cracks formed by this type of corrosion propagate over time and may lead to sudden, catastrophic failure. SCC often occurs in tanks, pipelines, and other structures that are exposed to applied or residual stresses, especially when there are aggressive chemicals such as chlorides. For example, SCC may lead to breaks in pipelines in aggressive soils. Reduction of tensile stress by proper design, by relieving stresses through certain processes, or by the use of corrosion-resistant metals is also part of the mitigation process [15]. In addition, the risk may be minimized by a change in environment, for instance by removing aggressive ions.

1.2.7   Erosion Corrosion

Material loss results from erosion corrosion, which happens when a fluid passes over a material's surface quickly removing oxide coatings or protective films. Such corrosion often takes place in systems such as pump impellers, pipe bends, and turbine blades due to the effect of high-velocity fluids that come into contact with material surfaces. For example, high velocity seawater exposes pump impellers in desalination plants to frequent erosion and corrosion. This type of corrosion is highly aggressive because of the combination of the chemical attack and mechanical wear. Wear-resistant materials, lower fluid velocity, and coatings that enhance surface resistance are just some of the preventive methods [16].

1.2.8   Microbiologically Influenced Corrosion (MIC)

MIC refers to the activity of microorganisms such as bacteria, algae, or fungi in saturated or waterlogged conditions. These microorganisms produce various compounds, which accelerate corrosion, with some examples being sulfurizing agents or acids. For example, hydrogen sulfide results from SRB, which reacts with metals to cause rapid failure in storage tanks and pipelines. Otherwise, corrosion-resistant systems can have unanticipated breakdowns due to MIC. The effective ways to mitigate MIC are biocides, protective coatings, and routine cleaning. Another practical way of lowering the danger is through proper design that minimizes the buildup of biofilm and stagnant water [17].

2.   Laboratory testing methods

Controlled laboratory testing is a must that will determine materials and coatings and their durability coupled with corrosion resistances under simulated conditions. That will help ascertain weaknesses and determine long-term exposure. It tends to mimic various real-life settings, usually at increased rates. The major techniques are basically divided into electrochemical advanced ones such as those including potentiodynamic polarization electrochemical impedance spectroscopy (EIS), LPR, whereas the physical testing of exposure relates to salt spray testing [18]. These methods help find the corrosion rate, understand why corrosion occurs, and assess how well inhibitors or protective coatings will work. Various types of testing in a laboratory can be applied by any industry for construction, automobile, aerospace, or marine application. They yield extremely useful data in failure analysis, quality control, and design of products. The tests ensure that the materials will pass through the various environmental conditions of temperatures, humidities, and exposure to chlorides for diverse applications.

Advances in Corrosion Science and Surface Engineering
Materials Research Forum LLC
Materials Research Foundations 188 (2026) 55-67
https://doi.org/10.21741/9781644903919-4

2.1   Salt Spray Testing (ASTM B117)

A very common accelerated corrosion test is the salt spray testing. This simulates chloride-rich environments, such as those found in ships, to evaluate the endurance of materials. Samples are subjected to a fine mist of a 5% sodium chloride solution in a controlled chamber at 35°C (±2°C). This technique tests coatings and materials for pitting, blistering, or rust over a given period of time, providing information on how well they function in corrosive environments [19]. Though it is easy to use and relatively inexpensive, its utility to chloride-induced corrosion scenarios is limited by its inability to reproduce complex real-world conditions. It is extensively used in the automobile, aerospace, marine, and construction industries.

2.2   Potentiodynamic Polarization Curves

This electrochemical method produces a polarization curve where it measures the current output based on an alternating voltage to be applied through the metal being held in a fluid electrolyte. The information contained in this curve is about crucial details with regards to how this material could potentially corrode - the potential (Ecorr) and corrosion current density (Icorr). Using these factors allows for calculating corrosion rates [20]. It is excellently used in examining inhibitors, pitting resistance, and the passivation behavior of materials in various conditions.

2.3   Electrochemical Impedance Spectroscopy (EIS):

EIS applies a small AC signal over a range of frequencies to measure the impedance of a material. This provides ample information about electrochemical reactions happening on the material surface, like diffusion, resistance in coatings, and resistance through pores. The application of this technique is most important while measuring the rates of corrosion, the performance over long time spans by protective coatings, and the ways materials under corrosive exposure break down [21].

2.4   Linear Polarization Resistance (LPR):

EIS applies a very small alternating current (AC) signal across a range of frequencies to determine a material's impedance. It provides detailed information regarding the electrochemical reactions that are occurring at the surface of the material, such as diffusion of ions, coating resistance, and pore resistance. The method is especially useful in monitoring the rate of corrosion, evaluating the long-term protection that coatings may offer, and understanding how materials that are exposed to corrosive environments deteriorate over time [19].

3.   Application techniques and Coating thickness

3.1   Importance of coating in corrosion prevention

Coatings, by acting as physical barriers, reduce the possibility of corrosion by preventing contact between the underlying metal and environmental factors such as oxygen, moisture, and contaminants. In infrastructure, automotive, and aerospace applications, successful coatings increase material life, extend service life, and lower maintenance costs.

Advances in Corrosion Science and Surface Engineering   Materials Research Forum LLC
Materials Research Foundations 188 (2026) 55-67   https://doi.org/10.21741/9781644903919-4

3.2 Methods of Measuring Coating Thickness

3.2.1 Magneto Gages

These are portable tools operating on magnetic field strength. This tool measures thickness of non-magnetic coatings, which may adhere to a ferromagnetic substrate. These gauges are portable and very useful for field evaluations.

3.2.2 Ultrasonic Thickness Gauges

This is a good method for non-destructive testing on thicker coatings because ultrasonic waves can measure the thickness of the coating on a substrate without damaging it.

3.2.3 X-ray fluorescence

XRF, is the technique that applies X-rays for the measurement of coating thickness and material composition in high accuracy. It is particularly suitable for multi-layer coatings as well as for quality control purposes in an industrial environment.

3.3 Application Techniques

3.3.1 Spray Coating

This is a method with consistent application suited for intricate surfaces. The coating is applied with a spray cannon. However, overspray and uneven coverage are likely to occur if the right tools and expertise are not used.

3.3.2 Dip Coating

Materials get submerged in the coating solution; then they are drawn out. Since it is fairly easy to work with and relatively affordable, it is a good way to coat even though it may show some disadvantages-such as non-uniform thickness on small geometries.

3.3.3 Powder Coating

This is a dry finishing method that applies powdered material using electrostatics and uses heat to cure it. Since it does not employ solvents, it provides greater durability, chip resistance, and environmental benefit.

3.3.4 Electroplating

This process utilizes an electrolytic solution and an electric current to deposit a layer of metal on a substrate. Although it is cumbersome and demands proper handling of chemicals and wastes, it increases corrosion resistance, wear resistance, and aesthetic appeal.

3.4 Monitoring and Inspection in the Field

3.4.1 Visual Inspection

Visual Examination A basic inspection may be made of the signs and symptoms of corrosion, like rust, pitting, cracking or discoloration, using direct visual means via instruments, magnifying lenses, mirrors, or borescopes. Regular checks help in identifying corrosion early in the process, and the knowledge gained helps perform maintenance on time.

3.4.2 Non-Destructive Testing (NDT)

This method ensures the structural integrity with no damage of the material used. Eddy current techniques recognize surface and near-surface corrosion through the alteration of electromagnetic fields, radiography testing reveals the presence of faults below the surface through X-rays, and ultrasonic testing determines the internal weakness by using the sound waves.

3.4.3 Real-Time Monitoring

Sensors of electrical resistance and electrochemical probes are used together with corrosion coupons exposed to surrounding environments for periodic inspection to monitor corrosion in real-time. Data generated through the application of such methods help determine corrosion rates; hence, wise decisions can be made regarding potential actions to mitigate corrosion.

3.5 Factors Affecting Environment Corrosion

3.5.1 Categories of Environment: Environmental and Industrial

Natural as well as industrial settings significantly increase the rates of corrosion. Since industrial settings normally expose materials at elevated temperatures plus chemical attack, degradation in them often accelerates faster. Similarly, in extreme natural settings, where there are sand storms or beach environments, chemicals like sulphur dioxide or chlorides cause accelerated material damage.

3.5.2 Role of Temperature, Humidity, and Pollutants

Although the contribution of pollutants, such as sulphates and chlorides, promotes material degradation, higher temperatures and humidity accelerate the corrosion process by allowing electrochemical activity. These parameters vary between industries and different geographical locations.

3.5.3 Influences of the Material-Specific Environment

Some materials corrode fast in saline or moist environments; on the other hand, others are more resistant under particular conditions because of passive oxide coatings, like in the case of stainless steel. So, selection based on environmentally friendly materials is significant for durability.

3.5.4 Examples of How the Environment Affects Corrosion

Examples that highlight the need to consider the environment in the selection of the material are the effects of industrial pollutants on aluminium structures and the rapid corrosion of steel in coastal areas due to chloride ions.

Advances in Corrosion Science and Surface Engineering   Materials Research Forum LLC
Materials Research Foundations 188 (2026) 55-67   https://doi.org/10.21741/9781644903919-4

3.6   Cost Implications for the Economy and the Environment

3.6.1   Global Cost Analysis

Corrosion is associated with both direct and indirect costs. Direct costs involve replacement and repair, while indirect costs include lost productivity and downtime. It represents 3-4% of the world's GDP and is a significant influence on industries like infrastructure, transportation, and oil and gas.

3.6.2   Sector-specific examples

Pipeline corrosion means costly pipeline repair and even probable environmental disaster in the oil and gas sector. Corrosion-related problems in bridges signify expensive maintenance work and safety hazard in infrastructure sectors.

3.6.3   Effects on the Environment

Corrosion failure releases dangerous materials into the soil, water, and eco-systems. Rusting pipelines may contaminate sources of fresh water, while leaching metals into the water can harm aquatic life.

3.6.4   Environmental standards, regulatory compliance

Regulations enforce a strict limit on material selection, waste management, and use of ecological coatings to reduce the influence on the environment. Examples include ISO and NACE environmental standards.

3.6.5   Prevention of corrosion through sustainability

The concept of sustainable corrosion prevention aims at minimizing resource utilization and environmental degradation using recyclable materials, eco-friendly coatings, and advanced technologies.

3.7   Assurance of Quality and Inspection

3.7.1   Value of Quality Control in Preventing Corrosion

Quality assurance reduces the likelihood of premature failure because it ensures that applied coatings and materials meet performance requirements. Reliability and longevity are assured through regular inspections by conformity to international standards [22].

3.7.2   Important international standards

This helps protect against corrosion using standards on the inspection and maintenance procedure; those include standards set by ISO 12944 in terms of the coating system to protect from corrosion, ASTM G48 regarding testing the pitting or crevice, and standards provided by NACE.

3.7.3   Routine maintenance operations

They look for indications of corrosion; thus, one routine maintenance entails a surface cleanup reapplication of coats, keeping up with conditions as far as environmental concerns go.

Advances in Corrosion Science and Surface Engineering   Materials Research Forum LLC
Materials Research Foundations 188 (2026) 55-67   https://doi.org/10.21741/9781644903919-4

3.8   New Developments in Corrosion Testing Technologies

3.8.1   Smart Coatings and Self-Healing Materials

Self-healing materials repair microcracks autonomously to lengthen their lifespan and reduce maintenance needs, whereas smart coatings detect early corrosion indicators and spill inhibitors once environmental changes are detected [23].

3.8.2   Cutting-edge sensor technologies include

Cutting-edge sensors such as wireless corrosion probes and nano-sensors that allow for highly precise real-time corrosion monitoring which aides predictive maintenance and timely interventions.

3.8.3   Prediction of Corrosion with the Help of AI and Machine Learning

It analyses large datasets from sensors and lab testing using AI and ML for forecasting corrosion trends, improving material selection procedures, and optimizing maintenance schedules-all of which revolutionize corrosion control tactics.

Conclusion

Corrosion affects the environment, infrastructure, and industry and still remains a ubiquitous and costly phenomenon. The effective testing, monitoring, and mitigation techniques of the industries can be understood with knowledge of the corrosion fundamentals, mechanisms, and types. Important understanding about corrosion behavior is given through advanced approaches involving real-time monitoring and AI-based predictions and in laboratory procedures through salt spray tests, potentiodynamic polarization, and EIS. Advanced sensors, self-healing materials, and intelligent coatings are few of the innovations that highly improved corrosion avoidance with increased material longevity and lessened negative effects on both the environment and economy. A good fight against corrosion requires proactive management and sustainability. Resource consumption as well as damage to the environment will be reduced in the use of predictive maintenance techniques, international standards, and an eco-friendly coating. Industry will reduce corrosion hazards, increase the lifespan of material, and also contribute to a safer and sustainable future by bringing together technological breakthroughs, quality assurance, and international standards.

References

[1]L. Veleva, Atmospheric Corrosion, 2003.
https://www.researchgate.net/publication/261176470

[2]Ferit Artkin, Platanus_Dr.FeritARTKIN_Galvanized_Coating, (2023).
https://doi.org/10.5281/zenodo.10060775

[3]L. De Arriba-Rodríguez, F. Ortega-Fernández, J.M. Villanueva-Balsera, V. Rodríguez-Montequín, Corrosion Predictive Model in Hot-Dip Galvanized Steel Buried in Soil, Complexity 2021 (2021). https://doi.org/10.1155/2021/9275779

[4] S.U. Ofoegbu, Comparative gravimetric studies on carbon steel corrosion in selected fruit juices and acidic chloride media (Hcl) at different ph, Materials 14 (2021). https://doi.org/10.3390/ma14164755

[5] Madhu Dowlapalli, Plameen Atanassov, Electrochemical Oxidation Resistance of Carbonaceous Materials, (2005). https://doi.org/10.13140/2.1.2720.3841

[6] N.F.E. Boraei, M.A.M. Ibrahim, S.S.A. El Rehim, I.H. Elshamy, Electrochemical corrosion behavior of β-Ti alloy in a physiological saline solution and the impact of H2O2 and albumin, Journal of Solid State Electrochemistry 28 (2024) 2243–2256. https://doi.org/10.1007/s10008-023-05751-z

[7] K.C. Manu, C. Madhushree, M.S. Chandini, N. Shree, S. Hemanth, T.P. Jeevan, Corrosion in Steel Structures: A Review, Journal of Mines, Metals and Fuels (2025) 189–198. https://doi.org/10.18311/jmmf/2025/46985

[8] M. Hanif, F. Abida, Comparison of Corrosion of metal connected to Zn anode for 1%, 3% and 5% NaCl solution SEE PROFILE, 2019. https://www.researchgate.net/publication/355737148

[9] D. Landolt, S. Mischler, M. Stemp, Electrochemical methods in tribocorrosion: A critical appraisal, Electrochim Acta 46 (2001) 3913–3929. https://doi.org/10.1016/S0013-4686(01)00679-X

[10] X. Li, Y. Zhao, W. Qi, J. Wang, J. Xie, H. Wang, L. Chang, B. Liu, G. Zeng, Q. Gao, H. Sun, T. Zhang, F. Wang, Modeling of Pitting Corrosion Damage Based on Electrochemical and Statistical Methods, J Electrochem Soc 166 (2019) C539–C549. https://doi.org/10.1149/2.0401915jes

[11] A.G. Adeniyi, O.O. Ogunleye, M.O. Durowoju, O. Eletta, O. Olaosebikan Ogunleye, O. Abosede, A. Eletta, Modelling of type 304 stainless steel crevice corrosion propagation in chloride environments, 2018. https://www.researchgate.net/publication/337678876

[12] Z. Zhang, Z. Li, F. Wu, J. Xia, K. Huang, B. Zhang, J. Wu, A comparison study of crevice corrosion on typical stainless steels under biofouling and artificial configurations, Npj Mater Degrad 6 (2022). https://doi.org/10.1038/s41529-022-00301-w

[13] Z. Ławrynowicz, Effect of The Degree of Cold Work and Sensitization Time on Intergranular Corrosion Behavior in Austenitic Stainless Steel, Advances in Materials Science 19 (2019) 32–43. https://doi.org/10.2478/adms-2019-0003

[14] E. Metin Tumer, T. Atıcı, M. Efe Tümer, N. Furkan Şahin, B. Akçalıoğlu, Ç. Çelik İmalat, M. ve Tesisat AŞ, EFFECTS OF BACKING GAS CHARACTERISTICS AND SENSITIZATION HEAT TREATMENT PARAMETERS ON INTERGRANULAR CORROSION BEHAVIOR OF 304L GRADE STAINLESS STEEL PIPE WELDING, 2024. https://www.researchgate.net/publication/384458019

[15] M. Hussain, Stress Corrosion Cracking issues in energy pipeline and Arab Countries, 2020. https://www.researchgate.net/publication/346523145

[16] S. Matthews, B. James, M. Hyland, High temperature erosion-oxidation of Cr3C2-NiCr thermal spray coatings under simulated turbine conditions, Corros Sci 70 (2013) 203–211. https://doi.org/10.1016/j.corsci.2013.01.030

[17]   M.A.A. Khan, M. Hussain, F. Djavanroodi, Microbiologically influenced corrosion in oil and gas industries: A review, International Journal of Corrosion and Scale Inhibition 10 (2021) 80–106. https://doi.org/10.17675/2305-6894-2021-10-1-5

[18]   M.G. Sohail, M. Salih, N. Al Nuaimi, R. Kahraman, Corrosion performance of mild steel and epoxy coated rebar in concrete under simulated harsh environment, International Journal of Building Pathology and Adaptation 37 (2019) 657–678. https://doi.org/10.1108/IJBPA-12-2018-0099

[19]   S. Al-Saadi, Thesis_Silane Coatings for Mitigation of Microbiologically Influenced Corrosion of Mild Steel, 1968. https://www.researchgate.net/publication/369830358

[20]   Z. Zhang, X. Zhong, X. Teng, Y. Huang, H. Han, T. Chen, Q. Zhang, X. Yang, Y. Gong, Effect of Annealing Temperature on Electrochemical Properties of Zr56Cu19Ni11Al9Nb5 in PBS Solution, Materials 16 (2023). https://doi.org/10.3390/ma16093389

[21]   P. Wang, D. Cai, Preparation of Graphene-Modified Anticorrosion Coating and Study on Its Corrosion Resistance Mechanism, International Journal of Photoenergy 2020 (2020). https://doi.org/10.1155/2020/8846644

[22]   E. Harrison, Machine Learning-Driven Testing Automation for Efficient Software Quality Assurance in Distributed Systems, (2024). https://doi.org/10.13140/RG.2.2.11467.81440

[23]   S. Sanyal, S.J. Park, R. Chelliah, S.J. Yeon, K. Barathikannan, S. Vijayalakshmi, Y.J. Jeong, M. Rubab, D.H. Oh, Emerging Trends in Smart Self-Healing Coatings: A Focus on Micro/Nanocontainer Technologies for Enhanced Corrosion Protection, Coatings 14 (2024). https://doi.org/10.3390/coatings14030324

Materials Research Forum LLC
https://doi.org/10.21741/9781644903919-5

Chapter 5

Corrosion Testing and Evaluation

M. Malekli[1], E. Famkar[1], H. Eivaz Mohammadloo[2]*

[1]Department of Polymer Engineering and Color Technology, Amirkabir University of Technology (Tehran Polytechnic), Tehran P.O. Box 15875-4413, Iran

[2]Iran Polymer and Petrochemical Institute, Tehran P.O. Box 14965-115, Iran

h.mohammadloo@ippi.ac.ir

Abstract

Coatings play a vital role in protecting and enhancing everyday materials and surfaces. They safeguard against corrosion, environmental damage, and wear, while also improving aesthetics and functionality. However, maintaining long-term coating performance is complex, requiring deep understanding of coating-substrate interactions and environmental factors. Different coating types, such as organic, inorganic, and metallic, offer unique properties for corrosion protection. Among them, Ni-P coatings stand out for their exceptional corrosion resistance, wear resistance, and hardness, making them crucial in sectors like automotive, aerospace, and electronics. To help with the evaluation of these coatings, this chapter will also cover the corrosion assessment techniques for Ni-P coatings.

Keywords

Ni-P, Coating, Corrosion, Testing, EIS

Contents

1.   Introduction

Nearly every object encountered, interacted with, and utilized in our daily lives undergoes a coating process. The applications of coatings and coating systems clearly and substantially impact various aspects of our lives and routines. At the same time, some operate subtly but permeate all facets of our activities [1, 2]. Coatings play a fundamental and essential role in fortifying a wide range of surfaces, augmenting their performance, and bolstering their overall effectiveness. These encompass diverse types, such as metal coatings, polymer films, and ceramic coatings. The term "coating" refers to the precise application of a thin layer of specialized substances onto a given surface, forming a continuous or discontinuous film after dyeing [2]. This meticulously applied layer is a vital protective shield, diligently safeguarding the underlying substrate against many environmental hazards and potential threats. By effectively acting as an impenetrable barrier, coatings offer formidable defense against a broad spectrum of corrosive agents, oxidation, moisture, UV radiation, chemicals, abrasion, and other detrimental factors. The paramount significance of coatings lies in their exceptional capacity to significantly extend the lifespan of materials while steadfastly preserving their innate structural integrity. Techniques such as spraying, dipping, brushing, and electroplating ensure adequate coverage and adhesion to the substrate, ensuring optimal application.

Coatings play a significant role in enhancing the aesthetics of surfaces, improving the overall performance of materials, and enabling the addition or enhancement of functional properties to the substrate. Moreover, they reinforce durability by establishing a protective layer and streamline maintenance and cleansing processes. Coatings can also integrate additional functionalities into substrates, including sensor properties. Coatings assume paramount importance during the modern era of technology due to their protective nature, aesthetic enhancements, performance improvements, durability reinforcements, ease of maintenance, and potential for integrating additional functionalities. Their applications span diverse industries, encompassing construction, automotive, electronics, healthcare, and military applications [2]. The significance of coatings derives from their capacity to augment the functionality, appearance, and longevity of various materials and surfaces [1].

Ensuring the continuous effectiveness of coatings in preventing corrosion and preserving material integrity is of utmost significance in coating areas, specifically across various industries, necessitating regular inspection and maintenance [3]. Corrosion, an inherent process characterized by the deterioration of materials through chemical reactions with the surrounding environment, poses substantial economic and safety risks and converts refined metals into stable chemical forms, like oxides, hydroxides, or sulfides [4]. Corrosion protection coatings play a vital role by acting as a protective barrier that stands between the material and corrosive elements, such as moisture, oxygen, chemicals, gases, and contaminants, effectively shielding the substrate from their detrimental impacts and ensuring good adhesion to its substrate [3, 5, 6]. Chemical or electrochemical reactions between a material and its surrounding environment give rise to its occurrence [7]. Applying coatings involves adding an additional material layer to the surface, which forms a barrier obstructing the penetration of corrosive substances. As a result, the longevity of the protected object is significantly prolonged. The significance of coatings in corrosion prevention becomes particularly evident in safeguarding several industries against the detrimental effects of corrosion, specifically within sectors such as construction, marine, automotive, and oil and gas [5].

Scientific estimates suggest that corrosion and its prevention incur substantial annual costs, forming a noteworthy portion of the gross national product in different countries [3, 7]. The financial consequences of corrosion are overwhelming, with estimations indicating that the global annual cost of corrosion reaches trillions of dollars. Beyond economic implications and technological setbacks, corrosion can cause structural problems, yielding dire repercussions for both humans and environment. Materials affected by corrosion release pollutants, leading to contamination of the environment and posing a threat to soil, air, and water. This pollution has detrimental effects on ecosystems and biodiversity, impacting aquatic life, vegetation, and wildlife. Additionally, corrosion-related leaks from pipelines or storage tanks can result in hazardous material spills, endangering human health and causing ecological harm [6]. Due to these substantial costs and potential hazards, it becomes imperative to implement effective measures and strategies to mitigate corrosion and its associated detrimental impacts.

Despite significant advancements in coating technologies, preserving long-term metal protection against threats and aggressive environments still poses challenges. The intricate nature of the coating-substrate system, coupled with multiple factors influencing the effectiveness and lifespan of corrosion resistant coatings, contributes to the limited availability of high-performance systems. Parameters such as the composition of the coating, type of substrate, pretreatment procedures, curing methods, coating thickness, adhesion properties, and external environmental conditions collectively impact the longevity and efficiency of corrosion resistant coatings. Assessing the overall performance and durability of a coating can be changed based on manipulating its chemical, mechanical, and physical characteristics through the selection of binder systems, pigmentation, solvents, and additives in their chemical formulas. A thorough comprehension of

Advances in Corrosion Science and Surface Engineering　　　　　Materials Research Forum LLC
Materials Research Foundations 188 (2026) 68-120　　　　　https://doi.org/10.21741/9781644903919-5

how components of coating interact with one another and the fundamental mechanisms responsible for anticorrosive coating failures is crucial for developing high-performance coating systems [7].

Corrosion protection coatings can be categorized into organic, inorganic, and metallic coatings, each with distinct properties and applications. The choice of coating system and coating methods depends on the desired characteristics [8]. Organic coatings, commonly used, are composed of resins, binders, pigments, and additives, forming a continuous film that acts as a physical barrier against corrosive agents. Inorganic coatings, such as ceramic coatings, offer high-temperature resistance and durability, providing excellent resistance to chemical attacks. Metallic coatings provide protection through sacrificial or barrier mechanisms, with sacrificial coatings sacrificing themselves to protect the substrate and barrier coatings, creating a physical barrier against corrosive agents. Using both organic and inorganic components, hybrid coatings combine their benefits. Every kind of coating has special properties and mechanisms for corrosion protection. These coatings play a crucial role in safeguarding against corrosion in various applications [6]. They also serve as viable alternatives to solvent-based counterparts.

Over the past few decades, both organic and inorganic coatings have found extensive application in safeguarding metals against corrosion. The field of coating technology continues to advance, with continuous research efforts focused on the development of innovative materials and techniques that offer improved durability, environmental sustainability, and superior performance in challenging environments. By utilizing coatings, metals/components can achieve enhanced efficiency, leading to more effective industrial processes, cost cutting, conservation of limited material resources, and reduced emissions of pollutants. Its development also allows for the incorporation of various functional properties into metal/component surfaces [9]. Remarkable strides have been made in the battle against corrosion within various realms of coatings technology. Therefore, addressing corrosion is of utmost importance not only for economic considerations but also for the mitigation of environmental degradation.

Ni-P (nickel-phosphorus) coatings stand out as one of the most important types within the coating group, particularly renowned for their significance in corrosion protection. These coatings are applied through electroless deposition and offer exceptional corrosion resistance, wear resistance, and hardness [10]. They are highly valued across various industries for their ability to provide uniform, defect-free layers, thereby enhancing the durability and lifespan of components. In the corrosion industry, Ni-P coatings play a pivotal role as a physical barrier against corrosive elements, surpassing sacrificial coatings like aluminum and zinc. This pivotal role is evident in critical sectors such as automotive, aerospace, electronics, oil and gas, and chemical processing, where Ni-P coatings safeguard vital components, ensuring longevity and optimal performance [10, 11].

In this chapter, we will delve into various aspects of Ni-P coatings. We will begin with an introduction to Ni-P coatings and their applications, emphasizing the importance of corrosion resistance. This will be followed by a definition of corrosion and an overview of corrosion assessment methods. We will then explore specific corrosion evaluation tests for Ni-P coatings, including destructive tests such as polarization tests, weight loss tests, salt spray tests, and non-destructive tests like electrochemical impedance spectroscopy and surface tests. Advanced corrosion tests will also be discussed. Finally, we will examine case studies, applications, future technology trends, and applications in this field.

2.   Ni-P coatings

In 1844, the nickel deposition process, a significant milestone in the history of materials science, was initiated through Wurtz's research [12, 13]. Over a century later, Brenner and Riddel [14] pioneered formulations and techniques for Nickel-phosphorous (Ni-P) deposition on carbon steel, a breakthrough that would revolutionize the industry. Since then, a series of significant research conducted by different scientists, including Duncan [15], Colaruotolo [16], Mainier et al. [17], Tallinn [18], Weil et al. [19], Mainier and Araújo [20], Delaunois et al. [21], Liu et al. [22], and Baudrand [23] and other famous scientists in several years (Fig.1) have indicated a significant rise in the adoption and advancement of Ni-P coatings across various industries since the 1980s, further cementing its historical importance [24].

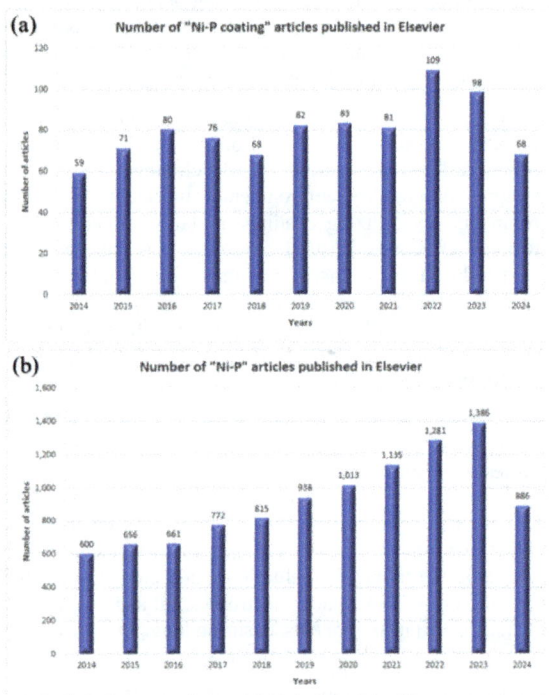

Figure 1. The Number of (a) "Ni-P coating" and (b) "Ni-P" articles published in recent years (data was extracted from the ScienceDirect website by searching the keywords "Ni-P coating" and "Ni-P")

The advancement of electroless nickel composite coatings stems from the concept of embedding diverse second-phase particles within the nickel deposits during the electroless plating process. This approach capitalizes on the uniformity, hardenability, wear resistance, and corrosion resistance inherently offered by these particles. Initial endeavors to integrate second-phase

Advances in Corrosion Science and Surface Engineering   Materials Research Forum LLC
Materials Research Foundations 188 (2026) 68-120   https://doi.org/10.21741/9781644903919-5

particles into the electroless nickel matrix commenced in the 1960s but encountered setbacks due to methodological similarities with traditional electroplating techniques, leading to bath decomposition. Nonetheless, electroless nickel composite coatings were successfully synthesized by introducing suitable stabilizers. An advantage of this method over electrodeposition lies in its ability to faithfully replicate the base geometry without requiring subsequent mechanical finishing [25].

Electroless nickel (EN) coatings find extensive applications across various industries, such as chemical, mechanical, and electronic industries, due to their inherent qualities like robust corrosion resistance, durability against wear, and consistent coating thickness [26]. Another advantageous characteristic is their capability to uniformly coat components of any size. Unlike electrodeposits, the properties of electroless nickel are largely influenced by its structure. However, given the substantial differences in structure and composition between electroless and electrodeposited nickel, their properties diverge as well [27].

Nickel-phosphorous (Ni-P) coatings, particularly when applied using electroless plating, stand out for their distinctive mechanical and chemical properties. Electroless nickel-phosphorus plating, also known as E-nickel, is a chemical method that uniformly deposits a layer of nickel-phosphorus alloy onto the surface of a solid material, such as metal or plastic, without the need for an electrical current. This method ensures an even distribution of the coating, even on intricate geometries and surfaces, making it highly adaptable for a range of industrial applications [27]. Crucially, these processes do not require an electric current to flow through the bath and the substrate, a unique characteristic distinguishing Ni-P coatings from other surface coatings [28].

Electroless nickel-phosphorus coatings are increasingly utilized as protective barriers against corrosion, wear resistance, and abrasion, along with a reduced coefficient of friction, forming a uniform layer adaptable to various product shapes [29, 30]. These coatings offer notable corrosion resistance, especially at high temperatures, and facilitate soldering and wetting processes for soldering materials. The conductivity and magnetic properties of nickel make it particularly valuable in the electronics industry. Incorporating phosphorus in the Ni-P structure results in a less porous and harder coat than electro-deposited metals. Heat treatment further enhances its hardness. Additionally, these coatings exhibit good wear resistance, low ductility, low porosity, excellent solderability, and traceability, all at a reduced cost and with improved corrosion resistance compared to magnesium substrates. Various methods, such as multi-layer coatings with higher phosphorus content, sol-gel deposition, or direct current electroplating following electroless deposition, enhance the performance of these coatings [29].

These coatings are categorized based on phosphorous content within the matrix, Ni-P coatings can be classified into three groups: low (LP, 3-5% of P), medium (MP, 6-9% of P), and high (HP, 10-14% of P). The metalloid composition dictates the coating's properties. The phosphorus content variation in electroless   Ni-P coatings affects the crystal structure, influencing properties such as corrosion resistance and wear. Widely employed for corrosion protection in harsh conditions, electroless Ni-P acts as a barrier shield, sealing substrates from corrosive elements rather than through sacrificial means. The corrosion resistance depends on the phosphorus content, which is higher in high-phosphorus deposits and lower in low-phosphorus ones. The quantity of phosphorus significantly impacts coating characteristics by modifying the metallic alloy's microstructure [11, 30, 31]. The as-deposited electroless plated Ni-P alloy comprises a blend of amorphous and microcrystalline nickel at low and medium phosphorus levels. Conversely, a fully amorphous

Advances in Corrosion Science and Surface Engineering   Materials Research Forum LLC
Materials Research Foundations 188 (2026) 68-120   https://doi.org/10.21741/9781644903919-5

microstructure can develop with high phosphorus content. Moreover, amorphous alloys tend to demonstrate inherently superior corrosion resistance compared to crystalline Ni due to their extreme uniformity, which eliminates defects and preferential corrosion paths like grain boundaries seen in crystalline materials. This enhanced corrosion resistance is also attributed to the absence of crystalline defects and their chemically homogeneous single-phase nature, facilitating the formation of a consistent and highly protective passive film [32].

The properties of Ni-P coatings are linked to the levels of nickel and phosphorus (as outlined in Table 1), and vary depending on their composition. During the critical inspection of the Ni-P deposition process, it's crucial to understand the various procedures involved, including preparation stages such as sandblasting and chemical cleaning, assessments within the process control laboratory, evaluation of electrochemical baths, and scrutiny of finishing processes to ensure quality and performance.

As previously mentioned, the protective ability of Ni-P coating against corrosion relies on the levels of nickel and phosphorus within the applied layer, with higher concentrations of phosphorus enhancing its resistance to corrosion. Chemical guidelines outlined in ISO 4527 (Table 2) [24] detail acceptable compositions for Ni-P coatings. Analytical methods such as wet chemical analysis, atomic absorption, x-ray fluorescence, and plasma techniques can determine the nickel and phosphorus content within the Ni-P deposit. In demanding industrial environments, the phosphorus concentration must exceed 10% by mass to ensure optimal performance under severe conditions.

Table 1. Nickel-phosphorus coatings' characteristics [24].

Properties	Phosphorus content (%)		
	Low	Medium	High
Nickel, %(mass)	96-99	92-95	88-91
Phosphorus, %(mass)	1-4	5-8	9-12
Vickers microhardness without heat treatment, HV	650 - 750	500 – 550	450 - 500
Vickers microhardness with heat treatment, HV	1000 - 1050	900 – 950	850 - 900
Melting point, °C	1200	890	870
Density. g/cm^3	8.5-8.7	8.1-8.3	7.7-7.8
Resistivity. μΩ/cm	50	70	90
Resistance to abrasion	Superior	Very good	Very good
Weldability	Good	Regular	Bad

Advances in Corrosion Science and Surface Engineering   Materials Research Forum LLC
Materials Research Foundations 188 (2026) 68-120   https://doi.org/10.21741/9781644903919-5

Table 2. Chemical composition of Ni-P deposit based on ISO 4527 Chemical guidelines [24].

Elements	Chemical composition (% mass)		
	Minimum	Maximum	Typical
Nickel	85	98	88-95
Phosphorus	2	15	5-12
Other (Al, As, B, Bi, C, Cd, Co, Cr, Cu, Fe, Mn, Nb, Pb, S, Sb, Se, Si, Sn, V, Zn)	0	2.0	0.05

The mechanism behind electroless Ni-P deposition has undergone extensive examination, yet certain aspects of the process remain enigmatic. Despite the elucidation of Ni deposition principles by scrutinizing the plating solution's constituents, such as the nickel source, reducing agent, and complexing agent/buffer solution, some mechanisms are still under debate. The autocatalytic nature of the ENP process, catalyzed by the substrate itself, underscores the need for chemical stabilization to prevent uncontrolled deposition and byproduct formation. Efforts to enhance ENP applicability and sustainability include developing lead-free stabilizers and tailoring the phosphorus content in coatings through adjustments in the plating solution's chemistry, fostering a deeper understanding of the interplay between reagents for improved deposition and coating properties [33].

Ni-P coatings are created by combining nickel ions with phosphorus-rich acids, leading to the formation of Ni-P compounds through an oxidation process [34]. The crystal structure of these coatings changes based on the phosphorus levels in the alloy, transitioning from crystalline to nanocrystalline and eventually becoming amorphous as phosphorus content increases. Higher phosphorus concentrations limit crystal growth by reducing nickel and co-depositing phosphorus, resulting in smaller grain sizes. Essentially, phosphorus acts as a grain refiner in this process. Using hypophosphite ions as the reducing agent in the Ni-P coating system, phosphorus co-deposition occurs alongside the nickel matrix, resulting in a coating characterized by its hardness, solderability, and corrosion resistance. Hence, Ni-P coatings have extensive applications in safeguarding engineering components against surface degradation in environments and harsh conditions with contaminants and deteriorating agents.

The hardness of these coatings varies depending on their phosphorus content and the heat treatments they undergo. Coatings containing a medium to high phosphorus concentration (typically 6-9 wt% phosphorus) are renowned for their remarkable hardness and corrosion resistance, positioning them as a premier choice for harsh environments. Additionally, Ni-P coatings demonstrate exceptional corrosion resistance, a critical attribute for safeguarding metal substrates in challenging conditions. An additional benefit of Ni-P electroless deposition is its ability to produce highly uniform coatings without requiring specific substrate geometries.

Electroless Ni-P composite coatings have gained popularity due to their notable enhancements in hardness and resistance to wear and abrasion. Some researchers suggest that the corrosion resistance of these composite coatings may be inferior to traditional electroless Ni-P coatings,

Advances in Corrosion Science and Surface Engineering   Materials Research Forum LLC
Materials Research Foundations 188 (2026) 68-120   https://doi.org/10.21741/9781644903919-5

attributing it to the presence of co-deposited second-phase particles that reduce passivity and corrosion resistance. However, studies by various researchers, including Hubbell, Hussain, Such, and Shoeib, have shown satisfactory corrosion performance in certain conditions [30]. Further investigation into the corrosion resistance of electroless Ni-P composite coatings is warranted, with evaluation techniques such as the neutral salt spray test, CASS test, and electrochemical impedance spectroscopy offering insights into their effectiveness in different environments.

3.   Ni-P coatings utilities

There has been an increasing trend over the past few decades in surface engineering towards increased utilization of coatings. Ni-P coatings, readily available commercially, offer a combination of strong wear resistance and adhesion to the substrate. These coatings maintain a consistent thickness and provide an alternative to hard chromium, mitigating health and environmental concerns. Their widespread adoption is driven by emerging opportunities to enhance their tribological properties through methods like thermal treatment and the integration of tough particles such as alumina [9].

Ni-P coatings hold significant value due to their outstanding coating properties, including durability, resistance to corrosion, favorable lubricity, satisfactory ductility, and high wear resistance, making them indispensable across diverse industries [11]. The excellent corrosion resistance makes Ni-P coatings particularly suitable for the automotive, aerospace, oil, and gas industries, where components often face corrosive agents.

Regarding oil and gas industries, grappling with high corrosiveness and the growing complexities of petroleum extraction in harsh environments, have been compelled to explore novel material solutions. Ni-P coatings have emerged as a notable choice due to their exceptional resistance to abrasion and corrosion, particularly in critical components such as valves, specialized tools, risers, pumps, and pipelines used in oil production and refining processes. This coating's efficacy is illustrated in various applications within the petroleum sector [24]. Furthermore, specialized variations of Ni-P coatings, such as Ni-P-PTFE, offer additional benefits like reduced friction and non-stick properties, expanding their potential uses and demonstrating their versatility in real-world applications [33].

Ni-P coatings' broad range of applications is also highlighted by their use in electronics and other precision fields where consistency and uniformity are key. Electroless nickel-phosphorous plating can produce coatings with consistent thickness and high durability, along with excellent resistance to wear and heat, making it a preferred choice for a variety of industrial applications [35]. The versatility of this coating has resulted in its application across diverse industrial sectors, including pulp and paper manufacturing, plastics production, petrochemicals, textiles, automotive, aerospace, electronics, and food processing. Furthermore, the oil industry, grappling with severe corrosive environments and the escalating difficulties in extracting petroleum under adverse conditions, has been compelled to explore novel material solutions. Ni-P coatings, renowned for their exceptional resistance to abrasion and corrosion, have demonstrated remarkable efficacy, particularly in applications involving valves, specialized tools, risers, pumps, and production pipelines [24].

Ni-P coatings are vital in electronics to improve durability and protection against wear and corrosion for certain metal parts and are increasingly utilized as protective barriers against corrosion and wear. Additionally, they have diverse applications, serving as barriers against diffusion, primers for gold-plated pieces, materials for creating thin film resistors, and coatings to facilitate hydrogen evolution reactions [34]. These coatings offer notable corrosion resistance, especially at high temperatures, and facilitate soldering and wetting processes for soldering materials. The conductivity and magnetic properties of nickel make it particularly valuable in the electronics.

4.   Corrosion

Corrosion, an electrochemical process, poses severe threats to material integrity and industries' financial stability. Metal protection involves anodic passivation or sacrificial anode defense mechanisms. Coatings offer further protection, shielding metal from corrosive environments. However, metal vulnerability persists in acidic or saline conditions, where surface degradation occurs rapidly, exposing the metal to corrosion risks. Damage to coated surfaces mainly stems from inadequate coating thickness or coating defects, facilitating corrosive media access to the substrate [36].

Corrosion, the deterioration of metals within specific environments, varies among metals due to factors like chemical composition and electrochemical reactions. The ability to endure harsh conditions determines a metal's corrosion resistance and operational longevity. Various definitions of corrosion exist, including one by the International Union of Pure and Applied Chemistry (IUPAC) describing it as an irreversible reaction between a material and its environment, often leading to detrimental effects on the material's utility. While corrosion is generally viewed as harmful, it also serves advantageous purposes. Degradation resulting from corrosion outweighs its benefits, although metals are predominantly chosen for their favorable properties [10, 36].

Understanding corrosion requires a grasp of thermodynamics and electrochemistry, with metallurgical factors significantly influencing corrosion resistance. Physical chemistry plays a crucial role in analyzing corrosion mechanisms and metal surface conditions. Classification of corrosion involves considering its mechanisms, such as physical, chemical, or electrochemical reactions based on their reaction mechanisms. The electromechanical reaction involving electron transfer, is particularly significant. Corrosive environments contain species adjacent to corroding metals, and materials prone to corrosion are termed corrodible. Factors affecting corrosion resistance include environmental conditions like oxygen levels, temperature, corrosive concentration, and metallurgical aspects. Preventing corrosion-induced failures is crucial due to their potential costliness. Engineers combat this by developing superior anticorrosion coatings, enhancing the lifespan and performance of metallic components while reducing resource consumption and emissions. These coatings, including advanced materials like composites and nanoparticles, find extensive use across various industries.

ASM (2000) identified three main factors for classifying corrosion: the nature of the corrosive agent, the corrosion mechanism, and the appearance of the corroded metal (Table 3). This classification system is used because it provides essential information about the failure mode linked to the corroded materials. Although the many types of corrosion are separate in principle, there are real-world situations when corrosion might fall into more than one category based on surface morphology. We can categorize the corroded metals into eight types of wet (or aqueous) corrosion: pitting, uniform or general corrosion, filiform, poultice, galvanic, erosion-corrosion (including cavitation erosion and fretting corrosion), intergranular (including sensitization and

Advances in Corrosion Science and Surface Engineering   Materials Research Forum LLC
Materials Research Foundations 188 (2026) 68-120   https://doi.org/10.21741/9781644903919-5

exfoliation), dealloying (including dezincification), and environmentally assisted cracking (including stress cracking corrosion, corrosion fatigue, and hydrogen damage).

Various coating types offer effective corrosion prevention, with electroless nickel-phosphorous (Ni-P) coatings renowned for their corrosion resistance. These coatings, applied without electricity, form a cost-effective barrier against corrosion and are widely adopted across industries. Incorporating different particles into Ni-P coatings significantly enhances their corrosion resistance. Despite advancements, research gaps exist in understanding the effects of coating parameters on corrosion resistance. Methods for corrosion rate measurement range from gravimetric determinations to electrochemical techniques like potentiodynamic polarization and impedance spectroscopy. While electrochemical methods are preferred for their relevance to corrosion origins, newer techniques like localized electrochemical impedance spectroscopy and scanning electrochemical microscopy provide insights into corrosion at high spatial resolutions [10].

Anti-corrosion coatings employ various mechanisms, primarily categorized into three: forming barriers between substrate materials and their environment, inhibiting corrosion processes, and serving as sacrificial layers. A recent advancement known as "active-passive" involves the coating functioning as barrier layers, preventing the penetration of corrosive agents to the metal surface passively. Meanwhile, the active aspect facilitates the creation of an effective passive layer, hindering corrosion half reactions and leading to the formation of a Schottky barrier at the interface, resulting in electron depletion.

In laboratory settings, sodium chloride solution serves as a prevalent electrolyte for corrosion testing. Comparative studies between various solutions, such as 3.5 wt% NaCl and 2.67 wt% H_2SO_4, indicate different corrosion resistances of electroless Ni-P and Ni-P-SiC-PTFE composite coatings. Varied results in different solutions underscore the significance of electrolyte choice, impacting corrosion rate and resistance. Distinct electrolytes yield diverse corrosion behaviors, and corrosion rates vary significantly across solutions. Electrochemical analysis reveals notable differences in charge transfer resistance between acidic and saline environments, with the latter exhibiting superior corrosion resistance. Different findings elucidate the complex interplay between electrolyte composition and corrosion behavior, informing corrosion mitigation strategies tailored to specific environmental conditions.

Corrosion imposes significant burdens on society through its exorbitant expense, excessive depletion of natural resources, and disruptions to daily life, often resulting in fatalities. It is the primary consumer of metals globally and presents one of the most enduring challenges to industrial development. The financial impact of corrosion is staggering, with costs reaching billions of dollars in developed nations alone, though precise national calculations remain elusive. Though harder to quantify, indirect costs likely surpass direct economic losses, estimated to be around $300 billion in the United States and billions more in developing countries like Nigeria and India. Furthermore, corrosion exacerbates resource depletion, intensifying economic competition for metal resources amidst rapid industrialization, thereby driving up prices. While corrosion can have beneficial applications, its detrimental effects, including aesthetic degradation, increased maintenance costs, and safety concerns, underscore the urgency of reducing its economic and environmental toll through preventive measures and innovative solutions.

Advances in Corrosion Science and Surface Engineering   Materials Research Forum LLC
Materials Research Foundations 188 (2026) 68-120   https://doi.org/10.21741/9781644903919-5

Table 3. Types of corrosion classified by ASM [37].

General Corrosion	Localized Corrosion	Metallurgical Influenced Corrosion	Mechanically Assisted Degradation	Environmentally Induced Cracking
Corrosive attack dominated by uniform thinning	High rates of metal penetration at specific sites	Affected by alloy chemistry and heat treatment	Corrosion with a mechanical component	Cracking produced by corrosion, in the presence of stress
Atmospheric corrosion	Crevice corrosion	Intergranular corrosion	Erosion corrosion	Stress-Corrosion cracking
Galvanic corrosion	Filiform corrosion	Dealloying corrosion	Fretting corrosion	Hydrogen Damage
General biological corrosion	Localized biological corrosion		Corrosion fatigue	Solid metal induced embrittlement
Molten salt corrosion				
Corrosion in liquid metals				

The significance of corrosion control cannot be overstated, considering the substantial financial resources allocated to it annually. Various methods are employed to safeguard materials from corrosion. Some strategies entail careful material selection and system design to eliminate opportunities for specific types of corrosion. Additionally, corrosion prevention or suppression often involves altering the environment, enhancing material properties, applying protective coatings, and implementing cathodic and anodic protection measures. The decision on whether to avoid, suppress, or simply manage corrosion is primarily dictated by economic factors. Coating, in particular, plays a crucial role as it encompasses various corrosion control and prevention

Advances in Corrosion Science and Surface Engineering   Materials Research Forum LLC
Materials Research Foundations 188 (2026) 68-120   https://doi.org/10.21741/9781644903919-5

methods, including sacrificial cathodic protection, environmental modification through barriers and inhibitors, and the development of corrosion-resistant materials like conductive polymers [10].

5. Corrosion assessment methods

There are two categories of corrosion assessment techniques: destructive techniques and non-destructive techniques. By directly measuring the material's electrochemical characteristics, destructive corrosion evaluation techniques frequently need the removal or alteration of the sample. The kinetics and mechanisms of the corrosion process are explained in depth by these techniques. Non-destructive electrochemical corrosion testing techniques enable the material to be evaluated without suffering irreversible harm. These methods are especially helpful for long-term surveillance and in-situ monitoring of structures and components. Although there are many ways to study corrosion, we will just cover a few of them in this chapter.

5.1 Destructive techniques

Potentiodynamic polarization is an electrochemical method that determines a material's potential-current response in a corrosive electrolyte and yields data on passivation behavior, corrosion potential, rate, and susceptibility to localized corrosion. A straightforward technique called weight loss can be used to directly calculate the average corrosion rate by measuring the mass loss of a specimen or coupon that has been pre-weighed and exposed to a corrosive environment. The comparative corrosion resistance of materials and coatings can be evaluated using the salt spray (fog) test, an accelerated corrosion test that exposes samples to a fine mist of salt solution in a controlled chamber. The hydrogen evolution test measures the amount of hydrogen gas evolved from a sample subjected to a corrosive environment in order to determine the materials' sensitivity to hydrogen-induced cracking or embrittlement. The integrity of the structure or component being assessed may be jeopardized by these destructive corrosion evaluation procedures, which offer in-depth insights into the corrosion behavior of materials but frequently require removing or altering the test sample.

5.2 Non-destructive methods

Electrochemical impedance spectroscopy (EIS) is a non-destructive technique that evaluates a material's impedance response throughout a frequency range. It offers details on the processes of diffusion, surface film characteristics, and corrosion mechanisms. A version of EIS called localized electrochemical impedance spectroscopy (LEIS) measures the impedance response at particular locations on the material's surface using a tiny, localized probe. Evaluation of isolated or heterogeneous corrosion events is made possible by this. By using a vibrating electrode to map the distribution of corrosion currents on a material's surface, the scanning vibrating electrode method (SVET) makes it possible to identify and see specific corrosion spots. Using ion-selective electrodes, the scanning ion-selective electrode technique (SIET) measures the concentration of particular ions, like hydrogen or chloride, at the material's surface to provide information about the local corrosion environment. The technique known as scanning electrochemical microscopy (SECM) allows the identification and mapping of corrosion processes by measuring the local electrochemical activity at the material's surface using a tiny, electrochemically active probe. With the use of a scanning kelvin probe (SKP), one can determine the potential for corrosion and the development of surface films by measuring the work function, or contact potential difference, between the surface of a material and the probe. A non-destructive method of visualizing and

Advances in Corrosion Science and Surface Engineering   Materials Research Forum LLC
Materials Research Foundations 188 (2026) 68-120   https://doi.org/10.21741/9781644903919-5

analyzing a material's surface topography, including the identification of alterations brought on by corrosion, is confocal laser scanning microscopy (CLSM). Early identification and monitoring of localized corrosion are made possible by electrochemical noise analysis (ENA), a non-destructive approach that tracks random variations in the electrical potential and current between two or more electrodes in a corrosive environment. These non-destructive corrosion assessment techniques are especially helpful for evaluating important structures and components since they allow for long-term monitoring and in-situ material evaluation without resulting in irreversible damage.

6. Corrosion assessment methods for Ni-P coatings

6.1 Destructive methods

6.1.1 Potentiodynamic polarization

Potentiostatic and potentiodynamic polarization observations make up the two sorts of the polarization investigation. It is possible to ascertain the material performance and its chemical behavior using either of these analyses. The evaluation's analysis can yield a significant amount of experimental data regarding the processes, rate, and performance of material degradation. Potentiodynamic polarization analysis, when combined with the other two methods, offers precise and comprehensive details regarding the rate and potential of corrosion of a particular material.

In a conventional potentiodynamic polarization analysis, the three electrodes in the system layout are a sample serving as the working electrode, a counter electrode, and a reference electrode. Using the potentiodynamic sweeping approach, which sweeps in a range of potentials from the passive to the active region on the potentiodynamic curve, the electrode performance in a potentiodynamic polarization test is determined by the largest region that can be recognized. The passivation potential, open circuit, rupture, and anodic charge are all plainly visible through potentiodynamic polarization. Information about the passive range and sensitivity to pitting corrosion is also provided. Potentiodynamic polarization curves distinguish between four zones: the passive, trans-passive, anode, and cathode areas.

The electrochemical characteristics of the coating materials under investigation determine the form of the potentiodynamic polarization curve. To determine the improvement from the coating, the coated samples are frequently assessed in contrast to their bare metal counterparts. While test parameters are used to assess coating performance, potentiodynamic polarization curves can be employed to examine material behavior. The corrosion current density and self-corrosion potentials are a couple of the parameters. Better material and coating performance can be inferred from the polarization curve by combining a greater corrosion potential with a lower passivation current.

Since the electrode voltage must be changed continuously owing to changes in the charge density stored at the electrode-solution interface, the charging current always passes through the electrode during potentiodynamic polarization. The charge density rises with voltage, and the difference between the two voltages which depends on the scan rate and material degradation rate reveals how much the polarization curve has been altered. Various scan rates have been employed by researchers to attain a specific voltage range that offers insights into the properties of distinct materials [38, 39].

Advances in Corrosion Science and Surface Engineering   Materials Research Forum LLC
Materials Research Foundations 188 (2026) 68-120   https://doi.org/10.21741/9781644903919-5

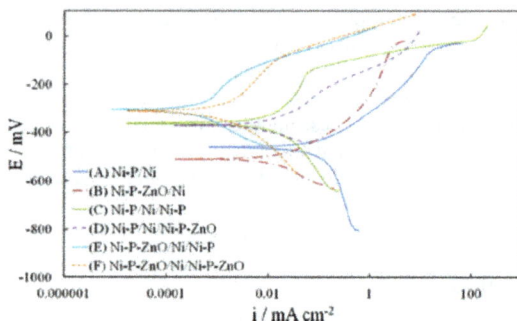

Figure 2. Tafel polarization curves in a 3.5 wt% NaCl solution for various multilayer coatings [40].

Potentiodynamic polarization was used by Varmazyar et al. [40] to assess the corrosion characteristics of six distinct monolayer and multilayer Ni, Ni-P, and Ni-P-Nano ZnO$_p$ coating specimens. An ambient temperature potentiodynamic polarization test employing a three-electrode cell setup was conducted in a 3.5 wt% NaCl solution to investigate the corrosion resistance of the coatings.

Fig. 2 displays the polarization curves of various monolayer and multilayer coatings. According to the polarization results, the Ni coating showed more resistance to corrosion than the Ni-P and Ni-P-ZnO$_p$ coatings. Furthermore, the Ni-P-ZnO$_p$ nano-composite coating showed a nobler corrosion potential than Ni-P coating and a corrosion current density that was almost half that of Ni-P film. The corrosion resistance of multilayer films was much higher than the monolayer films. Among all the samples, the multilayer coating with the Ni-P-ZnO$_p$/Ni/Ni-P layer arrangement showed the maximum resistance to corrosion. This sample's i$_{corr}$ was roughly 54 times lower than that of the Ni film, which had the best monolayer coating performance.

6.1.2 Weight loss

A quantitative method for tracking and measuring corrosion in materials interior or exterior is weight loss. It involves monitoring the weight disparity throughout the course of laps. Perhaps the simplest and fastest corrosion monitoring approach is this one, since it only requires two features to be obtained: the weight and the duration. It is employed to ascertain the kind of corrosion that will transpire under specific circumstances and the suitability of different materials for advancement. Using coupons is only one aspect of the weight reduction evaluation. It is a process that involves immersing coupons in a given medium, taking them out after a set amount of time, cleaning them to measure their net weight, and calculating the weight difference according to the immersion times.

Because of its simplicity, the weight loss method is unique in corrosion testing and is regarded by some as the "gold standard." Using an aqueous solution of chromic acid, the corrosion layers that developed on the specimens' surfaces following the immersion test are removed in this approach. The weight loss is then computed by comparing the specimen's pre- and post-corrosive weights. The weight loss calculations' clear experimental drawback is that they only offer a single measurement for every exposure period. When the measured values are less than the gravimetric device's resolution (usually 0.1 mg), the results become unreliable. Furthermore, it is probable that experimental errors, like insufficient or excessive corrosion product removal, will be incorporated into the final weight loss results.

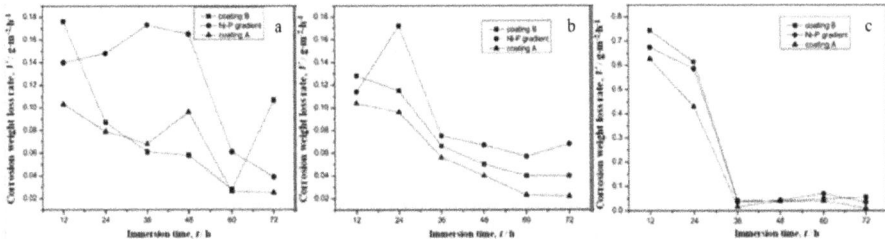

Figure 3. Corrosion weight loss rate of Ni-P coatings in a)10 wt% HCl solution, b)20 wt% NaOH solution and c)5 wt% NaCl solution [42].

This method's foundation is the Faraday relation, which establishes the relationship between electricity and reactant quantity so that the ions at the interface must react with the electrons in the metal to produce an electric current. As a result, the rate of corrosion reactions is attainable. 96,500 coulombs of electrons are required to either make or consume an equivalent. As a result, the mass expended in the reaction is equal to the product of the atomic weight, current, and duration divided by the Faraday number and the number of electrons transferred.

Determining the extent of corrosion in intricate structures is a useful method. The research results might be applied to ascertain not only the corrosion rates but also the thickness decrement. Although this approach is inexpensive and simple to use, it has the drawback that the corrosion products can occasionally adhere tightly to the substrate and be impossible to remove. Therefore, the measurement accuracy may be impacted by the mass of the uncleared corrosion products. Another problem with this method is that the sample mass must always be increased since the oxide layers need to be removed before the mass can be measured. However, in actuality, the oxide layer is not eliminated, so the removal of the oxide layer would not be responsible for the mass loss. As a result, the aforementioned issue affects this method's accuracy [41].

Sun et al. [42] prepared Ni-P gradient coatings, gradient Ni-P coatings with rare earth (RE) Yttrium (Y) (coating A) and non-gradient Ni-P coatings with RE Yttrium (coating B) separately on LY12 aluminum alloy by an electroless plating technique. By using the corrosion weight loss method, the corrosion resistance of the three types of coatings were assessed in various corrosive conditions. Based on the comparison of corrosion weight loss rates for the three types of coatings in alkaline (20 wt% NaOH and 5 wt% NaOH) and acidic (10 wt% HCl and 10 wt% H_2SO_4) environments, Fig. 3 describes the corrosion resistance as follows: coating A>coating B>Ni-P gradient coating, whereas in a neutral environment (5 wt% NaCl and 3.5 wt% NaCl), the characteristics are as follows: coating A>Ni-P gradient coating>coating B.

6.1.3 Salt spray

Spraying salt is a common corrosion analysis method for assessing a material's or surface coating's resistance to corrosion. This test simulates severely corrosive conditions because it is carried out in a high-salinity atmosphere (often NaCl 3.5wt% – the maximum amount of oxygen that can be dissolved). It is a high-moisture fog used for testing. Furthermore, the likelihood of increasing temperatures and surface scratches exacerbate the corrosion conditions. The materials that need to be tested are frequently metallic and have a surface coating applied to them in order to stop the metal underlying from oxidizing. The salt spray test is a type of accelerated corrosion experiment that evaluates whether a coating is suitable for use as a barrier layer by subjecting it to a corrosive attack. After the designated times, the development of corrosion products is evaluated. The corrosion resistance of the coating dictates how long the test will take to produce corrosion products; in general, the more corrosion resistance the coating exhibits, the longer the test will take. Instead of estimating protection against coating corrosion, the majority of salt spray chambers are currently used to monitor coatings processes, such as post- or pretreatment and painting, electroplating, galvanizing, and the like, on a conditional basis. For example, the coating, which consists of a pretreatment and painted sections, must pass a 96-hour salt spray test before it can be sold. The salt spray test is primarily used to compare the corrosion resistance of samples under real-world conditions. Predictions are typically contrasted with the times needed for oxides to appear on samples subjected to salt spray testing in order to assess if the testing effectively achieved its goals. Because of this, the salt spray test is often employed in quality assurance responsibilities, which include keeping an eye on the effectiveness of industrial processes like coating metal.

Because the salt spray test is so inexpensive, consistent, and repeatable, it is utilized extensively. For some coating materials, there is, however, little correlation between the anticipated durability of a coating and the duration of a salt spray test. Because the testing is inexpensive and produces results quickly, it has gained popularity. It is a serious misuse to compare or rate several materials or coatings with varied properties using the salt spray test. The test is very misleading when comparing coatings to metallic coatings. Additionally, comparing different metallic coatings might be deceptive. It may produce results that are radically at odds with behavior in the real world. Thus, in all likelihoods, the test results should not be interpreted as a clear indication of the metallic materials under investigation's ability to withstand corrosion. Salt spray testing is appropriate in the same way as quality assurance testing [43].

Rongjie et al. [44] investigated the corrosion resistance of electroless Ni-P deposits in sodium chloride solutions with phosphorous concentrations ranging from 12% to 14%. The deposits were subjected to a normal salt spray test for 15 days, with a pH of 6.5–7.2 and a solution of 5% NaCl. Six samples were used for the test. Following the conventional salt spray test, the corrosion rates of the Ni-P deposits were determined and are presented in Table 4. For every sample, the yearly average corrosion rates were roughly 0.023 mm. Fig. 4 shows an optical micrograph of the surface of the corroded Ni-P deposit. A few pinholes developed on the deposit following a 15-day standard salt spray test, and the weight content of phosphorus on the deposit's surface was higher than it was prior to the test) a development that helped the passivation film form (while the weight content of nickel was lower because the dissolved weight of nickel was larger than that of phosphorus.

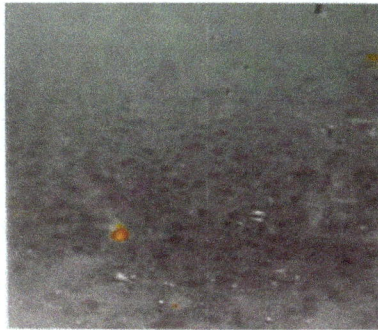

Figure 4. An optical micrograph of the surface of the Ni-P deposit following a typical salt spray test [44].

Table 4. Rates of deposit corrosion in the typical salt spray test [44].

Number	1	2	3	4	5	6
Corrosion rate (g m^{-2} d^{-1})	0.3638	0.4228	0.3702	0.4156	0.4421	0.5395
Average corrosion rate (g m^{-2} d^{-1})			0.4257			
Average Annual corrosion rate (mm year^{-1})			0.0233			

6.1.4 Hydrogen evolution

The hydrogen evolution approach is one of the easiest and most trustworthy ways to keep an eye on variations in the rates at which metals and alloys corrode. A technique for measuring corrosion according to stoichiometric ratios between the hydrogen gas and the concentrations of metals dissociated in the corrosive environment generated as a result is called the hydrogen evolution test (HET).

This method involves collecting evolved hydrogen in a burette above an inverted funnel positioned in the center above a specimen that is fully submerged in a beaker containing the test solution. A hydrogen molecule is produced during the metal corrosion process when a metal atom oxidizes. The drawbacks of this approach include an underestimate of the rate of corrosion because part of the molecular hydrogen produced by metal corrosion dissolves in the solution or is absorbed by the metal object. Furthermore, when the measured values are very little, the results are not very dependable [45].

Huang et al. [46] investigated the biocompatibility and corrosion behavior of a composite coating made of hydroxyapatite, magnesium phosphate, and zinc phosphate that was applied on AZ31 alloy. Hank's solution was soaked in uncoated and coated samples for 21 days at 37°C to study their corrosion behavior. Additionally, during the immersion, the hydrogen volume was checked every 24 hours. The drainage gas collecting method was used for this monitoring in order to get the volume of H$_2$. Every specimen was put into a beaker filled with Hank's solution. Every material was used five times. As the samples were submerged in Hank's solution, Fig. 5 depicted the hydrogen evolution of the uncoated, magnesium phosphate/hydroxyapatite (HM), zinc phosphate/hydroxyapatite (HZ), and hydroxyapatite/magnesium phosphate/zinc phosphate (HMZ) coated samples. As can be shown, after 21 days of immersion, the hydrogen evolution volume of the AZ31 disc reached approximately 4.5 mL/cm^2, which was the greatest compared to other tested samples. The H$_2$ evolution volumes of the uncoated and coated samples were approximately half, at 2.7 and 1.9 mL/cm^2, respectively. HZ produced 1.3 mL/cm^2, or around 25% of the uncoated sample, H$_2$. After 21 days of immersion, the hydrogen evolution of the HMZ coated AZ31 disc was also the lowest when compared to the uncoated, HA, and HM coated samples, coming in at 1.05 mL/cm^2. As a result, the HMZ composite coating showed the highest level of protection. Recently, Ni-P coatings have not been evaluated using HET alone, but it can be used in conjunction with other corrosion evaluation techniques to obtain more accurate results for Ni-P coatings.

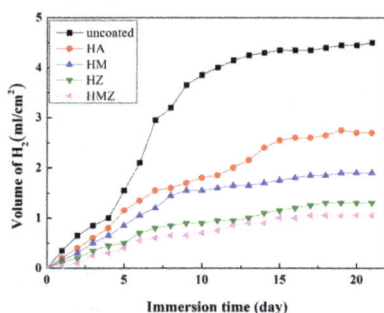

Figure 5. The amount of H$_2$ evolution for the uncoated, HA, HM, HZ, and HMZ coated samples [46].

6.2   Non-destructive methods

6.2.1   Electrochemical Impedance Spectroscopy

Crucially, EIS is a crucial instrument for tracking electrochemical changes in a system in situ with immersion time since it offers quantitative data on the procedures occurring inside the system. In the event of passive, active, or a mix of both protections, it accomplishes this by offering kinetic and mechanical details regarding the corrosion process.

The working electrode of this approach is typically subjected to a long-frequency (100 kHz to 10 mHz) classic three-electrode electrochemical cell applied with a small-amplitude sinusoidal potential perturbation (often in the range of 5–10 mV). The current response, with respect to the perturbation, is a sinusoidal signal with the same frequency that differs in phase and amplitude. Real (Zre) and imaginary (Zim) components of the impedance (Z), sometimes called modulus |Z| and phase shift θ, are calculated using the complex ratio between the applied voltage sinewave (V) and the response current (I).

$$Z(j\omega)=V(j\omega)/I(j\omega) \tag{1}$$

Advances in Corrosion Science and Surface Engineering   Materials Research Forum LLC
Materials Research Foundations 188 (2026) 68-120   https://doi.org/10.21741/9781644903919-5

where the angular frequency is represented by ω and j is the imaginary unit equal to $(-1)^{1/2}$.

he Nyquist plots usually show the imaginary impedance component (Z'') plotted against the real impedance component (Z') at each excitation frequency, while the Bode plots indicate the logarithm of the phase angle (θ) and impedance modulus $|Z|$ as a function of the applied frequency range. It is important to highlight that this analysis refers to Nyquist plots more often than Bode plots.

The process of modeling EIS data involves fitting them to an analogous electrical circuit consisting of series or parallel connections between resistors (R), capacitors (C) or constant phase elements (CPE), inductors (L), and Warburg impedance (W). In this way, corrosion properties like resistance and impedance can be obtained from frequency response data. One example of commercial software that can be used to match the impedance spectrum to the equivalent circuit is the Z View program (3.0a Scribner Associates, Inc., Southern Pines, NC, USA). A common method for evaluating fit quality is to use the chi-squared (x^2) error value. For the equivalent electrical circuit to be useful, its components have to be associated with the different corrosion processes taking place in the system under study.

One requirement prior to performing impedance analyses is validating the EIS data. The prerequisites of causality, linearity, stability, and finity value can be met by verifying electrochemical impedance data using Kramers-Kronig (K-K) transforms. The respectable agreement between the transformed and experimental data sets demonstrated that the system complied with these specifications.

Frequency-resolved EIS may differentiate between several electrochemical reactions in corrosion research and provide detailed information on each one's unique kinetics in the form of corrosion rates based on their individual relaxation times. Slow processes in the frequency domain include diffusion (transport phenomena) in solutions or the creation of adsorbed layers using intricate electrochemical reaction intermediates; fast processes in the frequency domain include the formation of the double electric layer, the presence of ohmic resistance, and the charge transfer resistance of electrochemical reactions. By determining the mechanism of corrosion (activation, concentration, adsorption, or diffusion), this may be used to assess the formation of passivation or corrosion layers on the metal surface and calculate the corrosion resistance, which is inversely related to the corrosion rate. Moreover, it offers details on the dielectric characteristics of the surface oxides and the double layer at the metal/electrolyte interface. EIS uses a small excitation amplitude, leading the corrosion potential to be perturbed less than with polarization curves. Since EIS is non-destructive, fewer samples are needed for measurements and real-time corrosion rate measurements can be made in situ during extended immersion times. Compared to corrosion rate values determined by non-electrochemical methods (weight loss or hydrogen evolution), EIS enables the measurement of substantially lower corrosion rate values. High-purity water, organic coating/metal systems, coatings and linings, and corrosion in a low-conductive solution are examples of high-impedance systems that can be studied using the EIS technique.

The main issue with EIS analysis is that it is frequently possible to fit the same EIS data using a variety of comparable circuit models, all of which have low chi-squared (x^2) error values. Data on resistance and capacitance depend on selecting the appropriate electric model. Therefore, building an appropriate equivalent circuit necessitates a thorough comprehension of the system's corrosion mechanisms, the application of supplementary characterization techniques, such as surface, optical, and/or physical chemical analysis, supported by models from publications. The low lateral

Advances in Corrosion Science and Surface Engineering   Materials Research Forum LLC
Materials Research Foundations 188 (2026) 68-120   https://doi.org/10.21741/9781644903919-5

resolution of global EIS measurements, which are the outcome of averaging the impedance throughout the whole macroscopic electrode surface, is another significant drawback. Studying localized electrochemical corrosion reactions (where the specific electrochemical properties of the micro-defect are averaged over the whole electrode surface) becomes more challenging as a result. One major limitation of AC impedance investigations is the inability to convert the polarization resistance into a corrosion rate without first knowing the values of the Tafel slopes and the Stern-Geary coefficient. Dispersion in the low-frequency measured impedance values can be caused by Mg corrosion process instabilities such as pitting corrosion or the relaxing of adsorbed species. Lastly, unlike polarization curves, EIS is unable to identify variations in solutions, microstructural characteristics (such as secondary phases), or alloying components that alter corrosion potentials and the relative kinetics of anodic and cathodic reactions [39].

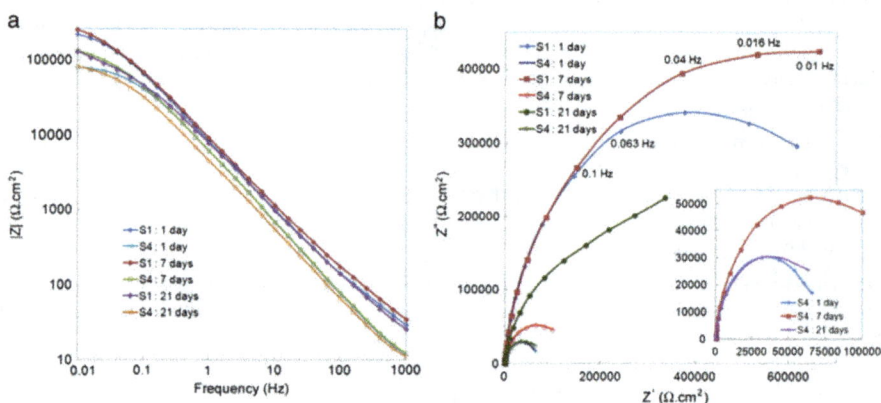

Figure 6. EIS data showing the (a) Bode diagram and (b) Nyquist plot for samples S_1 and S_4 after they were submerged in a 4 wt% NaCl solution for 1, 7, and 21 days. The Nyquist plot for S_4 is shown in greater detail in the inset in (b) [47].

Islam et al. [47] compared the corrosion resistance of pure and composite coatings in artificial seawater containing a 4 weight percent (wt%) NaCl solution using electrochemical impedance spectroscopy (EIS). Fig. 6 displays the Nyquist plot and Bode diagram (Log |Z| vs. log f) for pure Ni-P coatings (S_1 and S_4) at various immersion times. It is clear from comparing the Bode graphs of S_1 and S_4 that S_1 provides better corrosion resistance over all immersion durations. However, in both situations, the area impedance at low frequency (10 mHz) gradually drops, indicating that the coating in touch with the saline solution has deteriorated. After a single day of immersion, the Nyquist plots for both samples display a resistive loop, as seen in Fig.5b. Sample S_1 exhibits area impedance values between 0.5 and 1.5 MΩ·cm² at low frequency ($0.01 \leq f \leq 0.1$ Hz), which is ten times greater than that of sample S_4. Furthermore, sample S_1 exhibits greater stability in the electrolyte, with an impedance of approximately 1.3 MΩ·cm² at 10 mHz, while sample S_4's area impedance gradually drops below 0.1 MΩ·cm².

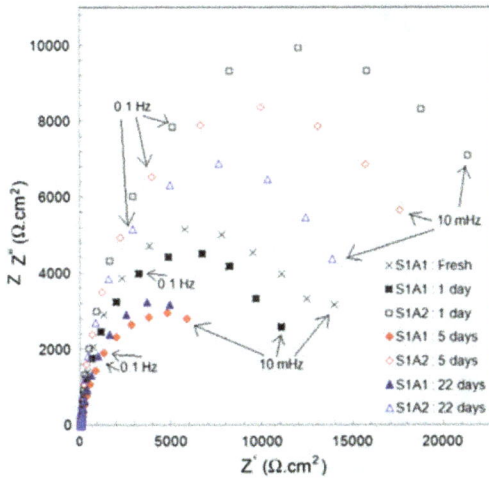

Figure 7. Nyquist plots for immersion in a 4 wt% NaCl solution for up to five days: (a) S_1A_1 and (b) S_1A_2 [47].

Fig. 7 compares the corrosion behavior of Ni-P-Al_2O_3 composite coatings using Nyquist plots for samples S_1A_1 and S_1A_2. Depending on the Al_2O_3 content and immersion period, the maximum area impedance value for Ni-P-Al_2O_3 coatings is around 10^4 $\Omega \cdot cm^2$ or less. At low and intermediate frequencies, the sample S_1A_2 shows larger area impedance values than S_1A_1 for all immersion periods. The S_1A_1 coating's corrosion behavior is represented by the data series with filled markers. When the sample S_1A_2 is fresh (that is, right after it has been submerged in a salt solution) its area impedance value is at its highest and gradually decreases as the immersion period increases. When comparing the S_1A_1 sample's behavior across five and twenty-two days of immersion, the same pattern is seen. Arrows represent the real (Z′) and imaginary (Z″) components of area impedance at 10 MHz and 0.1 Hz frequencies. The addition of Al_2O_3 nanoparticles has been shown to either improve or worsen corrosion characteristics, depending on the amount of the material in the bath. However, in contrast to other publications, the amount of Al_2O_3 nanoparticles in our instance is relatively modest (0.25 or 1.0 g/L), and as previously mentioned, corrosion resistance improves with an increase in Al_2O_3 content.

The EIS data for samples S_1C_1, S_4C_2, and S_1N_1 are displayed in Fig. 8, with filled and empty markers, respectively, denoting the data series for samples S_1C_1 and S_4C_2. After a day of immersion, all samples exhibit a resistive loop and a low frequency area impedance of 2 to 4 × $10^4 \Omega \cdot cm^2$. Among the three types of composite coatings, sample S_1C_1 has the best corrosion characteristics, with comparable Nyquist plots for immersion times of one and five days. The maximum area impedance decreases by a factor of two after 22 days of immersion, suggesting less corrosion resistance. After five days of immersion, the sample S_4C_2 has the lowest area impedance, exhibiting quicker degradation in the electrolyte and a low area impedance below $10^4 \Omega \cdot cm^2$.

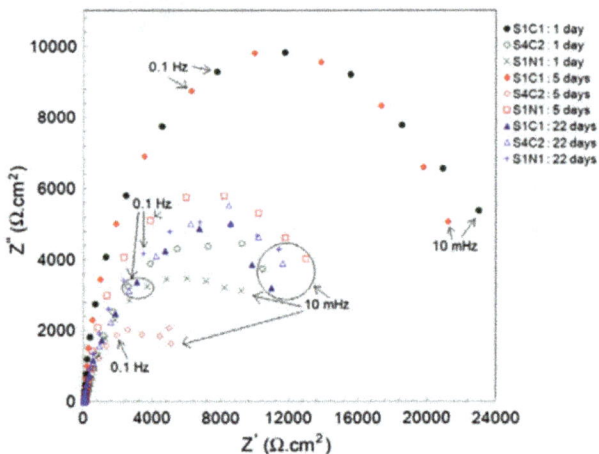

Figure 8. Nyquist plot for composite coatings S_1C_1, S_4C_2, and S_1N_1 following varying immersion periods in a 4 wt% NaCl solution [47].

6.2.2   Advanced surface analysis methods

6.2.2.1 Localized electrochemical impedance spectroscopy

Known also as localized electrochemical impedance spectroscopy (LEIS), Lillard et al. (1992) initiated the most recent advancement in electrochemical impedance spectroscopy (LEIS). The LEIS system can measure coated substances with low electrochemical activity. The LEIS system is subjected to an external voltage, similar to what was previously discussed in the EIS approach. The local potential difference can be measured with a moveable LEIS probe. Fig. 8 displays the schematic diagram for the LEIS measurement. The electronic components used in the LEIS system include a gamry potentiostat integrated with a frequency response analyzer (FRA), two differential amplifiers, an LEIS probe, and a computer display for data presentation. Global and local impedances are simultaneously recorded using a frequency response analyzer with four channels. Furthermore, frequency response analyzers are typically used to examine transfer functions. This arrangement includes a twin microprobe system that is contained within the moveable LEIS probe. 'd' stands for the distance between the two microprobes (micro reference electrodes). The tolerance of the most recent LEIS instruments is approximately 1 nanovolt (nV). Differential amplifiers are used for tracking data on current density and local potential by employing a high input impedance. The two microprobes sense the local potential difference, which is used to calculate the current density. The LEIS probe is controlled by an imposed digital signal and has three-dimensional mobility capabilities. An encoder can control the LEIS probe's positioning system by giving it location, velocity, and direction data. The potentiostat should be adjusted to produce the maximum sinusoidal voltage amplitude in order to optimize the signal-to-noise ratio. Using 50 acquisition cycles, seven points frequency per decade, and a 100 mV peak signal, certain LEIS experiments have been carried out.

Advances in Corrosion Science and Surface Engineering   Materials Research Forum LLC
Materials Research Foundations 188 (2026) 68-120   https://doi.org/10.21741/9781644903919-5

Metal wires make up the dual microprobe (microelectrodes). Every microprobe has a unique set of metallic wires that link to a unique differential amplifier. Micrometers, which are incredibly small in nature, can be used to measure the diameter of those metallic wires. Wires with extremely thin internal diameters are sealed with molten glasses using a nichrome wire equipped with a heater that controls resistance in response to current (Jaffe and Nuccitelli 1974). But in the case of Ag/AgCl microelectrodes, the silver wires are sealed using epoxy resin rather than melted glass. By anodizing the silver microelectrode in KCl solution, the electrochemical process that follows produces a thin layer of silver chloride:

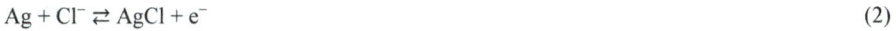

$$Ag + Cl^- \rightleftarrows AgCl + e^- \tag{2}$$

The silver microelectrode must first be cleaned at a rate of 100 mV/s in 2M KCl solution under conditions of 241.20 V_{SHE}. Potentiodynamic oxidation takes place in that electrolyte with 241.40 V_{SHE} after 5–10 minutes. Typically, a typical three-electrode cell is used for the electrochemical testing, which are carried out at room temperature. The potential value of Ag wires can vary depending on their dimensions. According to the schematic diagram (Fig. 9), a platinum grid (counter electrode) can be used to determine the potential difference with regard to a reference electrode. The position and size of the microprobe have a significant effect on the LEIS measurement. For the experiment, a tiny microprobe is always preferred. As per the experimental results for various probe placements above the working electrode (substrate), the resistance of electrolyte needs to be decreased to minimize the high frequencies' contribution.

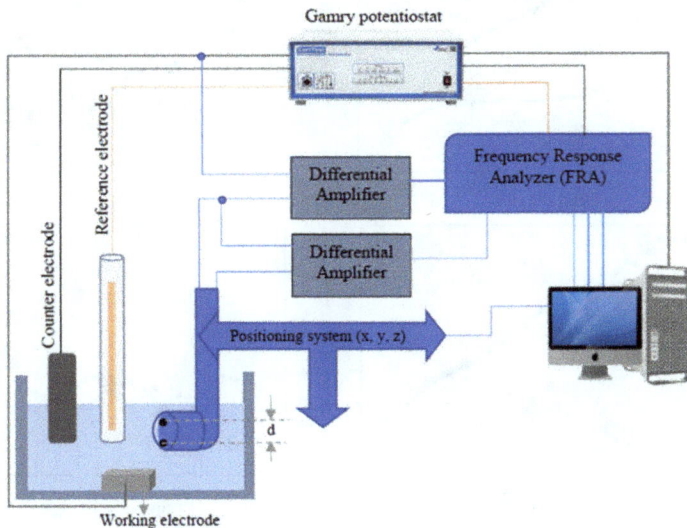

Figure 9. A schematic diagram of LEIS setup [36]

By adjusting the ohmic drop, the nearest microprobe may measure the local current density. Therefore, local interfacial impedance can be used to optimize the LEIS measurement and prevent high frequency. The results of a recent study suggest that the probe size has repercussions. The micro capillary electrochemical setup has also been proposed by several writers as a solution to the probe dimension problem [36, 48, 49].

Wan et al. [50] examined the corrosion process of aqueous acrylic coatings with zinc phosphate (0% and 8%) at the coating/metal interface using localized EIS (LEIS). The local alternating current impedance point measurements in 3.5% NaCl solution are shown in Fig. 10. The LEIS value of the 0% zinc phosphate coating decreased in 3.5% NaCl after immersion, however the 8% zinc phosphate coating remained constant for six hours. In the absence of zinc phosphate, the solution penetrated the coating and resulted in corrosion from below. However, the medium permeability of the coating was resisted by the coating that contained 8% zinc phosphate.

Fig.11 shows the map measurement of LEIS for the zinc phosphate coating with 0% zinc submerged in 3.5% NaCl. Compared to the undamaged coating, the coating defect's impedance value was significantly lower when it came into contact with the electrolyte solution at the start of immersion. After six hours of immersion, the coating defect's impedance modulus value rapidly decreased. The coating defect exhibited severe corrosion, and the surrounding area displayed exfoliation. Corrosion occurred after the solution medium reached the coating's bottom.

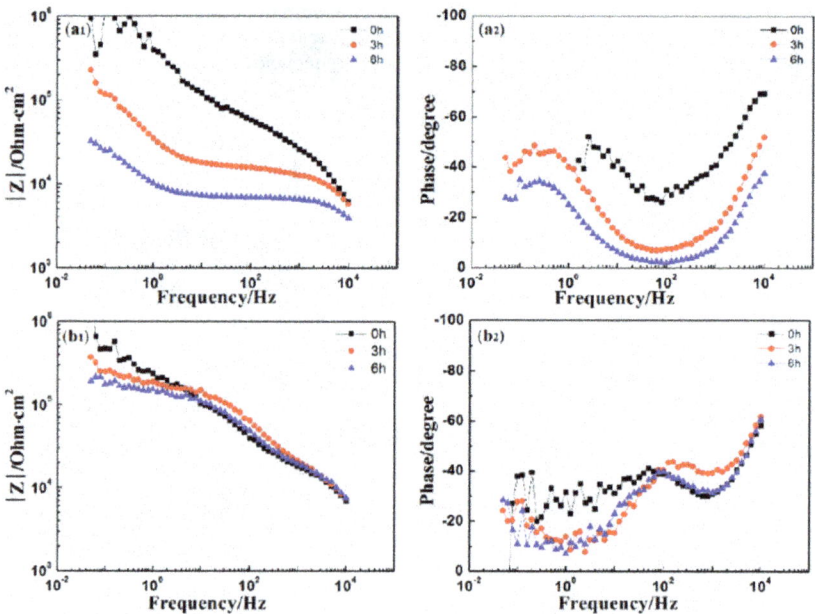

Figure 10. Point measurements of localized EIS (LEIS) of zinc phosphate coatings submerged in 3.5% NaCl solution at 0% (a1-Nquist, a2-Bode) and 8% (b1-Nquist, b2 Bode) [50].

Advances in Corrosion Science and Surface Engineering   Materials Research Forum LLC
Materials Research Foundations 188 (2026) 68-120   https://doi.org/10.21741/9781644903919-5

The LEIS map measurements of the of the 3.5% NaCl solution immersed in an 8% zinc phosphate layer are displayed in Fig.12. The coating defect's impedance value at the beginning of immersion was substantially lower than the intact coating, which was equal to the 0% zinc phosphate coating. The coating defect's impedance remained constant while the coating defect surrounding it marginally increased with extended immersion time. This was entirely different from the 0% zinc phosphate coating. The application of a zinc phosphate coating delayed the corrosion process and stopped corrosive media from diffusing in the coating/metal contacts.

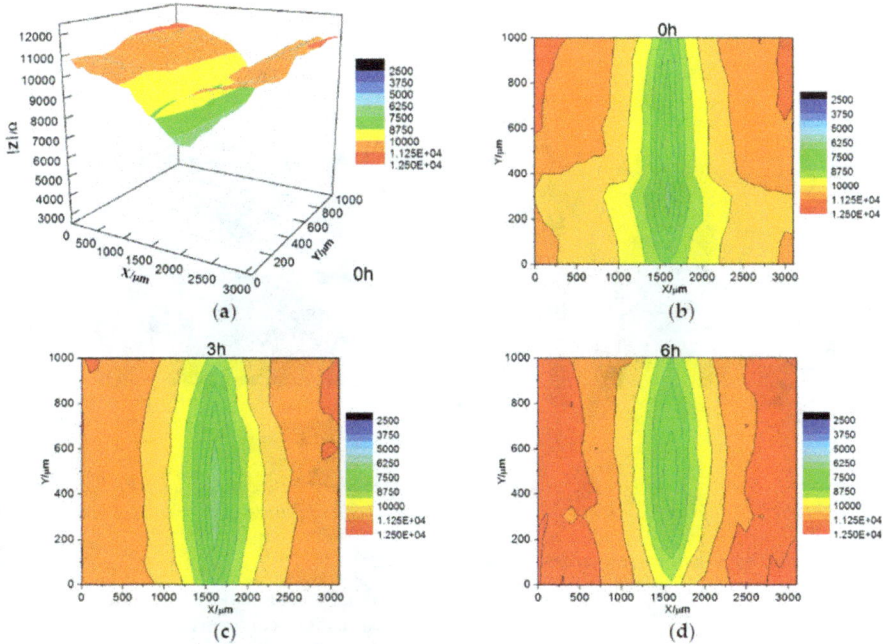

Figure 11. The LEIS map measurements of the 0% zinc phosphate coating immersed in 3.5% NaCl solution: (a) 3D of 0 h; (b) 2D of 0 h; (c) 2D of 3 h; (d) 2D of 6 h [50].

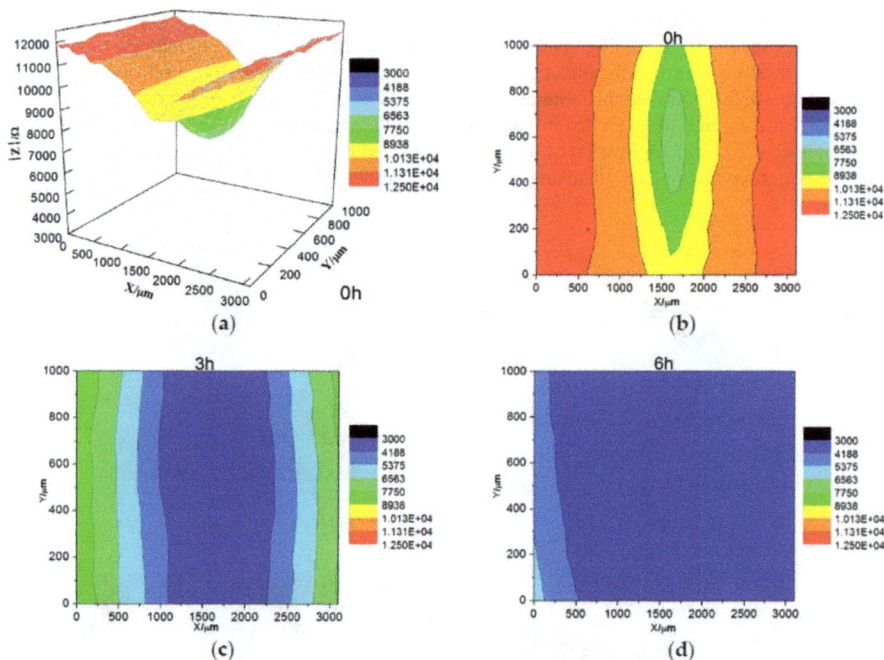

Figure 12. The LEIS map measurements of the 8% zinc phosphate coating immersed in 3.5% NaCl solution: (a) 3D of 0 h; (b) 2D of 0 h; (c) 2D of 3 h; (d) 2D of 6 h [50].

6.2.2.2 Scanning vibrating electrochemical technique

In the scanning vibrating electrochemical method (SVET) setup, the electrolyte's potential differential between the substrate and probe is measured using a conductive vibrating probe. A piezoelectric actuator holds the vibrating probe in place (Fig.13). It is possible to place a measurement probe just above the sample's surface. The platinum tip of the scanning probe equipment is extremely thin. The potential difference (ΔV) is measured by the tip of the SVET probe. The current source point, which ordinarily provides the system with 60 nA of current, is normally 150 µm away from the SVET probe. A thin, insulated wire made of platinum-iridium (Pt-Ir) is wrapped around the probe. The tip is coated with black platinum and has a spherical form with a diameter of 10 µm. The probe is placed 100 µm above the surface, and its vibration frequency typically stays at 398 Hz. Three-dimensional motion is possible using a piezoelectric motor. The probe's movement is controlled by the piezo ceramic device, which permits vibration amplitudes (d) ranging from 1 to 60 µmin in an angle that is perpendicular to the surface of the sample (Z axis). A 100 µA current is flowing between the B and C graphite electrodes. This current is externally controlled by a galvanostat and is produced naturally by electrochemical corrosion or biological processes. Calculating the potential difference at specific places (A to D) in the circuit is done. The ohmic drop in solution is measured using reference microelectrodes with low

Advances in Corrosion Science and Surface Engineering   Materials Research Forum LLC
Materials Research Foundations 188 (2026) 68-120   https://doi.org/10.21741/9781644903919-5

resistance. Two reference microelectrodes are moved in unison while remaining in the same spot. The potential difference is caused by the compact electric field that has generated in the electrolyte solution. Eq. 8 makes it simple to measure the potential difference (ΔV), the vibration amplitude (d), and the solution resistivity (ρ) in order to get the current density (I). To determine the current density value, the procedure is iterated within a predetermined framework, and the data is obtained by plotting the current density on a chosen sample surface. The rate of corrosion can be calculated from this current (I).

$$I = -(1/\rho)\,(\Delta V/2d) \tag{3}$$

A computer command that is created automatically controls the SVET probe position. The precise location of the probe can be determined by the close loop linear encoders. The probe's ability to vibrate perpendicular to the substrate surface can create an area where an AC signal can form. An excessively noisy environment's AC signal with its carrier wave is extracted using a lock-in amplifier (LIA) device, which uses the input phase angle to convert it to a DC signal. Larger signals can also be captured and demodulated by the LIA in addition to smaller ones. In this SVET configuration, the LIA serves as a rectifier. Typically, the lock-in amplifier's phase input is manually changed until no response is noticed in order to ascertain the input phase angle. Ninety degrees is the optimal phase angle that can be obtained. Certain commercial equipment has the capability to immediately obtain the reference phase angle. It is therefore possible to plot the converted DC signal from the LIA device to display the distribution of local activity. Fig. 14 displays the configuration of the SVET signal chain.

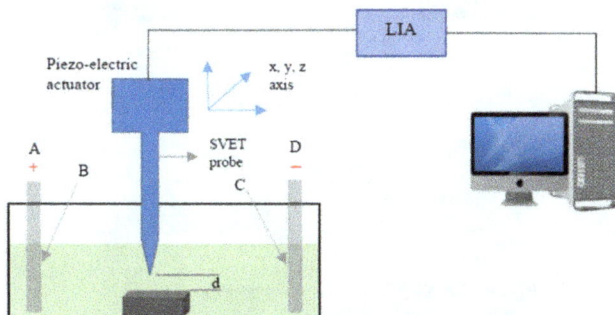

Figure 13. Schematic presentation of SVET [36].

There are significant experimental limitations to using SVET in corrosion investigations. In the matrix, data acquisition processing takes longer to complete. If the sample is not quite flat or is not positioned in line with the scanning probe, another problem may arise. Larger vibration amplitudes have the potential to harm the substrate and prevent the electrochemical corrosion response by causing the electrolytes to ripple, which can lead to an incorrect current density measurement. Galvanic corrosion rate, pitting corrosion, crevice corrosion, stress corrosion cracking, microorganism-related corrosion, inorganic coatings, coated materials, corrosion inhibitors, weldment corrosion, and polymer corrosion can all be determined using SVET methods [36, 49, 51].

Ni-P/alginate microgel coatings were investigated by Stankiewicz et al. [52] utilizing an electroless technique as possible metallic protective coatings with self-healing capabilities. NaOH and nickel chloride were present in the alginate microgels. It was demonstrated that the surface of the Ni-P coating and steel may reduce the nickel ions liberated from the microgels. SVET, or scanning vibrating electrode method, was used to investigate the system's capacity for self-healing. Because the coatings made of Ni-P alginate microgels produced less current density at the sample surface after immersion than did Ni-P coatings, they demonstrated improved corrosion protection. Furthermore, unlike the Ni-P sample, the Ni-P/alginate microgels coating did not exhibit the same intense, highly localized anodic site development. This suggests that the microgels' presence impedes any major deterioration of the Ni-P layer that would have otherwise exposed steel to a corrosive environment. Therefore, the Ni-P/alginate microgels covering has self-healing characteristics, as shown by the SVET analysis.

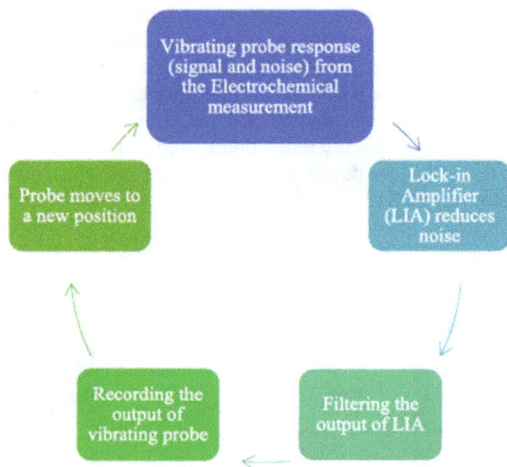

Figure 14. SVET signal chain configuration [36].

Fig. 15 shows the electrolyte's variations in current density above the coated surfaces after it has been submerged in 50 mM NaCl for 0, 5, 12, 18, and 20 hours. In this design, the positive current values indicated the anodic current direction, while the negative current values indicated the cathodic current direction. At the beginning (0 h), the entire sample surface appeared neutral and there was no activity seen above it because the Ni-P coating layer was acting as a continuous barrier of protection (Fig. 15a). After a 5-hour immersion, however, a broad anodic reaction was seen to be developing on the right side of the map, signifying the start of corrosion. A few hours later (12 h), the first signs of cathodic activity appeared on the left side of the map, some distance distant from the anodic zone. Concurrently, the anodic region shrank and became more limited, but the severity did not lessen. It's likely that the increased current density around the anodic region

Advances in Corrosion Science and Surface Engineering   Materials Research Forum LLC
Materials Research Foundations 188 (2026) 68-120   https://doi.org/10.21741/9781644903919-5

caused the coating to be removed, allowing steel corrosion to occur instead. It is advisable to observe the cathodic nature of the Ni-P coating only after the substrate's steel corrosion has started, as it is cathodic to steel.

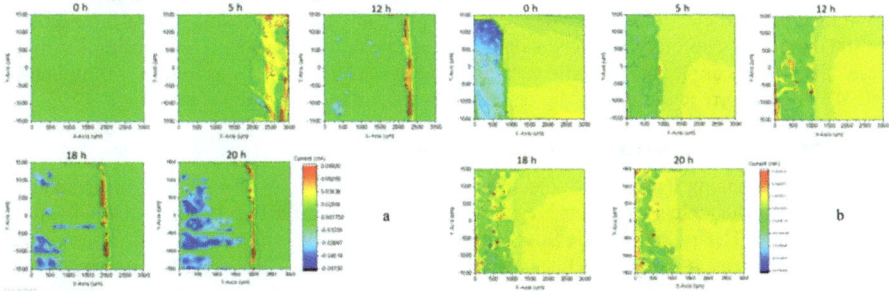

Figure 15. Current density changes above a) Ni-P coating surface and b) Ni-P\alginate microgels coating surface following 0, 5, 12, 18 and 20 h immersion in 50 mM NaCl measured by SVET [52].

The current densities on the Ni-P alginate microgels sample were an order of magnitude lower at the start of immersion than the current density on the Ni-P coating, indicating that the alginate microgels slowed down the rate of corrosion on the coating (Fig. 15b). In this case, the initial reaction observed at the onset of immersion was a widely distributed cathodic response. This cathodic current appeared to start as soon as the microgels were submerged in the electrolyte, and it may have been brought on by the nickel reduction that takes place following the discharge of nickel chloride and sodium hypophosphate. This behavior could indicate that the coating has the capacity to self-heal due to the autocatalytic reaction of phosphorous and nickel deposition in the alginate microgels. Anodic and cathodic current imbalance was observed at the start of the immersion in the NaCl solution. The following is an explanation:

- Because SVET can only detect net ionic fluxes, it cannot determine whether a given activity is cathodic or anodic based on the dominance of one over the other.

- When one activity is readily recognized due to its extreme localization, while the other activity is dispersed throughout the entire surface, the current density may drop below the detection limit.

- By reducing nickel and phosphorous, the cathodic process produces the nickel-phosphorous coating, which adheres to an area where the microgels are present. Ionic compounds are created via the anodic reaction, which is the reducer's oxidation. These compounds are easily soluble.

There was no maintenance of the cathodic activity. Longer immersion reduced the first robust cathodic reaction (5 h) because the microgel content in that region was leaching. The sample was primarily neutral after 12 hours of immersion, and there was some indication of slight anodic/cathodic activity interchanging at the early cathodic region. This behavior doesn't change for another 20 hours. When all of the substrates in the microgels were used up, the autocatalytic deposition of the Ni-P coating was complete. Nonetheless, the microgels themselves may work as a powerful corrosion inhibitor for steel substrates. The alginate derivatives have a twofold inhibitory mechanism. The cathodic hydrogen evolution reaction is inhibited by the possibility of alginate being adsorbed on the cathodic sites of carbon steel. Nevertheless, the anodic metal dissolution reaction is hindered by –OH groups, which can be adsorbed on the anodic sites and have lone pairs of electrons. There did not seem to be any more corrosion during the exposure period due to the extremely low corrosion current above the active spots. This implies that in the case of the Ni-P sample, the galvanic current between steel and nickel was more harmful, but the initial cathodic current was created by reactions on the surface of the nickel covering.

6.2.2.3 Scanning ion-selective electrode technique

Noninvasive scanning ion-selective electrode technique (SIET) provides accurate measurements of pH and specific ionic species just above the electrolyte's surface; Fig. 16A schematic diagrams the SIET setup. The microelectrode (Fig. 16B) of the SIET probe is a glass or plastic micropipette that is used to measure ions and pH changes at the electrode's tip; a glass capillary microelectrode has an oil-like selective ionophore membrane inside of it. The metallic tip and the ion-selective membrane are separated by an ion-to-electron transducer made of conductive polymer poly (3-octylthiophene-2,5-diyl). The electrode's ion-sensitive tip can range in length from 10 to 100 μm, depending on the application. About 10 μm is its length at this time. The base of the microelectrode is made up of an open-ended Pt-Ir wire that resembles a needle. Glass-capillary microelectrode that is selective for ions and has an aperture width ranging from 0.1 to 5 μm is used to achieve a competitive lateral resolution.

Using an active solution surface, the scanning ion-selective electrode approach measures particular ions at a quasiconstant micro-distance, serving as a micro-potentiometric tool. Potentiometric measurements are made in a two-electrode galvanic cell with no current flowing through it. An ion-specific microelectrode and a reference electrode comprise a potentiometric cell. One of the previously mentioned essential components of a SIET system, an ionselective microelectrode, is inserted and transferred above the specimen using a 3D computer-controlled stepper-motor system. A long-distance lens is used with a 400x magnification video camera that is positioned above the sample's surface. The potential difference measurement of this potentiometric cell is improved and digitalized. The reference electrode, which consists of Ag/AgCl wires, is placed above the computer-controlled 3D stepper motor, and both electrodes are moved in close proximity to it once the material has dissolved in the electrolyte. While taking images and helping to position the probe, video-assisted camera optics maintain the space between the probe and sample. To calibrate the microelectrode, a large amount of SIET pre- and post-assessment is required. A computer-generated command can regulate the micropipette's motion and speed as it moves across the sample's surface. The distribution of the sample's pH or ionic content is represented by a 3D data set. Potentiometry is the context in which SIET tests are performed. The possible variations causing fluctuation in ionic concentration are assessed during this control mode. Ionophore membranes are utilized in ion selective electrodes due to their capacity to permit particular ions to pass through the membrane. Through the use of internal and external reference electrodes, potential differences are noticed as a result of the chemical potential changing as a result of the ions diffusing through the membrane.

A reference electrode is often a wire that has been dipped in the electrolytes and coated with silver. Selective microelectrodes are usually positioned 50 μm above the tracked surface in SIET research. The H^+-selective microelectrode is calibrated using the buffer solution and the Nernst Equation.

Advances in Corrosion Science and Surface Engineering   Materials Research Forum LLC
Materials Research Foundations 188 (2026) 68-120   https://doi.org/10.21741/9781644903919-5

You can utilize the ASET-2 (Science Wares) programming tools to search the area in question. Potential can be computed using a preamplifier with an input impedance rating of 1 PΩ. The ion selective electrodes' physical design has a number of drawbacks. Glass micropipettes are fragile and brittle due to this. Because glass is transparent, it can be difficult to gauge the exact distance throughout the calibration and analysis procedure between the sample and the probe [36, 49, 53, 54].

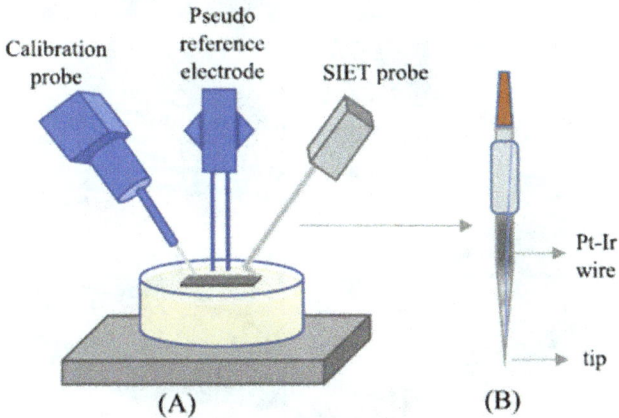

Figure 16. SIET technique setup in (A) standard form; glass-capillary ion-selective microelectrode with liquid membrane in (B) [36].

Localized corrosion of magnesium alloys coated with inhibitors was investigated by Gnedenkov et al. [55]. using the scanning ion-selective electrode method (SIET). The kinetics and mechanism of the self-healing process were determined by applying the SIET. The inhibitor action helped to slow down the corrosion process after immersion in the corrosion-active environment. This was shown by analyzing the dynamics of the pH values distribution at the surface of the composite coating over a seven-day sample exposure period (Fig. 17). In comparison to the undamaged region of the sample, the defect zone (center zone of the image) stayed more alkaline (red in color version and deeper grey in black-and-white), indicating that an inhibitor was preventing the anodic area from corroding. Seven days after exposure, the created defect zone's ΔpH value of 1.2 indicated that the self-healing process was still ongoing.

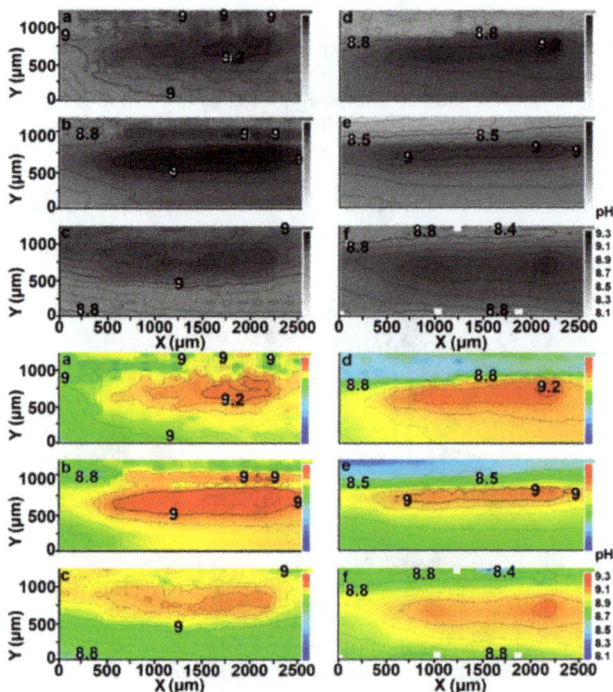

Figure 17. The pH distribution showed by SIET mapping on the surface of the composite inhibitor-containing coating with defect upon exposure of the sample to 0.05 M NaCl (in days): a) 2, b) 3, c) 4, d) 5, e) 6, and f) 7. Compared to the portion of the sample that was undamaged, the defect zone (the central zone of the figure) was more alkaline (red in the color version and deeper grey in the black-and-white version) [55].

6.2.2.4 Scanning droplet cell

A confined electrochemical cell is scanned across the sample's surface in a classic three electrode experiment called a scanning droplet cell (SDC). The SDC approach uses a positive displacement pump to push electrolyte through a tiny diameter tube from a reservoir that has been gas-purged. Pump rate can be used to control the convection profile created by the narrow tube. The single droplet on the substrate's surface that serves as a working electrode is subsequently filled with the electrolyte. This droplet is held in place on the surface by the cohesive forces between liquid molecules, and capillary action subsequently allows it to be pulled throughout the sample. The sample's contact area with the droplet is monitored as the electrochemical interaction between the electrolyte and electrode occurs. There are two possible configurations for SDC experiments. First, a data map is created by increasing the droplet's position and adding a constant bias, such as potential or current. Secondly, the droplet can be kept in a fixed position while doing tests like

100

Advances in Corrosion Science and Surface Engineering   Materials Research Forum LLC
Materials Research Foundations 188 (2026) 68-120   https://doi.org/10.21741/9781644903919-5

Tafel or EIS by supplying a static or dynamic electrochemical signal. Because the droplet is a chemical cell with three electrodes that is self-contained. The force sensor, fixedly mounted on the scanning flow cell (SFC), measures the force applied by the droplet as it presses on the working electrode. In addition to inlet and exit channels that cross in a polycarbonate block, the SFC has an elliptical aperture at the bottom of the cell body (Fig. 18). While the working electrode is positioned directly below the elliptical entrance onto an automated three-dimensional stage, the reference and counter electrodes are carefully positioned within the channels of the SFC cell. The reference electrode is Ag/AgCl, and the counter electrode is platinum wire. A 150 μm thick silicon gasket is affixed to the electrode's aperture to prevent electrolyte leakage. The distance between the working electrode and the cell bottom is 500 μm. The minimum capillary diameter and droplet contact area are necessary. The induced current range is as small as picoamperes due to the extremely small sample measuring area. This extremely little quantity of current can be measured by the high resolution potentiostat. Following the application of a continuous polarization and a single-direction scan across the surface of the SDC head and droplet, the responses) current, voltage, and open circuit potential (are monitored. Despite the dynamic nature of droplet mobility on the surface, the ohmic drop must be taken into account. Because it is a flow-type cell, it may also be easily integrated with downstream analysis or spectrometry. When measuring DC and AC simultaneously with the same configuration, the SDC approach is applicable [36, 49, 56].

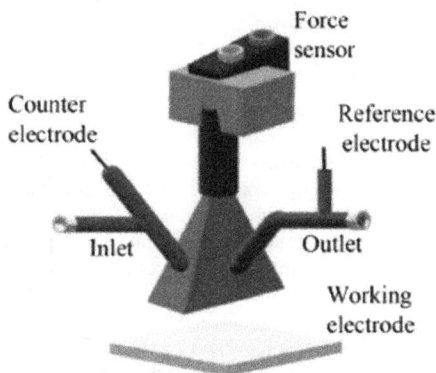

Figure 18. Scanning droplet cell 3D diagram [36].

The car heat exchanger's cross-sectional profiles of the corrosion potential for aluminum-alloy brazing sheets were obtained by Fushimi et al. [57] by the successful application of the flowing scanning droplet cell (f-SDC) technique (Fig. 19). Three layers make up the brazing sheets: an Al-Mn-Cu core layer, an Al-Si brazing filler, and a sacrificial anode layer made of zinc alloy. To assess the sacrificial anode's effectiveness in preventing corrosion, the profiles of its potential for corrosion are crucial. It was discovered that because sequential surface cleaning is not required, the determination of the corrosion-potential profile using f-SDC takes a lot less time than it does using the traditional method.

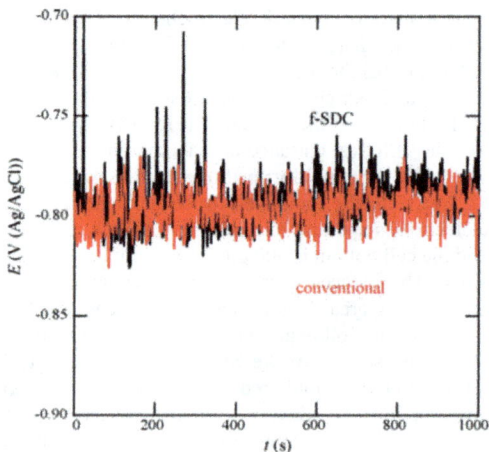

Figure 19. Time fluctuation of the sacrificial anode layer surface potential measured using f-SDC or a traditional technique [57].

6.2.2.5 Scanning electrochemical microscopy

The working electrode's (substrate's) electric potential can be more accurately analyzed using scanning electrochemical microscopy (SECM) techniques. This method makes it simple to scan the surface reactions at a macroscopic level. An ultra-microelectrode (UME), which serves as an SECM probe, a biopotentiostat, and a piezoelectric actuator are combined in the experimental setup (Fig. 20) of SECM. The noble metal used to make SECM probes typically makes them stiff. The topographic image of the substrate's surface can be captured by placing the ultra-microelectrode tip vertically above the substrate's surface. While platinum electrode can be utilized as a reference electrode, substrate and UME tip are considered working electrodes. To measure the current response, the SECM probe is typically subjected to DC voltage. The bipotentiostat can detect currents and polarizes the substrate and UME tip separately. The probe signal is amplified using a potentiostat before being transformed to a digital signal. While the other electrode is used to assess the response, the first electrode is in charge of carrying out the electrochemical reactions between the electrolytes and the substrate. Data on tip creation and collecting can be observed by varying the polarization level. A computer-generated program uses the position controller's signal to trigger a stepper motor. By creating a data map, the positioning system can determine the electrochemical parameters by scanning the location of the measuring tip. Once the electric signal has been transformed into mechanical energy, the piezoelectric actuator can drive the UME probe. The two-dimensional (X and Y) movements of the SECM probe are controlled by a position controller, while its vertical movement with respect to the substrate surface is done by hand. Depending on the position of the SECM probe, current is scanned in either the X or Y direction and recorded. The SECM has a sensitivity of 100 μV [36, 49, 58, 59].

In order to investigate the effect of multiwalled carbon nanotube loading on the local electrochemical behavior of the films as seen by scanning electrochemical microscopy (SECM),

Advances in Corrosion Science and Surface Engineering
Materials Research Foundations 188 (2026) 68-120

Materials Research Forum LLC
https://doi.org/10.21741/9781644903919-5

Oliveira et al. [60] employed electroless deposition to construct composite Ni-P-multiwalled carbon nanotube films.

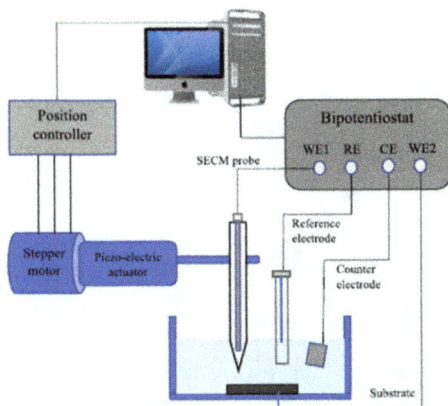

Figure 20. The SECM experimental setup's schematic diagram [36].

The uncoated API5LX80 steel, Ni-P, and composite Ni-P/MWCNT coatings are shown in SECM 2D maps in Fig. 21. This kind of material corrodes by oxidizing Fe^{2+} to Fe^{3+} at the anodic sites, where the Fe^{2+} ions originate from the material breakdown. Fe^{2+} ions produced on the surface as a result of corrosion are thus reachable. Therefore, when the Pt tip (polarized at $+600$ $mV_{Ag/AgCl}$) travels over the anodic sites, Fe^{2+} ions are oxidized to Fe^{3+} by Eq. 4. Elevated values of oxidation current indicate heightened surface electrochemical activity. The SECM maps show current spikes where the examined surface exhibits higher levels of electrochemical activity. The uncoated substrate showed the greatest currents throughout the entire region when compared to the coated material, indicating a higher level of electrochemical activity. The significantly lower currents for the Ni-P film, which also led to a decrease in the electrochemical activity that the tip could probe in close proximity to the sample surface, suggested that the electrolessly deposited layer was protective. The cathodic values of the observed currents confirm the minimal activity for Fe^{2+} oxidation.

$$Fe^{2+} \leftrightarrow Fe^{3+} + e \tag{4}$$

The tendency of these species to be easily oxidized to Fe^{3+} and then diffuse to the bulk electrolyte once the concentration of Fe^{2+} ions is low at the metallic surface may give rise to cathodic currents. The cathodic current would then be probed at the sites where Fe^{3+} ions are produced as a result of reaction (4). Such areas would imply that although the film shows little electrochemical activity, it is prone to corrode at its weak points. The SECM map shown in Fig. 21B shows how the current values vary over the surface, suggesting that the surface is not uniform in the places where Fe^{2+} oxidation occurs.

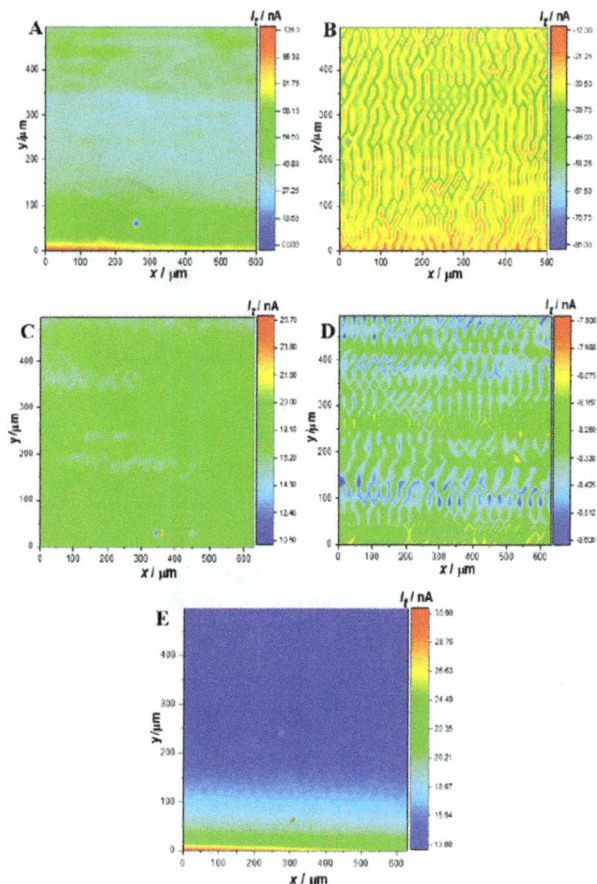

Figure 21. SECM 2D maps acquired with the sample at the open circuit voltage and the tip biased at +600 mV$_{Ag/AgCl}$: Substrate without coating (A); layer of Ni-P (B); CNT-0.25 (C), CNT-0.50 (D); and CNT-1.0 (E). Electrolyte: ambient temperature 0.1 M NaCl solution [60].

The investigated anodic currents are lower on the surface of the CNT-0.25 sample (Fig. 21C) than on the uncoated substrate, suggesting that carbon nanotubes have a beneficial effect on decreasing the metallic substrate's electrochemical activity. The values of current are constant throughout the examined region. As a result, the SECM map of the CNT-0.50 film showed numerous current oscillations and values that are slightly cathodic (Fig. 21D). These low cathodic currents are most likely caused by the absence of the electroactive species under investigation (Fe^{2+} ions), indicating that the CNT-0.50 film is less active than the CNT-0.25 film.

The surface of the CNT-0.25 sample (Fig. 21C) shows lower studied anodic currents than the uncoated substrate, indicating that carbon nanotubes are helpful in lowering the metallic substrate's electrochemical activity. All across the area under investigation, the existing values remain constant. Consequently, several current oscillations and values that are slightly cathodic were displayed on the SECM map of the CNT-0.50 film (Fig. 21D). The absence of the electroactive species under examination (Fe^{2+} ions) is most likely the origin of these low cathodic currents, suggesting that the CNT-0.50 film is less active than the CNT-0.25 film. The global electrochemical behavior discussed in the preceding section is in good agreement with this result.

6.2.2.6 Scanning kelvin probe

There is no need for a conducting route between the kelvin probe and the substrate when using the nondestructive scanning Kelvin probe (SKP) technique. The Kelvin probe is capable of measuring the difference in work function in various media, such as open air, humid air, vacuum, or by employing a non-destructive capacitance approach with a drop of electrolyte on the surface. Typically, the metallic probe is composed of tungsten and has a brass container around it. The needed speed and predetermined height can be used to regulate the probe's vibration frequency and maximum extent. The external potential stifles the current produced by the probe vibration. The SKP setup's particular configuration enables the measurement of capacitive height during data collection, hence enabling the Kelvin probe to detect the precise profile. The capacitive height is measured using two parallel capacitor plates. One plate is referred to as the Kelvin probe surface and the other as the substrate surface when two parallel plates are present. The electrons' energy variations are seen when the Kelvin probe's tip is close to the substrate. The idea of work function is defined as the amount of energy needed to remove an electron from a conductor's surface; variations in work function can be seen in the potential of various conductive materials. Depending on the characteristics of the conductors, the electron flow can change from a higher to a lower work function. The value of corrosion potential (E_{corr}) can be related to the relative work function. An induced potential difference occurs when two conducting surfaces make electrical contact with one another. Consequently, the two distinct metals produce positive and negative charges.

The respective potentials are denoted by E_{F1} and E_{F2}, and the work functions of the associated plates are represented by Φ_1 and Φ_2. If there is no electrical connection between two parallel wires, there won't be a potential difference (VC). Furthermore, the differences in the charge and energy level of two plates are shown in Fig. 22A. On the other hand, charge and energy level seem in an equilibrium state when the plates are electrically connected. Fig. 22B depicts the situation of the connections between two plates. In this case, a potential difference (VC) is produced by the subsequent charge flow that results from the electrical connection between the two conductors. The work function fluctuations can be equalized by multiplying the electron charge (e) and potential difference (VC), as shown in Eq. 5.

$$-eV_C = \Phi_1 - \Phi_2 \qquad\qquad (5)$$

The plates are given an equal and opposing backup potential (VB) in order to balance the potential difference (VC) during the vibration of the probe. The neutralize condition is essential to determining the work function differential value throughout the measurements. The corrosion potential, or this voltage potential, can be computed using Eq. 6.

Advances in Corrosion Science and Surface Engineering   Materials Research Forum LLC
Materials Research Foundations 188 (2026) 68-120   https://doi.org/10.21741/9781644903919-5

Corrosion Potential = $K + (\Phi_1 - \Phi_2)/e$ (6)

where $(\Phi_1 - \Phi_2)$ represents the difference between the substrate and Kelvin probe work functions. The value of a constant, K, varies according on the kind of probe being utilized.

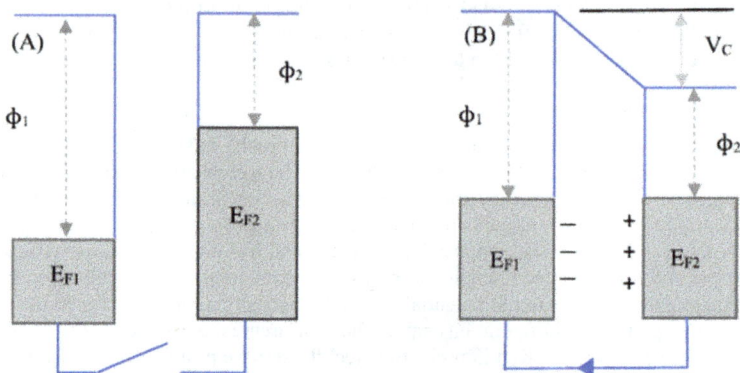

Figure 22. The plates' energy and charge levels when they are (A) not connected electrically and (B) connected electrically [36].

The HR-SKP system was designed to evaluate possible differences in rough, uneven, and curved surfaces by controlling the scanning Kelvin probe height. The experimental setup for the HR-SKP is shown in Fig. 23. Many electromechanical parts, including a needle, speaker, isolator, stepper motor, and piezo electric actuator, make up this type of Kelvin probe. An auditory signal with a frequency of 1-2 kHz starts the Kelvin probe's normal operation. The sample chamber can be positioned using a computer-generated command to establish the distance between the substrate surface and the needle. A tiny needle made of nickel/chromium wire is used. Height is controlled by a piezo-electric actuator, which can move fast and accurately in the Z direction and change the position vertically from the substrate surface. A function generator generates sinusoidal current at a frequency of 10 Hz and maintains a 300 mV voltage for the distance control circuit. Two computer screens display the two internal control circuits that are connected to each other. X and Y directed stepper motors are first driven by a computer command to locate the matching spot on the substrate surface. The piezoelectric actuator is moved by a circuit that regulates height. If a precise height is needed and the piezoelectric actuator isn't able to provide it, an additional Z-stepper motor can be utilized in tandem with it. The needle distance is the only command that needs to be explicitly set; all other orders are fulfilled automatically by a computer program. A LIA is utilized for the potential measurement. Once the probe is properly positioned on the substrate surface, the voltage potential is measured by the potential control circuit. The commands from the LIA to the probe then move the needle to the next location.

This task function may be related to the open circuit potential or corrosion of bare or coated metals. One application of the SKP is to examine the surface beneath a covering of a material to identify areas that are actively corroding or that may eventually become corroded. This technique has also

Advances in Corrosion Science and Surface Engineering   Materials Research Forum LLC
Materials Research Foundations 188 (2026) 68-120   https://doi.org/10.21741/9781644903919-5

been applied to solar cell research, forensics, and the manufacturing of organic light-emitting diodes. By letting the user set and maintain the probe-to-sample distance, SKP which relies on the idea of a capacitor allows the researcher to eliminate topographical contribution over the surface of reality. Because SKP allows electrochemical analysis to be performed even in the absence of electrolytes, it is a non-destructive technique (NDT) that saves the sample for additional analysis. Another option is the SKP non-contacting optical surface profile (OSP), which generates relative height information that might potentially be the basis for constant distance mode operation. Topographic changes on a sample surface can also be mapped using the OSP technique to estimate the size, amount, and depth of corrosion pits [36, 49, 61].

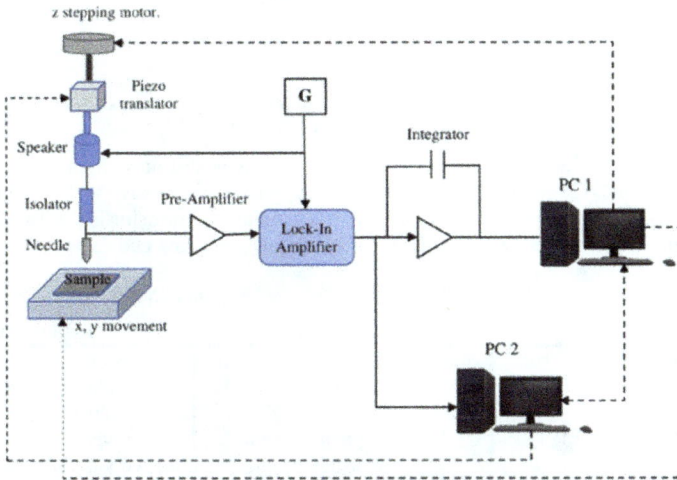

Figure 23. The HR-SKP Experimental setup [36].

Wang et al. [62] created an electroless Ni-P coating for the surface of carbon steel (Ni-P-G) to increase its resistance to corrosion. Simultaneously, a Ni-P electroless coating (Ni-P) was heated (Ni-P-H) under the same conditions as Ni-P-G for comparison purposes. The Ni-P and Ni-P-G surface potential maps for determining corrosion propensity are shown in Fig. 24. The Ni-P-G and Ni-P surfaces have differing surface potentials of around 375–750 mV, where Ni-P-G has a greater positive surface potential (400–600 mV) than Ni-P (150~25 mV). This implies that Ni-P-G has a higher corrosion resistance than Ni-P. Moreover, the surface potential distribution on the surfaces of the two specimens shows no appreciable local variation, suggesting a continuous graphene sheet on Ni-P-G and a homogeneous Ni-P on the Q235 steel substrate.

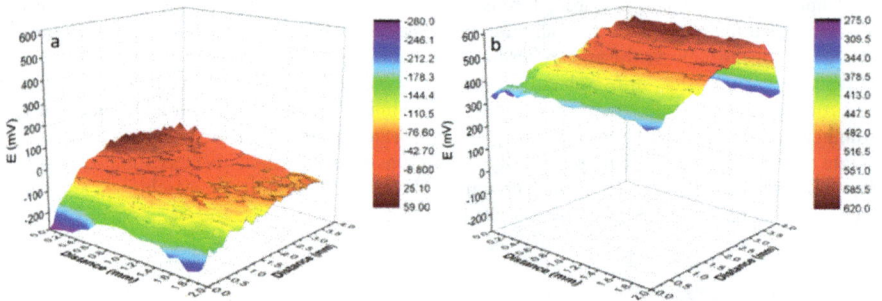

Figure 24. The Ni-P (a) and Ni-P-G (b) surface potential maps obtained with the scanning Kelvin probe (SKP) [62].

The application of LEIS, SDC, and SIET to Ni-P coatings is still uncommon because they are relatively new methods for evaluating corrosion. It goes without saying that applying these techniques in the future will be highly beneficial for assessing the corrosion issue. Table 5 provides a summary of the interpreted methodologies for easier understanding and comparison.

Table 5. An overview of the methods, characteristics measured, and applications [49].

Technique	Measurement	Example of uses
Scanning vibrating electrode Technique (SVET)	Potential Difference/Current Density	Location of oxidation and reduction half-cells and corrosion rate
Localized EIS (LEIS)	Local Capacitance and Resistance	Local variations in EIS and related variables such as CPE
Scanning Electrochemical Microscopy (SECM)	Current	Localized electrochemical activity, O_2 concentration, etc
Scanning Kelvin Probe (SKP)	Work Function	Delamination studies, impact of intermetallic particles on corrosion
Scanning Ion-Selective Electrode Technique (SIET)	Local pH, ion concentration	Corrosion inhibitor mechanisms, ion migration
Scanning Droplet Cell (SDC) Technique	Potential, Current	Alloying element impact on corrosion

6.2.2.7 Confocal laser scanning microscopy

In terms of surface measurement and quantification, confocal laser scanning microscopy performs better than traditional optical microscopy. In a typical CLSM configuration, a laser beam scans the whole x-y plane of the sample surface. Tiny pinholes are positioned in front of the sensor at a position that corresponds visually to the sample surface's focus point. For every x, y location, the detector measures the amount of light that the specimen on the focus surface returns. The majority

Advances in Corrosion Science and Surface Engineering
Materials Research Foundations 188 (2026) 68-120

Materials Research Forum LLC
https://doi.org/10.21741/9781644903919-5

of the light scattered by out-of-focus planes passes through the pinhole and is invisible. Precise imaging of a slice, or single sample surface, is made possible by this phenomenon. By combining the slices obtained at each z-step and gradually moving the sample surface over the focal plane with the aid of a z scanning stage, the three-dimensional surface topography may be computed. This is known as a stack. It is possible to ascertain surface profiles, pit depths, and surface roughness. It is also possible to compute a two-dimensional image with a greater depth of focus. This technique assigns a gray value (corresponding to the maximum brightness level across all slices at the x-y coordinate) to each pixel in the image. The CLSM operates with a 1 mm working distance between the lenses and specimen and is a contact-free method. One benefit of technologies like CLSM is that they eliminate the requirement for suction-based drying assessment and serial extraction, allowing for real-time observations to be made as corrosion occurs. It is no longer absolutely necessary to observe corrosion evolution at the same surface inhomogeneity after the extraction and drying processes. While they can be carried out in situ, other widely used methods for examining the surface topographies in corrosion, like AFM, need contact with or close proximity to the surface to be studied. Therefore, it is not suitable for examining corrosion events occurring beneath the thick layers of organic matter. This method's requirement for a transparent covering is its drawback [63].

Domínguez et al. [64] investigated the corrosion evaluation of a double black layer with two electroless Ni-P coatings in an acidic bath. The three-dimensional images obtained from confocal laser scanning microscopy for the single- and double-layer Ni-P coatings are presented in Fig. 24. With the aid of confocal laser scanning microscopy technology, we were able to determine the internal and external structures of one- and two-layer black Ni-P coatings through their fluorescence, which is dependent on the composition and structure of the surface. The CLSM pictures for black Ni-P coatings applied in single and double layers are displayed in Fig. 25. One possible explanation for the green fluorescence was the presence of a phosphorus molecule, maybe P203. The red-colored fluorescence was ascribed to nickel, which exhibited a monochromatic metal-like emission. The information above allows for the identification of the phosphorus and nickel constituents of the material as well as the size and distribution of the phosphorus-containing particles. Fig. 25a and 25b depict the three-dimensional images of a black Ni-P coating on a single layer. The three-dimensional photographs of the double layer black Ni-P coating in Fig. 25c and 25d provide a clear picture of the dispersion of phosphorus compounds after the coating has been applied to the steel surface. The levels of phosphorus in Figs. 25a and 25b are higher than those in a double layer black Ni-P coating. When black Ni-P layers are split or fractured, the phosphorus-containing particles' fluorescence can show through from below the layer represented by nickel red fluorescence.

6.2.2.8 Electrochemical noise analysis

The electrode dynamics and processes that result in current and potential changes on their own are explained by the ENA. The term ENA refers to the random fluctuations in current and potential observed in corrosion studies at the free corrosion potential. Thus, potential noise and current noise are the two categories into which ENA can be separated. The two elements that are commonly referred to as "working electrodes" and are used to measure noise in corrosion systems are potential noise and current noise. Additionally, a noiseless reference electrode (RE) is utilized to provide

Advances in Corrosion Science and Surface Engineering   Materials Research Forum LLC
Materials Research Foundations 188 (2026) 68-120   https://doi.org/10.21741/9781644903919-5

the most accurate measurements of corrosion system noise. With a zero-resistance ammeter, the current flowing between the two working electrodes is measured. A voltmeter records the potential of free corrosion events with respect to the reference electrode. Two working electrodes with a bias potential and a micro-counter electrode are among the other options. Analyzing the temporal domain noise signal is one of the key mathematical methods. Under typical corrosion conditions, the system operates as intended, and fortunately, external polarizing is not required for the noise analysis. It is now possible to notice important changes in an electrochemical process without jeopardizing or needlessly changing the structure itself. This technique is effective for studying fissures, stress-corrosion cracking, fretting, localized corrosion monitoring, and general corrosion detection. In laboratory settings, it is also very helpful for live corrosion monitoring. Alternative concepts have been established to promote the advancement of electrochemical noise measurements when measuring in extremely low conductivity processes, such as those where the impedance approach frequently fails because of signal loss in the high resistance of the solution [65].

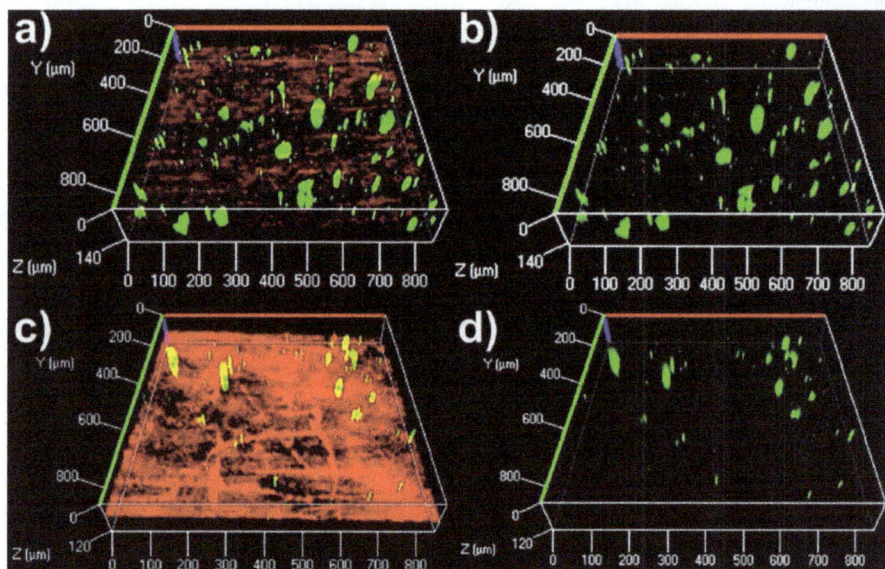

Figure 25. CLSM 3D fluorescence images of a black Ni-P coating red and green section a) one layer, c) two layers and green section b) one layer, d) two layers [64].

The study conducted by Ashassi-Sorkhabi et al. [66] examined the improvement of electroless Ni-P coating corrosion resistance with the use of ultrasonically dispersed diamond nanoparticles. Ultrasonically dispersed diamond nanoparticles (DNP) in a nickel electroless solution were used to sono-deposit the Ni-P/DNP nanocomposite coatings on mild steel. Fig. 26 displays typical raw potential and current noise data. Noise resistance was calculated by analyzing the noise data in time domains. The following formula defines the noise resistance, or R_n:

Materials Research Forum LLC
https://doi.org/10.21741/9781644903919-5

$$R_n = \frac{\sigma v}{\sigma i} \tag{7}$$

where σv and σi represent for the standard population deviations of the noise data related to potential and current, respectively. Results regarding electrochemical noise may be impacted by the test electrode's instability during the testing time. Data on electrochemical noise is typically thought of as random oscillations around a mean value. The corrosion potential represents the mean value in the electrochemical potential noise scenario. During an experiment, it is frequently noticed that the corrosion potential tends to drift. The method of eliminating this phenomenon is known as trend elimination, and it is also known as the direct current (DC) trend. The standard deviations of the potential and current fluctuation have been acknowledged to be impacted by DC trend. Consequently, before conducting additional statistical analysis, the DC trend of the time data was eliminated.

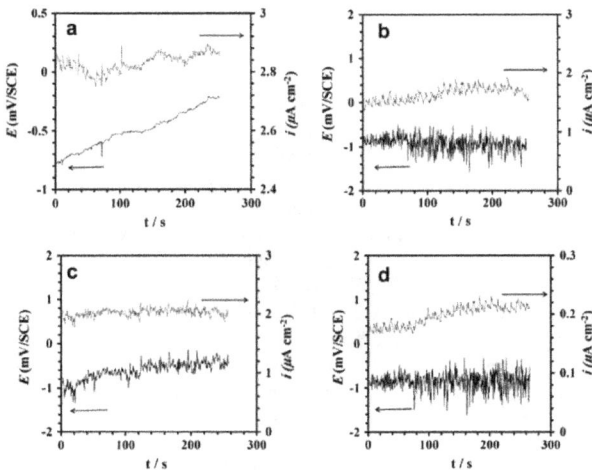

Figure 26. Typical raw electrochemical potential (—) and current (...) noise records for Ni-P coatings with varying DNP concentrations in 3.5% NaCl solution: (a) Ni-P; (b) Ni-P/DNP (20 mg/l); (c) Ni-P/DNP (100 mg/l) and (d) Ni-P/DNP (200 mg/l) [66].

Table 6 provides the electrochemical noise derived characteristics, such as the noise resistances and standard deviations of the potential and current. The values of the potential noise standard deviation in the corrosive solution are shown in Table 6-3 as follows: E_{Ni-P} (0.16 mV) > $E_{Ni-P/DNP(20\ mg/l)}$ (0.18 mV) > $E_{Ni-P/DNP(200\ mg/l)}$ (0.19 mV) > $E_{Ni-P/DNP(100\ mg/l)}$ (0.20 mV), indicating a significant difference in the systems' metal/electrolyte interfaces. Fig. 26 shows that, in contrast to Ni-P coating, the current noise for the nanocomposite coating has somewhat fluctuated, indicating minimal contact between the aggressive solution and the surface of the Ni-P/DNP coating. However, the magnitude of current noise has varied in the Ni-P coating case. Pitting corrosion is the cause of this phenomenon. The mild steel's exposure to pitting corrosion in the saline environment was significantly decreased by the nanocomposite coatings. Furthermore, the mild steel with the highest noise resistance was coated with Ni-P/DNP (100 mg/l).

Table 6. The EN data obtained for Ni-P/DNP coatings with different concentrations of DNP in 3.5% NaCl solution [66].

Specimen	σ_V (mV)	σ_i ($\mu A \cdot cm^{-2}$)	R_n ($\Omega \cdot cm^2$)
Ni-P	0.163	0.0312	5224 ± 97
Ni-P/DNP (20 mg/l)	0.180	0.0177	$10,169 \pm 134$
Ni-P/DNP (100 mg/l)	0.201	0.0145	$13,862 \pm 152$
Ni-P/DNP (200 mg/l)	0.191	0.0157	$12,165 \pm 134$

7.   Technology and trends for the future

7.1   Novel contactless technique

The contactless approach for corrosion system, developed here, is a novel technique that reduces the requirement for direct physical contact between the potentiostat and substrate in order to measure the polarization resistance (R_p). Typically, direct contact between each electrode is necessary in order to polarize the substrate, which serves as a working electrode, and then use the appropriate wires to detect the polarization resistance and potential difference. Electrostatic separation is produced when a current passes between the electrolyte and the substrate, polarizing it. To achieve the desired result, the substrate can be exposed to a magnetic field or either direct current (DC) or alternating current (AC). The methodology relies on the finding that the electrode can be electrically polarized by placing it in the path of an external electrical field created by passing a current between two external graphite electrodes, A and C. When an external voltage or current is introduced to the substrate (M), the electrolyte becomes electrostatically polarized. As a result, the applied current can pass the substrate in parallel field lines, allowing the response to be measured. To record the potential differences, utilize the reference electrodes (REFC and REFA). A schematic of the experimental setup for a four-electrode system, with each electrode connected to a galvanostat, is shown in Fig. 27. The tank is filled with electrolyte, and substrate 'M' serves as the working electrode of the system. The most inventive aspect of this inventive method is that the substrate (M) is positioned vertically, parallel to the reference and counter electrodes, rather than being electrically connected to the galvanostat. To observe the response, current is supplied between two graphite electrodes (A and C) using a galvanostat or potentiostat. The voltage fluctuation at the substrate (M) before and after the applied current is monitored using two saturated calomel reference electrodes (REFA and REFC) next to the bar. As soluble corrosive compounds are formed during the corrosion process, the concentration of electrolyte may drop.

Advances in Corrosion Science and Surface Engineering   Materials Research Forum LLC
Materials Research Foundations 188 (2026) 68-120   https://doi.org/10.21741/9781644903919-5

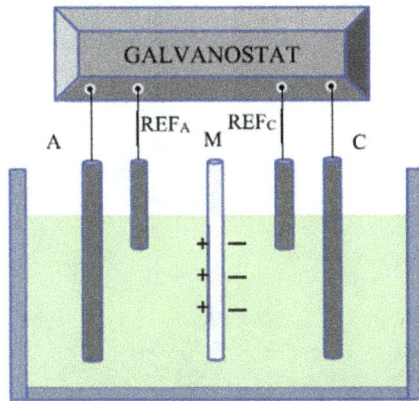

Figure 27. Diagrammatic illustration of the electrode configurations for the novel contactless method [36].

This novel contactless technique uses a multi-phase corrosion rate measurement process. The formula $R_e = V_{Re}/I_{ap}$ can be used to calculate the electrolyte resistance. An external current (I_{ap}) is applied to the substrate (M) immersed in the electrolyte without coming into touch with it. The applied current polarizes the substrate by changing the potential and producing an external electric field in the surrounding electrolyte. In cases where the substrate (M) is not present in the electrolyte, V_{Re} is measured with a four-electrode setup. The substrate (M) is then immersed in the electrolyte using the same configuration, and V_{e+M} is seen. The voltage differences between the reference electrodes (REFA and REFC) are referred to as the value of V_{e+M}, where $V_{e+M} = V_2 - V_0$. The reference electrode voltage drop (V_0) is measured before testing. After applying I_{ap} current for a predetermined period of time, a second voltage drop (V_2) between the reference electrodes is then measured; it typically takes less than a minute to establish a constant value of V_2. It is possible to compare the substrate resistance (R_M) with the conventional polarization resistance (R_p). V_{e+M} is then retested after a few hours or days to obtain R_M values and determine whether or not it is comparable to R_p. Fig. 28 depicts the electrical circuit for the contactless method, in which current passes through the electrolyte and substrate bar in parallel. It is believed that the current takes two different routes: one polarizes the substrate bar (M), while the other passes via the electrolyte. The corrosion resistance of the substrate bar can be assessed using the following expression:

$$1/(R_{e+M}) = 1/R_e + 1/R_M \tag{8}$$

where Re is the electrolyte resistance in the absence of substrate, RM is the substrate resistance, and Re+M is the resistance in the presence of substrate in the electrolyte. Using the Stern–Geary relationship, the corrosion current (icorr) can be computed from RM (Stern and Geary 1957). Because of the separation of charges, the substrate produces currents that are approximately proportionate to the voltage generated, meeting a fundamental need of the linear polarization technique. If the currents are directly proportionate to the electric potential, which is contingent upon the substrate condition, then the ratio of the current to the electric potential (induced voltage) is referred to as the polarization resistance. Non-destructive testing, or NDT, is the capacity to determine the rate of corrosion without the working electrode coming into direct contact with the calculation. This could pave the way for the creation of clever applications in new ways [36].

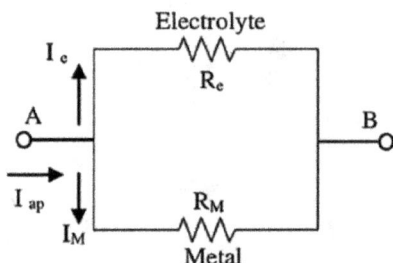

Figure 28. An electrical model that depicts the non-contact method's design [36].

This is a novel method in the field of corrosion evaluation and may be widely used in the future. Corrosion monitoring is a critical component that needs to be carried out using the particular methodologies, as per the ways previously discussed. Due to the fact that the kind of corrosion and the circumstances around the corroding samples affect the test's results or prohibit the use of specific methodologies. Many highly accurate approaches have been established to date, such as methods for determining local impedance, impedance, and polarization resistance. Furthermore, precise microscopy of the corroding systems is unavailable, despite the fact that it may yield a wealth of useful data. It appears that the direction of corrosion microscopy going forward will be to combine electron and atomic microscopy with extremely sensitive electrodes. Furthermore, corrosion products can be characterized; however, the accuracy of data can be improved by combining corrosion monitoring with material characterizations.

References

[1] C. R. Hegedus, "A holistic perspective of coatings technology," JCT research, vol. 1, no. 1, pp. 5-20, 2004. https://doi.org/10.1007/s11998-004-0020-4

[2] S. K. Ghosh, "Functional coatings and microencapsulation: a general perspective," Functional Coatings: by polymer microencapsulation, pp. 1-28, 2006. https://doi.org/10.1002/3527608478.ch1

[3] P. A. Sørensen, S. Kiil, K. Dam-Johansen, and C. E. Weinell, "Anticorrosive coatings: a review," Journal of coatings technology and research, vol. 6, pp. 135-176, 2009. https://doi.org/10.1007/s11998-008-9144-2

[4] O. Ali, Corrosion is a natural process that converts a refined metal into a more chemically-stable form such as oxide, hydroxide, or sulfide. 2020.

[5] H. Aljibori, A. Alamiery, and A. Kadhum, "Advances in corrosion protection coatings: A comprehensive review," Int. J. Corros. Scale Inhib, vol. 12, no. 4, pp. 1476-1520, 2023. https://doi.org/10.17675/2305-6894-2023-12-4-6

Advances in Corrosion Science and Surface Engineering   Materials Research Forum LLC
Materials Research Foundations 188 (2026) 68-120   https://doi.org/10.21741/9781644903919-5

[6] G. H. Koch, M. P. Brongers, N. G. Thompson, Y. P. Virmani, and J. H. Payer, "Corrosion cost and preventive strategies in the United States," United States. Federal Highway Administration, 2002.

[7] C. M. Hussain et al., Sustainable corrosion inhibitors I: fundamentals, methodologies, and industrial applications. ACS Publications, 2021. https://doi.org/10.1021/bk-2021-1403

[8] B. Fotovvati, N. Namdari, and A. Dehghanghadikolaei, "On coating techniques for surface protection: A review," Journal of Manufacturing and Materials processing, vol. 3, no. 1, p. 28, 2019. https://doi.org/10.3390/jmmp3010028

[9] W. Sade, R. T. Proença, T. D. d. O. Moura, and J. R. T. Branco, "Electroless Ni-P coatings: preparation and evaluation of fracture toughness and scratch hardness," International Scholarly Research Notices, vol. 2011, 2011. https://doi.org/10.5402/2011/693046

[10] A. Popoola, O. Olorunniwo, and O. Ige, "Corrosion resistance through the application of anti-corrosion coatings," Developments in corrosion protection, vol. 13, no. 4, pp. 241-270, 2014. https://doi.org/10.5772/57420

[11] C. Gu, J. Lian, G. Li, L. Niu, and Z. Jiang, "High corrosion-resistant Ni-P/Ni/Ni-P multilayer coatings on steel," Surface and coatings technology, vol. 197, no. 1, pp. 61-67, 2005. https://doi.org/10.1016/j.surfcoat.2004.11.004

[12] A. C. R. Wurtz, "On Copper Hydride," Hebdomadaires des Séances de l'Académie des Sciences, Vol. 18, 1844, pp. 702-704.

[13] A. C. R. Wurtz, "On Copper Hydride," Hebdomadaires des Séances de l'Académie des Sciences, Vol. 21, 1845, p. 149.

[14] K. H. Krishnan, S. John, K. Srinivasan, J. Praveen, M. Ganesan, and P. Kavimani, "An overall aspect of electroless Ni-P depositions-A review article," Metallurgical and materials transactions A, vol. 37, pp. 1917-1926, 2006. https://doi.org/10.1007/s11661-006-0134-7

[15] R. Duncan, "Performance of electroless nickel coated steel in oil field environments," Mater. Performance;(United States), vol. 22, no. 1, 1983. https://doi.org/10.5006/C1982-82136

[16] F. Colaruoto, B. V. Tilak and R. S. Jasinki, "Corrosion Charactheristcs of Electroless Nickel Coating of Oil Field Environments," Proceedings of Electroless Nickel Conference IV, Chicago, 22-24 April 1985

[17] F. B. Mainier, I. M. R. A. Brüning and E. F. Pamplona, "Desenvolvimento de Recipientes para Acondicionamento de Gás Natural Contendo Gases Corrosivos," In: VI Encontro Brasileiro de Tratamento de Superfícies (EBRAT-1989), ABTS, Sao Paulo, 1987, pp. 66-81.

[18] V. T. Talinn, "In the World of Electroless Nickel," Finishing, Vol. 12, 1988, p. 26.

[19] R. Weil, J. Lee, I. Kim, and K. Parker, "Comparison of Some Mechanical and Corrosion Properties of Electroless and Electroplated Nickel--Phosphorus Alloys," Plat. Surf. Finish., vol. 76, no. 2, pp. 62-66, 1989.

[20] F. B. Mainier and M. M. de Araújo, "On the effect of the electroless nickel-phosphorus (Ni-P) coating defects on the performance of this type of coating in oilfield environments," SPE

Advanced Technology Series, vol. 2, no. 01, pp. 63-67, 1994. https://doi.org/10.2118/23635-PA

[21] F. Delaunois, J. Petitjean, P. Lienard, and M. Jacob-Duliere, "Autocatalytic electroless nickel-boron plating on light alloys," Surface and Coatings Technology, vol. 124, no. 2-3, pp. 201-209, 2000. https://doi.org/10.1016/S0257-8972(99)00621-0

[22] X. Liu, J.-Q. Gao and W.-B. Hu, "Application of Electroless Ni-P Alloys in Electronic Industry," Plating & Finishing, Vol. 28, No. 1, 2006, pp. 30-34.

[23] D. Baudrand, "Adhesion of Electroless Nickel Deposits to Aluminum Alloys-We Now Have a Better Understanding of the Factors Influencing Adhesion," Products Finishing, Vol. 63, No. 10, 2009, pp. 80-87.

[24] F. B. Mainier, M. P. C. Fonseca, S. S. Tavares, and J. M. Pardal, "Quality of electroless Ni-P (nickel-phosphorus) coatings applied in oil production equipment with salinity," Journal of Materials Science and Chemical Engineering, vol. 1, no. 6, pp. 1-8, 2013. https://doi.org/10.4236/msce.2013.16001

[25] J. Balaraju, T. Sankara Narayanan, and S. Seshadri, "Electroless Ni-P composite coatings," Journal of applied electrochemistry, vol. 33, pp. 807-816, 2003. https://doi.org/10.1023/A:1025572410205

[26] F. Bigdeli and S. R. Allahkaram, "An investigation on corrosion resistance of as-applied and heat treated Ni-P/nanoSiC coatings," Materials & Design, vol. 30, no. 10, pp. 4450-4453, 2009. https://doi.org/10.1016/j.matdes.2009.04.020

[27] G. O. Mallory and J. B. Hajdu, Electroless plating: fundamentals and applications. William Andrew, 1990.

[28] Thomas Publishing Company (2020): "The Electro Nickel Plating Process". Online article at the Thomasnet.com website. Accessed on 2020-07-11.

[29] N. El Mahallawy, A. Bakkar, M. Shoeib, H. Palkowski, and V. Neubert, "Electroless Ni-P coating of different magnesium alloys," Surface and Coatings Technology, vol. 202, no. 21, pp. 5151-5157, 2008. https://doi.org/10.1016/j.surfcoat.2008.05.037

[30] J. Balaraju, T. Sankara Narayanan, and S. Seshadri, "Evaluation of the corrosion resistance of electroless Ni-P and Ni-P composite coatings by electrochemical impedance spectroscopy," Journal of solid state electrochemistry, vol. 5, pp. 334-338, 2001. https://doi.org/10.1007/s100080000159

[31] M. Yan, H. Ying, and T. Ma, "Improved microhardness and wear resistance of the as-deposited electroless Ni-P coating," Surface and Coatings Technology, vol. 202, no. 24, pp. 5909-5913, 2008. https://doi.org/10.1016/j.surfcoat.2008.06.180

[32] M. Alvarez, S. Vazquez, F. Audebert, and H. Sirkin, "Corrosion behaviour of Ni-B-Sn amorphous alloys," Scripta materialia, vol. 39, no. 6, pp. 661-668, 1998. https://doi.org/10.1016/S1359-6462(98)00227-9

[33] V. Genova et al., "Medium and High Phosphorous Ni-P Coatings Obtained via an Electroless Approach: Optimization of Solution Formulation and Characterization of Coatings," Coatings, vol. 13, no. 9, p. 1490, 2023. https://doi.org/10.3390/coatings13091490

[34] N. Kothanam, K. Harachai, J. Qin, Y. Boonyongmaneerat, N. Triroj, and P. Jaroenapibal, "Hardness and tribological properties of electrodeposited Ni-P multilayer coatings fabricated through a stirring time-controlled technique," journal of materials research and technology, vol. 19, pp. 1884-1896, 2022. https://doi.org/10.1016/j.jmrt.2022.05.145

[35] Don Baudrand, Brad Durkin, "Automotive applications of electroless nickel", Metal Finishing,

Volume 96, Issue 5, 1998, Pages 20-24, ISSN 0026-0576, https://doi.org/10.1016/S0026-0576(98)80080-9. https://doi.org/10.1016/S0026-0576(98)80080-9

[36] I. A. Shozib, A. Ahmad, A. M. Abdul-Rani, M. Beheshti, and A. A. A. Aliyu, "A review on the corrosion resistance of electroless Ni-P based composite coatings and electrochemical corrosion testing methods," Corrosion Reviews, vol. 40, no. 1, pp. 1-37, 2022. https://doi.org/10.1515/corrrev-2020-0091

[37] J. Scully and D. Taylor, "Electrochemical methods of corrosion testing," Metals Handbook, vol. 13, pp. 212-228, 1987.

[38] P. C. Okonkwo et al., "Potentiodynamic polarization test as a versatile tool for bipolar plates materials at start-up and shut-down environments: a review," International Journal of Green Energy, vol. 18, no. 11, pp. 1193-1202, 2021. https://doi.org/10.1080/15435075.2021.1904948

[39] S. Feliu Jr, "Electrochemical impedance spectroscopy for the measurement of the corrosion rate of magnesium alloys: Brief review and challenges," Metals, vol. 10, no. 6, p. 775, 2020. https://doi.org/10.3390/met10060775

[40] A. Varmazyar, S. R. Allahkaram, and S. Mahdavi, "Deposition, Characterization and Evaluation of Monolayer and Multilayer Ni, Ni-P and Ni-P-Nano ZnO p Coatings," Transactions of the Indian institute of metals, vol. 71, pp. 1301-1309, 2018. https://doi.org/10.1007/s12666-018-1279-y

[41] M. Esmaily et al., "Fundamentals and advances in magnesium alloy corrosion," Progress in Materials Science, vol. 89, pp. 92-193, 2017. https://doi.org/10.1016/j.pmatsci.2017.04.011

[42] W. Sun, P. Zhang, F. Zhang, W. Hou, and K. Zhao, "Microstructure and corrosion resistance of Ni-P gradient coatings," Transactions of the IMF, vol. 93, no. 4, pp. 180-185, 2015. https://doi.org/10.1179/0020296715Z.000000000249

[43] C. Suryanarayana, K. C. Rao, and D. Kumar, "Preparation and characterization of microcapsules containing linseed oil and its use in self-healing coatings," Progress in organic coatings, vol. 63, no. 1, pp. 72-78, 2008. https://doi.org/10.1016/j.porgcoat.2008.04.008

[44] R. Gao, M. Du, X. Sun, and Y. Pu, "Study of the corrosion resistance of electroless Ni-P deposits in a sodium chloride medium," Journal of Ocean University of China, vol. 6, pp. 349-354, 2007. https://doi.org/10.1007/s11802-007-0349-2

[45] S. S. Jamali, S. E. Moulton, D. E. Tallman, M. Forsyth, J. Weber, and G. G. Wallace, "Evaluating the corrosion behaviour of Magnesium alloy in simulated biological fluid by using SECM to detect hydrogen evolution," Electrochimica Acta, vol. 152, pp. 294-301, 2015. https://doi.org/10.1016/j.electacta.2014.11.012

[46] W. Huang et al., "Corrosion behavior and biocompatibility of hydroxyapatite/magnesium phosphate/zinc phosphate composite coating deposited on AZ31 alloy," Surface and Coatings Technology, vol. 326, pp. 270-280, 2017. https://doi.org/10.1016/j.surfcoat.2017.07.066

[47] M. Islam, M. R. Azhar, N. Fredj, and T. D. Burleigh, "Electrochemical impedance spectroscopy and indentation studies of pure and composite electroless Ni-P coatings," Surface and Coatings Technology, vol. 236, pp. 262-268, 2013. https://doi.org/10.1016/j.surfcoat.2013.09.057

[48] V. M. Huang, S.-L. Wu, M. E. Orazem, N. Pébère, B. Tribollet, and V. Vivier, "Local electrochemical impedance spectroscopy: A review and some recent developments," Electrochimica Acta, vol. 56, no. 23, pp. 8048-8057, 2011. https://doi.org/10.1016/j.electacta.2011.03.018

[49] N. Jadhav and V. J. Gelling, "The use of localized electrochemical techniques for corrosion studies," Journal of The Electrochemical Society, vol. 166, no. 11, p. C3461, 2019. https://doi.org/10.1149/2.0541911jes

[50] H. Wan, D. Song, X. Li, D. Zhang, J. Gao, and C. Du, "Effect of zinc phosphate on the corrosion behavior of waterborne acrylic coating/metal interface," Materials, vol. 10, no. 6, p. 654, 2017. https://doi.org/10.3390/ma10060654

[51] S. Rossi, M. Fedel, F. Deflorian, and M. del Carmen Vadillo, "Localized electrochemical techniques: Theory and practical examples in corrosion studies," Comptes Rendus Chimie, vol. 11, no. 9, pp. 984-994, 2008. https://doi.org/10.1016/j.crci.2008.06.011

[52] A. Stankiewicz, Z. Kefallinou, G. Mordarski, Z. Jagoda, and B. Spencer, "Surface functionalisation by the introduction of self-healing properties into electroless Ni-P coatings," Electrochimica Acta, vol. 297, pp. 427-434, 2019. https://doi.org/10.1016/j.electacta.2018.12.026

[53] D. Fix, E. V. Skorb, D. G. Shchukin, and H. Möhwald, "Quantitative analysis of scanning electric current density and pH-value observations in corrosion studies," Measurement science and technology, vol. 22, no. 7, p. 075704, 2011. https://doi.org/10.1088/0957-0233/22/7/075704

[54] S. Lamaka, R. M. Souto, and M. G. Ferreira, "In-situ visualization of local corrosion by Scanning Ion-selective Electrode Technique (SIET)," Microscopy: Science, technology, applications and education, vol. 3, pp. 2162-2173, 2010.

[55] A. Gnedenkov, S. Sinebryukhov, D. Mashtalyar, and S. Gnedenkov, "Localized corrosion of the Mg alloys with inhibitor-containing coatings: SVET and SIET studies," Corrosion Science, vol. 102, pp. 269-278, 2016. https://doi.org/10.1016/j.corsci.2015.10.015

[56] M. Lohrengel, A. Moehring, and M. Pilaski, "Electrochemical surface analysis with the scanning droplet cell," Fresenius' journal of analytical chemistry, vol. 367, pp. 334-339, 2000. https://doi.org/10.1007/s002160000402

[57] K. Fushimi, S. Yamamoto, R. Ozaki, and H. Habazaki, "Cross-section corrosion-potential profiles of aluminum-alloy brazing sheets observed by the flowing electrolyte scanning-droplet-cell technique," Electrochimica Acta, vol. 53, no. 5, pp. 2529-2537, 2008. https://doi.org/10.1016/j.electacta.2007.10.044

[58] M. Beheshti, S. Kakooei, M. C. Ismail, and S. Shahrestani, "Design and implementation of scanning electrochemical microscopy (SECM) for scanning open circuit potential system on damaged coated surfaces," J Eng Appl Sci, vol. 11, pp. 14299-14302, 2016.

[59] C. G. Zoski, "advances in scanning electrochemical microscopy (SECM)," Journal of The Electrochemical Society, vol. 163, no. 4, p. H3088, 2015. https://doi.org/10.1149/2.0141604jes

[60] M. C. L. de Oliveira et al., "Structural characterization, global and local electrochemical activity of electroless Ni-P-multiwalled carbon nanotube composite coatings on pipeline steel," Metals, vol. 11, no. 6, p. 982, 2021. https://doi.org/10.3390/met11060982

[61] M. Wicinski, W. Burgstaller, and A. W. Hassel, "Lateral resolution in scanning Kelvin probe microscopy," Corrosion Science, vol. 104, pp. 1-8, 2016. https://doi.org/10.1016/j.corsci.2015.09.008

[62] Q.-Y. Wang et al., "Realization of graphene on the surface of electroless Ni-P coating for short-term corrosion prevention," Coatings, vol. 8, no. 4, p. 130, 2018. https://doi.org/10.3390/coatings8040130

[63] P. Jakupi, J. Noël, and D. Shoesmith, "The evolution of crevice corrosion damage on the Ni-Cr-Mo-W alloy-22 determined by confocal laser scanning microscopy," Corrosion science, vol. 54, pp. 260-269, 2012. https://doi.org/10.1016/j.corsci.2011.09.028

[64] A. S. Domínguez, J. P. Bueno, I. Z. Torres, and M. M. López, "Characterization and corrosion resistance of electroless black Ni-P coatings of double black layer on carbon steel," Surface and coatings Technology, vol. 326, pp. 192-199, 2017. https://doi.org/10.1016/j.surfcoat.2017.07.044

[65] J. Lv, Q. x. Yue, R. Ding, X. Wang, T. j. Gui, and X. d. Zhao, "The application of electrochemical noise for the study of metal corrosion and organic anticorrosion coatings: a review," ChemElectroChem, vol. 8, no. 2, pp. 337-351, 2021. https://doi.org/10.1002/celc.202001342

[66] H. Ashassi-Sorkhabi and M. Es, "Corrosion resistance enhancement of electroless Ni-P coating by incorporation of ultrasonically dispersed diamond nanoparticles," Corrosion science, vol. 77, pp. 185-193, 2013. https://doi.org/10.1016/j.corsci.2013.07.046

Advances in Corrosion Science and Surface Engineering   Materials Research Forum LLC
Materials Research Foundations 188 (2026) 120-138   https://doi.org/10.21741/9781644903919-6

Chapter 6

Corrosion Influencing Factors: Environmental, Material and Electrochemical Perspectives

N. Muthukumaran[1]*, B. Arulmurugan[1], V. Manoj Mohan Prasath[2], G. Kausalya Sasikumar[3,] Rajender Boddula[4]

[1]Department of Mechanical Engineering, KPR Institute of Engineering and Technology, Coimbatore, India

[2]Department of Mechanical Engineering, Karpagam Academy of Higher Education, Coimbatore, India

[3]Centre for Research and Development,KPR Institute of Engineering and Technology, Coimbatore, India

[4]School of Sciences, Woxsen University, Hyderabad - 502345, Telangana, India

nmuthu1966@gmail.com

Abstract

Corrosion compromises the structural integrity of metals, presenting difficulties across various industries. This chapter examines environmental elements that affect corrosion, emphasizing exposure conditions, temperature, humidity, and chemical environments. Marine and industrial environments with elevated chloride levels and contaminants exhibit heightened corrosion rates, whereas rising temperatures and humidity intensify material degradation. Chemical exposures induce localized and uniform degradation through complex reactions. Case studies emphasize these repercussions, illustrating the importance of understanding these factors to develop effective prevention strategies. This chapter focuses on the fundamental principle of metal preservation for future use, highlighting the essential importance of corrosion science in maintaining durability.

Keywords

Corrosion, Environmental Factors, Temperature, Chemical Exposure, Marine Environment, Material Degradation

Contents

1. Introduction

Corrosion refers to the progressive degradation or destruction of materials, primarily metals, due to chemical or electrochemical interactions with their surroundings. This process typically involves the material's exposure to elements such as oxygen, moisture, acids, or other chemicals, leading to the formation of oxides, hydroxides, or other compounds on the metal surface [1]. The following flowchart (Fig. 1) illustrates the key factors influencing corrosion, categorized into environmental, material-related, and electrochemical aspects.

Environmental exposure plays a critical role in determining the rate and nature of corrosion. Marine environments, characterized by high salinity and chlorides, are particularly aggressive, significantly accelerating corrosion in metals. Industrial environments introduce additional challenges due to the presence of pollutants, chemicals, and acidic emissions, which further degrade metallic structures. Urban environments, though less corrosive than marine and industrial settings, still experience corrosion from air pollution and moisture. Rural environments tend to exhibit slower corrosion rates; however, exposure to agricultural chemicals such as fertilizers and pesticides may still contribute to corrosion [2].

Temperature is a key factor influencing the rate and type of corrosion. High temperatures generally accelerate corrosion by increasing the rates of electrochemical reactions. As temperature rises, the diffusion of ions and electrons within the electrolyte intensifies, leading to faster oxidation and reduction reactions. In humid environments with high temperatures, metals like steel and aluminum are more prone to uniform and pitting corrosion. Additionally, high temperatures can destabilize protective oxide layers, leaving the metal surface more susceptible to further attack. Conversely, low temperatures may slow corrosion rates due to reduced electrochemical activity, though other challenges such as frost-induced stress and moisture condensation can still occur. In

Advances in Corrosion Science and Surface Engineering   Materials Research Forum LLC
Materials Research Foundations 188 (2026) 120-138   https://doi.org/10.21741/9781644903919-6

marine environments, freezing temperatures can cause saltwater to concentrate on metal surfaces, intensifying localized corrosion when the ice melts [3].

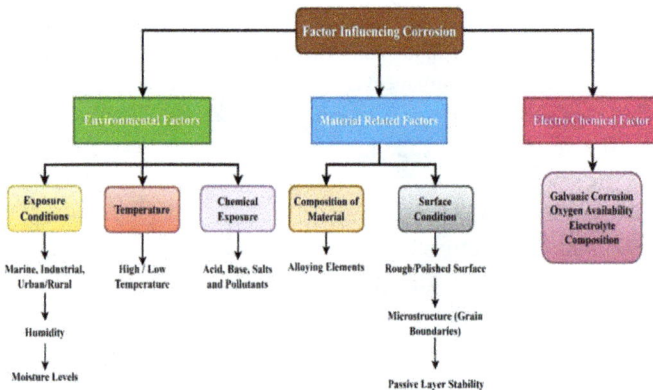

Figure 1. Categorization of Factors Influencing Corrosion in Metals

Chemical exposure plays a significant role in corrosion, with acids, bases, salts, and pollutants all affecting metal integrity. Acidic environments lower the pH, accelerating metal dissolution and promoting corrosion, particularly in metals like steel and aluminum. Although less aggressive, bases can also cause corrosion, especially in concrete structures where steel reinforcement may corrode if the protective oxide layer is compromised. Saltwater, particularly in marine environments, is a major contributor to corrosion, as chloride ions break down protective oxide layers, leading to pitting and crevice corrosion. Pollutants such as sulfur dioxide and nitrogen oxides can combine with moisture to form acids, further accelerating corrosion, especially on exposed metal surfaces. These chemical exposures underline the importance of implementing effective corrosion protection strategies, particularly in environments with aggressive chemical conditions [4].

Material-related factors significantly influence the corrosion behavior of metals. The environment in which a metal is placed, such as marine, industrial, or urban settings, directly impacts its corrosion rate. For instance, metals in marine environments with high humidity and saltwater exposure undergo more aggressive corrosion. Temperature variations, both high and low, can accelerate or slow corrosion depending on the material and environmental conditions.

The composition of a material, including the presence of alloying elements, is crucial in determining its corrosion resistance. Alloys containing elements like chromium, nickel, and molybdenum generally exhibit better resistance to corrosion. In contrast, metals with lower alloy content are more susceptible to corrosion under certain conditions. Surface condition is another critical factor; rough surfaces tend to accumulate moisture and chemicals, increasing the likelihood of corrosion. Polished surfaces, on the other hand, offer better resistance by reducing the accumulation of corrosive agents. Additionally, the material's microstructure, such as grain boundaries and the stability of the passive layer, influences corrosion resistance. A stable passive

Advances in Corrosion Science and Surface Engineering   Materials Research Forum LLC
Materials Research Foundations 188 (2026) 120-138   https://doi.org/10.21741/9781644903919-6

layer, like that found on stainless steel, can protect the metal from further degradation. However, disruption of this layer can significantly lower the material's resistance to corrosion [4].

Electrochemical factors play a vital role in corrosion, as corrosion is primarily an electrochemical process. Galvanic corrosion is a common phenomenon where two dissimilar metals are in electrical contact within an electrolyte. The more active metal corrodes, while the less active metal is protected, leading to accelerated degradation of the anode material. This process is often observed in environments where different metals, such as steel and copper, are used in proximity, such as in marine applications [5].

Oxygen availability also influences corrosion. In environments with abundant oxygen, such as open air, corrosion rates increase as oxygen readily reacts with the metal surface. In oxygen-deprived environments, such as underwater or buried structures, corrosion may slow, but other forms, such as anaerobic corrosion, can occur. The composition of the electrolyte, including factors like pH and ion concentration, plays a pivotal role in corrosion. Electrolytes with high conductivity, like salty water, accelerate corrosion by facilitating ion flow between the anodic and cathodic areas of the metal surface. The presence of ions such as chloride can exacerbate localized corrosion by breaking down protective oxide films, leading to pitting and crevice corrosion. Thus, understanding the electrochemical environment is essential for predicting and mitigating corrosion in various applications [6]. Fig 2 gives inference the corrosion losses in GDP of various countries.

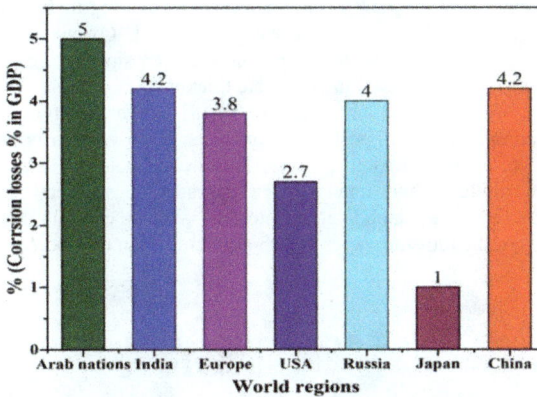

Figure 2. % in GDP losses by corrosion in the world regions.

This chapter has outlined the various factors influencing corrosion, including environmental conditions, material properties, and electrochemical interactions, which collectively contribute to significant global losses. As industries face the challenge of mitigating these losses, understanding these factors becomes crucial in addressing corrosion-related issues. The following sections will delve deeper into these influencing factors, providing a comprehensive analysis of their impact on corrosion, with a focus on how they shape the development of effective corrosion prevention strategies, coatings, and other technological advancements.

Advances in Corrosion Science and Surface Engineering   Materials Research Forum LLC
Materials Research Foundations 188 (2026) 120-138   https://doi.org/10.21741/9781644903919-6

2.   Environmental Factors

2.1   Marine environments

The corrosive nature of marine environments is driven not only by the high concentration of chloride ions but also by the electrochemical reactions occurring on the surface of metals. In marine atmospheres, a thin aqueous layer forms on the metal, acting as an electrolyte, which is essential for corrosion to take place. This moisture can be as thin as a few monolayers or as thick as several hundred microns, depending on factors such as rainfall, humidity, and temperature changes, which promote condensation [7].The process of atmospheric corrosion (AC) is a complex electrochemical reaction involving both oxidation and reduction. The anodic reaction involves the oxidation of the metal, typically iron, as described by the Equ (1):

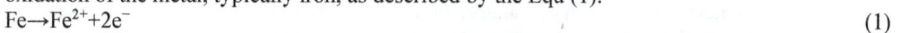

$$Fe \rightarrow Fe^{2+} + 2e^- \tag{1}$$

In the presence of oxygen, which is highly soluble in the aqueous layer, the reduction reaction

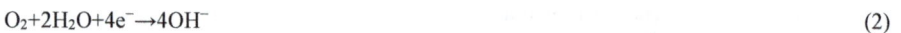

occurs as Equ (2):

$$O_2 + 2H_2O + 4e^- \rightarrow 4OH^- \tag{2}$$

The hydroxide ions produced migrate to the anodic areas, leading to the formation of ferrous hydroxide [$Fe(OH)_2$], which is the initial corrosion product. Oxygen diffusion through the moisture layer plays a significant role in controlling the corrosion rate. This rate can reach its maximum when the aqueous layer is of intermediate thickness, as excessively thick or thin layers hinder the movement of oxygen and ions, respectively [8]. In marine environments, factors such as the presence of sodium chloride (NaCl) and other salts, along with air pollutants like SO_2 and NO_x, further contribute to the corrosion process. These chemicals can lead to the formation of additional corrosion products and intensify the degradation of metals. As the electrolyte evaporates, the process becomes cathodically controlled, with the corrosion rate increasing due to the reduced thickness of the moisture layer, which accelerates ion transport [9].

Figure 3. Marine corrosion case study model

Fig 3 illustrated the model boat to conduct the real time case study. The study highlights that corrosion in the Spanish fishing fleet, particularly in boats made from aluminum, wood, and glass fiber reinforced polyester (GFRP), has led to significant repair and maintenance costs. The corrosion rate varied depending on environmental factors such as moisture, chloride concentration, and exposure time. Although specific cost losses or exact corrosion rates were not quantified in the study, it was noted that corrosion-related issues significantly increased maintenance expenses. In aluminum boats, the corrosion problems, especially due to galvanic corrosion and poor paint adhesion, were linked to increased costs for repairs and protective treatments. Similarly, in steel boats, the galvanic corrosion due to immersion in seawater contributed to increased costs for repainting and cathodic protection. The research concluded that proactive measures, such as employing corrosion protection systems and conducting regular inspections, could help reduce these costs over time [10].

2.2   Industrial Corrosion

Industrial corrosion refers to the degradation of materials, particularly metals, caused by their exposure to industrial environments containing chemicals, gases, and contaminants. Aggressive agents such as sulfur dioxide (SO_2), nitrogen oxides (NO_x), and chlorine compounds, commonly released from manufacturing operations, power plants, and refineries, accelerate corrosion processes. Metals like steel and aluminum exposed to these pollutants rapidly form oxides, sulfides, and chlorides, leading to material deterioration and structural failures. The consequences of industrial corrosion include significant material loss, safety risks, and high repair costs, making corrosion control essential for preserving the integrity of industrial equipment, pipelines, and structures [11].

Figure 4. Corroded pipe line in Oil refinery industries (Courtesy: NACE).

In the oil and gas industry, pipelines play a crucial role in transporting crude oil and natural gas but are especially vulnerable to corrosion, particularly in coastal areas. Fig 4 shows the corroded pipe line fitting. A major oil company reported substantial corrosion issues in their underground pipelines due to environmental pollutants, including high levels of SO_2 and chloride ions, originating from nearby industrial activities and the marine environment. Environmental factors such as high humidity, salt spray, and seasonal temperature fluctuations further accelerated corrosion. The pipelines, made of carbon steel, suffered from rust and galvanic corrosion due to their connections with dissimilar metals. Over a period of five years, the company faced frequent leaks, ruptures, unplanned shutdowns, and substantial financial losses, including costs related to environmental remediation. In response, they implemented several strategies: corrosion-resistant coatings, cathodic protection systems, regular monitoring technologies, and collaboration with

Advances in Corrosion Science and Surface Engineering   Materials Research Forum LLC
Materials Research Foundations 188 (2026) 120-138   https://doi.org/10.21741/9781644903919-6

local industries to reduce pollutant emissions. These measures significantly reduced corrosion rates, minimized downtime, and lowered repair expenses, leading to enhanced operational efficiency and cost savings. This case study underscores the importance of addressing environmental factors, material properties, and proactive corrosion mitigation techniques to ensure the safety and longevity of industrial infrastructure [12].

2.3   Humidity Corrosion

Humidity corrosion, or moisture-induced corrosion, occurs when metals are exposed to environments with high levels of atmospheric moisture, especially in the form of water vapor. In such conditions, a thin aqueous film forms on metal surfaces, providing the electrolyte necessary for electrochemical reactions. This film facilitates oxidation, with the rate of corrosion increasing as relative humidity (RH) rises due to enhanced ion mobility within the water layer [2]. High RH is particularly detrimental in coastal or industrial regions, where persistent humidity accelerates metal degradation. The presence of environmental pollutants like sulfur dioxide, chloride ions, and carbon dioxide lowers the pH of the water film, intensifying the corrosion process by making the film more acidic [15]. Fig 5 shows the pipe line carrying water was corroded due to humidity.

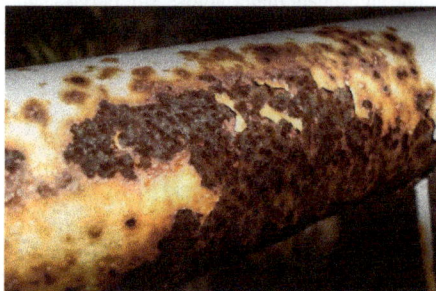

Figure 5. Water line pipe corrosion (Courtesy: Fractory)

Critical relative humidity (CRH) thresholds further influence corrosion rates. Studies indicate that for clean metal surfaces, CRH values are around 80–85% for carbon steel, 70% for zinc and nickel, and 76% for aluminum. However, salts and pollutants significantly reduce these thresholds, with deliquescent relative humidity (DRH) as low as 35% for some salts like MgCl. The effects of humidity are further compounded by temperature fluctuations and dew formation. When the surface temperature drops below the dew point, condensation forms, creating a conductive moisture layer that facilitates oxidation and accelerates rust formation, particularly in ferrous metals such as carbon steel.

Advances in Corrosion Science and Surface Engineering   Materials Research Forum LLC
Materials Research Foundations 188 (2026) 120-138   https://doi.org/10.21741/9781644903919-6

Figure 6. Moisture corrosion due to Chain links immersed in water body (Courtesy Test Labs)

In environments where RH frequently exceeds CRH, corrosion rates increase exponentially, as shown in magnesium alloys, where RH increases from 75% to 95% caused a substantial rise in corrosion current density [2]. Predictive models like the Peck and Vernon's exponential models estimate the acceleration factor (AF) of humidity-induced corrosion, although their effectiveness diminishes under near-saturation conditions.

Mitigation strategies include the use of hydrophobic coatings, corrosion-resistant materials, and environmental controls like dehumidification. Regular maintenance to remove salts and pollutants is also effective in reducing corrosion rates. Understanding the interaction between humidity, contaminants, and environmental factors is critical for developing effective protective measures to ensure the longevity of materials exposed to high-humidity conditions [13].

2.4   Moisture Corrosion

Moisture corrosion is closely related to humidity corrosion but specifically refers to the corrosion that takes place when metals are exposed to water (Fig 6) droplets or liquid moisture. This type of corrosion is common in areas subject to direct contact with water or where condensation occurs. In the presence of liquid water, corrosion can occur more rapidly due to the direct contact between the metal surface and the electrolyte. Water can also facilitate the movement of aggressive ions such as chloride (Cl^-), which is especially harmful in coastal environments. In cases of stagnant water, such as in industrial equipment or pipes, corrosion can cause localized damage, including pitting and crevice corrosion.

The rate of moisture corrosion is dependent on factors like water temperature, pH, and the presence of corrosive contaminants. In some cases, the corrosion rate can reach alarming levels if moisture is trapped between metal parts, preventing proper drying or ventilation [14].

2.5   Temperature

Temperature plays a pivotal role in the corrosion process by influencing the rate of electrochemical reactions. Elevated temperatures accelerate corrosion by enhancing ion mobility, reducing activation energy, and increasing the solubility of reactive gases like oxygen in aqueous solutions, all of which intensify metal oxidation and reduction reactions [15].

Figure 7. Sleeve of the burner corroded by high temperature.

For example, carbon steel experiences rapid iron oxide formation at high temperatures, particularly in acidic or saline conditions. Similarly, elevated temperatures can compromise the protective chromium oxide layer on stainless steel, leading to localized corrosion such as pitting or crevice corrosion. Aluminum also becomes more prone to oxidation at higher temperatures, especially in chloride-rich environments [16]. Fig 7 shows the example of the high temperature corroded sleeve of the boiler burner.

Temperature fluctuations further complicate corrosion behavior. Thermal cycling, caused by repeated expansion and contraction of metals, weakens protective oxide layers or coatings, increasing susceptibility to corrosion. Additionally, temperature affects the adhesion and structure of corrosion products. At lower temperatures, these products are often less adherent and protective, whereas higher temperatures may promote denser oxide layer formation, though thermal stress can cause these layers to crack.

Figure 8: Various acid container corroded by their pH levels.

The relationship between temperature and corrosion is highly influenced by other environmental factors. When relative humidity (RH) is below the critical relative humidity (CRH), temperature has little effect on corrosion due to the absence of an electrolyte film. However, when RH exceeds the CRH, elevated temperatures amplify corrosion rates by increasing reaction activation energy. This is especially pronounced in tropical regions with high humidity, where diurnal temperature

Advances in Corrosion Science and Surface Engineering   Materials Research Forum LLC
Materials Research Foundations 188 (2026) 120-138   https://doi.org/10.21741/9781644903919-6

fluctuations promote moisture condensation at night and evaporation during the day. Such conditions accelerate the formation and breakdown of electrolyte films, intensifying corrosion.

In freezing conditions, corrosion halts as the electrolyte film freezes, preventing oxygen from reaching the metal surface. Conversely, in marine or polluted environments, increased temperature can accelerate the diffusion of oxygen and chloride ions while simultaneously reducing oxygen solubility, potentially leading to compact but less protective corrosion products [17]. In certain settings, like those with high concentrations of carbon dioxide or nitric acid, the effect of temperature on corrosion rates may be negligible, as shown for zinc and copper.

Overall, temperature influences corrosion in complex and multifaceted ways. Its impact varies depending on the material, environmental conditions, and interactions with other factors such as humidity and pollutants. Understanding these dynamics is critical for selecting appropriate materials and implementing effective corrosion mitigation strategies in industries like oil and gas, marine applications, and power generation. Accurate analysis of temperature's effects allows for better design and maintenance of materials exposed to diverse thermal environments.

2.6   Chemical Exposure

2.6.1   Acidic and Basic Environments

Acidic and basic environments significantly influence the corrosion behavior of metals. Acids, such as sulfuric acid (H_2SO_4) and hydrochloric acid (HCl), accelerate corrosion by lowering the pH of the material surface (Fig 8), promoting metal dissolution. In acidic solutions, the high concentration of hydrogen ions (H^+) destabilizes protective oxide layers, exposing the bare metal to further attack. This phenomenon is particularly severe in reactive metals like carbon steel and aluminum, where the interaction between the acidic medium and the metal surface leads to the formation of soluble metal ions and hydrogen gas evolution. Conversely, bases, though generally less aggressive, can also induce corrosion under certain conditions. For example, in highly alkaline environments, such as those found in concrete structures, steel can corrode due to chloride ions or excessive alkalinity that destabilizes the protective oxide layer [18].

The behavior of stainless-steel grades SS304 and SS316 in acidic and basic environments exemplifies the differences in corrosion resistance under such conditions. Potentio dynamic and potentio static polarization studies revealed that SS316, due to its molybdenum content, exhibited superior corrosion resistance compared to SS304. In neutral chloride media (1M NaCl), the corrosion current density of SS316 was nearly four orders of magnitude lower than that of SS304, reflecting its enhanced resistance to chloride-induced corrosion. SEM analysis confirmed that SS304 developed pits even at passivation potentials, whereas SS316 maintained a smoother surface with minimal pit formation.

In acidic environments (1M NaCl and 1M HCl), SS304 suffered significant intergranular corrosion, with pits connecting to form larger corroded regions. In contrast, SS316 experienced localized deposition of corrosion products, such as hydroxides, which protected the surface. The corrosion current density of SS316 in acidic conditions remained 94% lower than that of SS304, underscoring its stability.

Under alkaline conditions (1M NaCl and 1M NaOH), both grades showed improved corrosion performance due to the passivation effects of hydroxide ions. However, SS316 demonstrated more consistent passivation with negligible current density increases, while SS304 exhibited passive layer breakdown and transpassivation under certain conditions. EDX analysis further revealed higher oxygen-to-iron (O/Fe) ratios on SS316 surfaces, indicating effective protective oxide layer formation, whereas SS304's lower ratios correlated with pit formation and localized corrosion.

This study emphasizes that SS316 outperforms SS304 in acidic and basic environments due to its superior passivation behavior and resistance to pitting corrosion. The findings highlight the importance of selecting appropriate stainless-steel grades based on environmental conditions to ensure long-term structural integrity [19].

2.6.2 Saltwater and Pollutants

Saltwater, particularly in marine environments, is one of the most aggressive accelerants of corrosion. The high concentration of chloride ions in saltwater destabilizes protective oxide films on metal surfaces, leading to localized forms of corrosion such as pitting and crevice corrosion. The high conductivity of saltwater solutions further accelerates electrochemical reactions, causing rapid material degradation. Metals like carbon steel, which lack inherent corrosion resistance, are particularly vulnerable, but even stainless steels are susceptible when exposed to high chloride concentrations over extended periods [20]. Pollutants such as sulfur dioxide (SO_2) and nitrogen oxides (NO_x) exacerbate this corrosive environment by dissolving in atmospheric water vapor to form sulfuric acid (H_2SO_4) and nitric acid (HNO_3). These acidic compounds deposit on metal surfaces, lowering the pH and accelerating corrosion. This combination of salt and acidic pollutants creates highly aggressive conditions, often seen in structural damage to bridges, pipelines, and industrial facilities [21].

Figure 9: Carbon steel flange coupling near marine port exposes to sea water and air pollutants (Courtesy Ulbrich)

Real-world case studies highlight the destructive effects of saltwater and pollutants. Coastal bridges frequently suffer from steel reinforcement corrosion due to salt-laden air and chloride deposition, resulting in concrete spalling and structural degradation. Offshore oil and gas pipelines are also at high risk of corrosion from saltwater exposure, necessitating advanced mitigation measures such as protective coatings and cathodic protection systems. Similarly, industrial cooling towers exposed to polluted atmospheres experience severe corrosion of metal components, reducing their operational efficiency and lifespan [22].

The study of seawater and pollutant impacts near the Kutubdia Channel, Bangladesh, further underscores the interplay between environmental pollutants and marine conditions. Seasonal

Advances in Corrosion Science and Surface Engineering   Materials Research Forum LLC
Materials Research Foundations 188 (2026) 120-138   https://doi.org/10.21741/9781644903919-6

sampling revealed elevated concentrations of heavy metals like Cu, Zn, Mn, and Pb, particularly in sediments, which acted as primary pollutant sinks, storing over 99% of the contaminants compared to water. Pollution indices such as the Contamination Factor (CF) and Pollution Load Index (PLI) categorized the contamination as low to moderate for these trace metals. However, enrichment factor (EF) analysis indicated severe to extremely severe enrichment of Cu and Zn, suggesting both natural and anthropogenic pollution sources. Seasonal variations were attributed to riverine inputs during monsoons and evaporation rates during pre-monsoon periods [21].

Although the potential ecological risk index (PER) suggested a low risk to aquatic ecosystems, human health risk analysis identified non-carcinogenic risks, particularly through inhalation, with children being the most vulnerable. These findings emphasize the importance of continuous monitoring and effective mitigation strategies to manage corrosion and pollution risks in coastal and industrial regions [22].

3.   Material Factors

3.1   Composition of Alloys

The chemical composition of a metal, including the presence of alloying elements such as chromium, nickel, and molybdenum, significantly affects its corrosion resistance. Alloys with higher concentrations of these elements exhibit superior resistance to corrosion. For example, stainless steel contains chromium, which forms a passive oxide layer that prevents further oxidation. In contrast, carbon steel, with lower alloy content, is more susceptible to corrosion under harsh conditions [23].

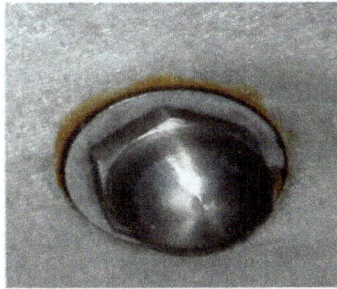

Figure 10: Stainless steel alloy related crevice region corrosion

Chromium plays a crucial role in enhancing corrosion resistance. When exposed to oxygen, chromium reacts to form a thin, adherent, and self-healing chromium oxide layer (Cr_2O_3) on the metal's surface shows in Fig 10. This passive layer serves as a protective shield against environmental factors like moisture and chloride ions. The effectiveness of this layer increases with higher chromium content, making the alloy more robust and durable under corrosive conditions [21].

Nickel further enhances corrosion resistance by stabilizing the alloy's passive layer and improving its adherence, particularly in acidic and high-temperature environments. In stainless steel, nickel contributes to an austenitic microstructure, which offers superior resistance to stress corrosion cracking and pitting compared to other phases. Nickel also provides improved resistance to reducing agents, such as hydrochloric and sulfuric acids, making alloys like Inconel 59 and AISI 904L highly suitable for chemical processing applications [8]. Together, chromium and nickel create alloys that resist both uniform corrosion and localized degradation, such as pitting, crevice corrosion, and intergranular corrosion.

Passivation is a key process in preventing corrosion by forming a protective oxide layer on the metal's surface, which acts as a barrier against environmental factors like oxygen and moisture. This process naturally occurs in metals like stainless steel, titanium, and aluminum due to their high content of alloying elements such as chromium. In stainless steel, the presence of at least 10.5% chromium ensures the formation and self-repair of the passive film if it becomes damaged. This feature is particularly important in high-humidity or chloride-rich environments, where unprotected metals are prone to rust and pitting corrosion [24].

Industrial applications often enhance passivation through chemical treatments, such as immersion in oxidizing acids, to remove contaminants and create a uniform, stable oxide layer. These treatments strengthen the protective barrier, significantly extending the durability and reliability of metals in challenging environments. Alloys such as Inconel 59 and AISI 904L, with high chromium and nickel content, are widely used in marine environments, chemical plants, and flue gas desulfurization systems due to their ability to maintain integrity even under aggressive conditions. This ensures long-term performance and minimizes the risk of structural degradation.

3.2   Surface Condition and Microstructure

The surface condition of a material has a profound influence on its susceptibility to corrosion. Rough surfaces are more prone to corrosion because they trap moisture, contaminants, and corrosive agents, creating localized environments that accelerate degradation [25]. The increased surface area of rough textures also disrupts the uniform formation of protective passive films, leaving the underlying metal exposed to corrosive attacks shows in Fig 11. In contrast, smooth, polished surfaces resist corrosion more effectively as they reduce the likelihood of crevices and promote uniform passive film coverage. Techniques such as polishing and electropolishing have been widely adopted in industries to enhance surface smoothness, thereby improving corrosion resistance.

The microstructure of a material is another critical factor affecting its corrosion behavior. Grain boundaries, phase distributions, and the presence of secondary phases can influence how corrosion initiates and propagates. Grain boundaries often act as initiation sites for corrosion, particularly in metals with heterogeneous microstructures or intergranular impurities. Conversely, refined grains and uniform microstructures stabilize passive films, distribute stresses more evenly, and reduce susceptibility to localized corrosion such as pitting or intergranular corrosion. For instance, in stainless steel, chromium-enriched grain boundaries promote the formation of a self-healing chromium oxide layer, contributing to enhanced corrosion resistance [15].

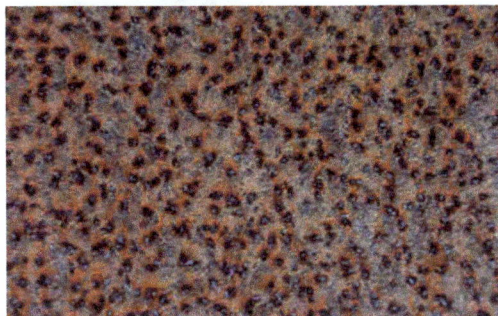

Figure 11: Surface corrosion

Industrial applications illustrate the importance of surface treatments in mitigating corrosion. Processes like shot peening introduce compressive stresses on metal surfaces, reducing the risk of crack propagation and stress corrosion cracking. Chemical passivation, especially in stainless steel, enhances the formation of protective passive films, while anodizing in aluminum alloys improves corrosion resistance in marine environments. Electropolishing of reactor vessels and pipelines in chemical plants has been shown to reduce the accumulation of corrosive chemicals, thereby extending their service life. Similarly, surface nitriding of steel components used in oil rigs increases their resistance to wear and corrosion, providing a robust solution in harsh, salt-laden environments.

These examples underscore how surface condition and microstructural optimization can significantly enhance a material's durability in corrosive environments. By addressing surface roughness and refining microstructures, industries can improve the longevity and reliability of materials in applications ranging from marine to chemical processing and oil and gas industries.

4.   Electrochemical Factors

4.1   Galvanic Corrosion

Galvanic corrosion occurs when two dissimilar metals are in electrical contact in the presence of an electrolyte, such as saltwater. In this process, the metal with the lower electrochemical potential (anode) corrodes at an accelerated rate, while the metal with the higher potential (cathode) is protected illustrated in Fig 12.

This electrochemical reaction is driven by the potential difference between the metals, creating a galvanic cell. Such corrosion is particularly problematic in environments where moisture or electrolytes are present, such as marine or industrial settings.To mitigate galvanic corrosion, it is essential to use materials with similar electrochemical properties. Metals that are close to each other on the galvanic series generate minimal potential differences, reducing the risk of galvanic reactions [26]. When dissimilar metals must be used, insulating layers such as coatings or gaskets can be applied to physically separate them, breaking the electrochemical pathway necessary for corrosion. This approach is highly effective in preventing galvanic interactions.

Advances in Corrosion Science and Surface Engineering   Materials Research Forum LLC
Materials Research Foundations 188 (2026) 120-138   https://doi.org/10.21741/9781644903919-6

Figure 12. Copper alloy and iron Galvanic corrosion

Examples of galvanic corrosion are commonly observed in maritime and offshore environments. For instance, in ships, steel or aluminum hulls often come into contact with copper-based materials like brass in propellers. In this case, the steel or aluminum acts as the anode, corroding faster due to the galvanic reaction. Offshore oil rigs also experience similar issues, with connections between steel and copper alloys in electrical wiring or piping. Such corrosion can lead to structural failures, increased maintenance costs, and safety hazards [27].

In pipelines, galvanic corrosion frequently occurs at connections between steel pipes and brass or copper fittings, particularly in moist or wet environments. This poses a significant challenge to the structural integrity of the pipeline and necessitates protective measures. Addressing galvanic corrosion through careful material selection, insulating barriers, and regular monitoring is crucial to extending the service life of these structures while minimizing maintenance expenses and safety risks.

4.2 Oxygen Availability

Oxygen availability is a critical factor influencing corrosion rates, as oxygen facilitates the electrochemical reactions that lead to metal degradation. In oxygen-rich environments, such as open air or oxygenated water, oxygen readily reacts with metal surfaces to form oxides. This reaction drives the cathodic process, accelerating the transformation of metals into their corroded state [28]. The presence of abundant oxygen promotes the formation of stable oxide layers, which may initially provide some protection. However, if these layers are porous or uneven, they fail to prevent further oxidation, allowing corrosion to progress illustrated in Fig 13. In marine environments, where structures like ships, offshore rigs, and subsea pipelines are exposed to oxygenated saltwater, corrosion occurs rapidly due to the combined effects of oxygen, salt, and humidity. The presence of chloride ions further exacerbates the corrosion process, often leading to localized damage such as pitting corrosion. Mitigating underwater corrosion in such conditions requires the application of protective coatings, cathodic protection systems, and consistent maintenance to safeguard metal components.

Figure 13 Oxygen pitting corrosion on a boiler tube

Conversely, in oxygen-deprived environments, such as buried pipelines, storage tanks, or deep-sea settings, corrosion rates may initially be slower. However, these environments often experience other localized forms of corrosion. Differential aeration cells, caused by oxygen concentration gradients, can accelerate corrosion in areas with higher oxygen exposure. Furthermore, anaerobic conditions often foster microbial-induced corrosion (MIC), where bacteria produce corrosive by-products, such as hydrogen sulfide, which contribute to metal degradation.

For example, in buried pipelines or enclosed spaces, microbial activity under oxygen-depleted conditions can lead to significant structural damage. Similarly, in deep-sea environments, sulfates act as electron acceptors, causing sulfate-reducing bacteria to produce hydrogen sulfide, which induces localized corrosion. These conditions highlight the importance of understanding oxygen availability in corrosion control strategies. Employing effective measures such as oxygen barriers, coatings, and chemical treatments can significantly reduce the risk of oxygen-driven or microbial-induced corrosion. Regular monitoring and maintenance remain critical in preventing corrosion in both oxygen-rich and oxygen-poor environments [24].

Conclusion

Corrosion is a complex phenomenon influenced by environmental, material, and electrochemical factors. Marine, industrial, and urban environments pose unique challenges, with factors such as chloride ions, acidic pollutants, and oxygen availability accelerating corrosion processes. Temperature, humidity, and chemical exposure further amplify the degradation of materials in these settings.

Material composition plays a critical role in corrosion resistance. Alloys containing chromium, nickel, and molybdenum enhance passivation, forming stable, protective oxide layers. For instance, stainless steels like SS316 demonstrate superior resistance to acidic and chloride-rich environments compared to SS304 due to their molybdenum content. Surface treatments, including electropolishing, passivation, and anodizing, further improve corrosion resistance by creating smooth and uniform protective barriers.

Mechanisms like galvanic corrosion and oxygen-driven degradation present additional challenges. Galvanic corrosion occurs when dissimilar metals are in contact within an electrolyte, while oxygen-rich environments accelerate corrosion through oxide formation. Conversely, oxygen-deprived conditions often result in microbial-induced corrosion (MIC), which leads to localized damage in buried pipelines or deep-sea structures.

Effective corrosion prevention requires a combination of material selection, surface treatment, and environmental control. Strategies such as cathodic protection, advanced coatings, and regular maintenance are essential to mitigate corrosion and ensure the longevity of structures in harsh environments. Understanding the interplay of these factors allows for improved durability, reduced maintenance costs, and enhanced safety across diverse applications.

References

[1]Zaki Ahmad. 2006. "Basic Concepts in Corrosion." Principles of Corrosion Engineering and Corrosion Control, 9–56.

[2]Wei, Chao, George Wang, Marcus Cridland, David L. Olson, and Stephen Liu. 2018. Corrosion Protection of Ships. Handbook of Environmental Degradation Of Materials: Third Edition. Third Edit. Elsevier Inc. https://doi.org/10.1016/B978-0-323-52472-8.00026-5

[3]Tortorelli, P. F., and K. Natesan. 1998. "Critical Factors Affecting the High-Temperature Corrosion Performance of Iron Aluminides." Materials Science and Engineering: A 258 (1–2): 115–25. https://doi.org/10.1016/S0921-5093(98)00924-1

[4]Solomon, Moses M., Ikenna B. Onyeachu, Demian I. Njoku, Simeon C. Nwanonenyi, and Emeka E. Oguzie. 2021. "Adsorption and Corrosion Inhibition Characteristics of 2–(Chloromethyl)Benzimidazole for C1018 Carbon Steel in a Typical Sweet Corrosion Environment: Effect of Chloride Ion Concentration and Temperature." Colloids and Surfaces A: Physicochemical and Engineering Aspects 610 (July): 125638. https://doi.org/10.1016/j.colsurfa.2020.125638

[5]Cwalina, B. 2014. Biodeterioration of Concrete, Brick and Other Mineral-Based Building Materials. Understanding Biocorrosion: Fundamentals and Applications. Woodhead Publishing Limited. https://doi.org/10.1533/9781782421252.3.281

[6]Al-Moubaraki, Aisha H., and Ime Bassey Obot. 2021. "Corrosion Challenges in Petroleum Refinery Operations: Sources, Mechanisms, Mitigation, and Future Outlook." Journal of Saudi Chemical Society 25 (12): 101370. https://doi.org/10.1016/j.jscs.2021.101370.

[7]Alcántara, Jenifer, Daniel de la Fuente, Belén Chico, Joaquín Simancas, Iván Díaz, and Manuel Morcillo. 2017. "Marine Atmospheric Corrosion of Carbon Steel: A Review." Materials 10 (4). https://doi.org/10.3390/ma10040406

[8]Cole, I. S., W. D. Ganther, J. D. Sinclair, D. Lau, and D. A. Paterson. 2004. "A Study of the Wetting of Metal Surfaces in Order to Understand the Processes Controlling Atmospheric Corrosion." Journal of The Electrochemical Society 151 (12): B627. https://doi.org/10.1149/1.1809596

[9]Schindelholz, E., B. E. Risteen, and R. G. Kelly. 2014. "Effect of Relative Humidity on Corrosion of Steel under Sea Salt Aerosol Proxies." Journal of The Electrochemical Society 161 (10): C450–59. https://doi.org/10.1149/2.0221410jes

[10]   Orosa, José A., Juan José Galán, and Mar Toledano. 2013. "A Real Case Study on the State of Corrosion of the Spanish Boats." Journal of Marine Science and Technology (Taiwan) 21 (4): 391–99. https://doi.org/10.6119/JMST-012-0518-4

[11]   Dong, Baojun, Wei Liu, Tianyi Zhang, Longjun Chen, Yueming Fan, Yonggang Zhao, Weijian Yang, and Wongpat Banthukul. 2021. "Corrosion Failure Analysis of Low Alloy Steel and Carbon Steel Rebar in Tropical Marine Atmospheric Environment: Outdoor Exposure and Indoor Test." Engineering Failure Analysis 129 (September): 105720. https://doi.org/10.1016/j.engfailanal.2021.105720

[12]   Xia, Dahai, Shizhe Song, Weixian Jin, Jian Li, Zhiming Gao, Jihui Wang, and Wenbin Hu. "Atmospheric corrosion monitoring of field-exposed Q235B and T91 steels in Zhoushan offshore environment using electrochemical probes." Journal of Wuhan University of Technology-Mater. Sci. Ed. 32, no. 6 (2017): 1433-1440.

[13]   Cai, Yikun, Yuanming Xu, Yu Zhao, and Xiaobing Ma. "Atmospheric corrosion prediction: a review." Corrosion Reviews 38, no. 4 (2020): 299-321.

[14]   Popov, Branko N. 2015. "Evaluation of Corrosion." Corrosion Engineering, 1–28. https://doi.org/10.1016/b978-0-444-62722-3.00001-x

[15]   Mohammadi Zerankeshi, Meysam, Reza Alizadeh, Ehsan Gerashi, Mohammad Asadollahi, and Terence G. Langdon. 2022. "Effects of Heat Treatment on the Corrosion Behavior and Mechanical Properties of Biodegradable Mg Alloys." Journal of Magnesium and Alloys 10 (7): 1737–85. https://doi.org/10.1016/j.jma.2022.04.010

[16]   Di, X. U., Y. A. N. G. Xiaojia, L. I. Qing, C. H. E. N. G. Xuequn, and L. I. Xiaogang. "Review on corrosion test methods and evaluation techniques for materials in atmospheric Environment." Journal of Chinese Society for Corrosion and protection 42, no. 3 (2021): 447-457.

[17]   Tian, Yuwan, Chaofang Dong, Xuequn Cheng, Yingqi Wan, Gui Wang, Kui Xiao, and Xiaogang Li. "The micro-solution electrochemical method to evaluate rebar corrosion in reinforced concrete structures." Construction and Building Materials 151 (2017): 607-614.

[18]   Chawla, S. K., T. Anguish, and J. H. Payer. "Microsensors for corrosion control." Corrosion 45, no. 7 (1989): 595-601.

[19]   Maharajan, S., F. Michael Thomas Rex, D. Ravindran, and S. Rajakarunakaran. "Surface morphology studies and corrosion behaviour of plasma sprayed Cr3C2/8YSZ composite coating on SS316." Surface Topography: Metrology and Properties 11, no. 2 (2023): 025003.

[20]   Little, Brenda J., and Jason S. Lee. "Microbiologically influenced corrosion." Oil and Gas Pipelines (2015): 387-398.

[21]   Fu, Yu, Jun Li, Hong Luo, Cuiwei Du, and Xiaogang Li. 2021. "Recent Advances on Environmental Corrosion Behavior and Mechanism of High-Entropy Alloys." Journal of Materials Science and Technology 80: 217–33. https://doi.org/10.1016/j.jmst.2020.11.044

[22]   Fanijo, Ebenezer O., Joseph G. Thomas, Yizheng Zhu, Wenjun Cai, and Alexander S. Brand. "Surface characterization techniques: a systematic review of their principles, applications, and perspectives in corrosion studies." Journal of The Electrochemical Society 169, no. 11 (2022): 111502.

[23]   Ezuber, Hosni M., Abdulla Alshater, SM Zakir Hossain, and Ali El-Basir. "Impact of soil characteristics and moisture content on the corrosion of underground steel pipelines." Arabian Journal for Science and Engineering 46 (2021): 6177-6188.

[24]   Gu, Ji Dong. 2018. Microbial Biofilms, Fouling, Corrosion, and Biodeterioration of Materials. Handbook of Environmental Degradation Of Materials: Third Edition. Third Edit. Elsevier Inc. https://doi.org/10.1016/B978-0-323-52472-8.00014-9

[25]   Kim, Changkyu, Lin Chen, Hui Wang, and Homero Castaneda. 2021. "Global and Local Parameters for Characterizing and Modeling External Corrosion in Underground Coated Steel Pipelines: A Review of Critical Factors." Journal of Pipeline Science and Engineering 1 (1): 17–35. https://doi.org/10.1016/j.jpse.2021.01.010

[26]   Kawahara, Yuuzou. 2002. "High Temperature Corrosion Mechanisms and Effect of Alloying Elements for Materials Used in Waste Incineration Environment." Corrosion Science 44 (2): 223–45. https://doi.org/10.1016/S0010-938X(01)00058-0

[27]   Karimi, Shabnam, Iman Taji, Tarlan Hajilou, Simona Palencsár, Arne Dugstad, Afrooz Barnoush, Kim Verbeken, Tom Depover, and Roy Johnsen. 2023. "Role of Cementite Morphology on Corrosion Layer Formation of High-Strength Carbon Steels in Sweet and Sour Environments." Corrosion Science 214 (September 2022). https://doi.org/10.1016/j.corsci.2023.111031

[28]   Pei, Zibo, Xuequn Cheng, Xiaojia Yang, Qing Li, Chenhan Xia, Dawei Zhang, and Xiaogang Li. 2021. "Understanding Environmental Impacts on Initial Atmospheric Corrosion Based on Corrosion Monitoring Sensors." Journal of Materials Science and Technology 64: 214–21. https://doi.org/10.1016/j.jmst.2020.01.023

Advances in Corrosion Science and Surface Engineering
Materials Research Foundations 188 (2026) 139-146

Materials Research Forum LLC
https://doi.org/10.21741/9781644903919-7

Chapter 7

Factors Influencing Corrosion Resistance in Ni-P Coatings

Isaac Otu[1]*, Samuel Osei-Amponsah[2]

[1]Materials Engineer-Corrosion Control. Quantum Terminals PLC

[2]Research Assistant. Department of Materials Science. University of Ghana, Ghana

Isaacotu982@gmail.com

Abstract

This review focuses on factors influencing the corrosion resistance of Ni-P coatings. The study explores the significance of protective coatings in mitigating corrosion of metals. Electroless nickel-phosphorus (Ni-P) coatings offer excellent corrosion resistance, with phosphorus content playing a crucial role. Factors such as pH, temperature, deposition time, chemicals, and residual stress also impact corrosion resistance. Understanding these factors is vital for optimizing Ni-P coatings in diverse industrial applications.

Keywords

Ni-P Coatings, Electroless Deposition, Corrosion Resistance, Phosphorus Content, pH, Temperature, Deposition Time

Contents

Advances in Corrosion Science and Surface Engineering   Materials Research Forum LLC
Materials Research Foundations 188 (2026) 139-146   https://doi.org/10.21741/9781644903919-7

1. Introduction

Anticorrosion refers to the methods employed to mitigate the degradation of materials, particularly metals, caused by chemical or electrochemical reactions with their environment [1]. These methods include the application of protective coatings [2], and the use of corrosion inhibitors [3].

Nickel coatings offer exceptional resistance to dry gases such as ammonia, hydrogen, and carbon dioxide, making them ideal for applications in corrosive environments. They also provide a barrier against substances like fuel, oil, soap, and carbon tetrachloride, ensuring longevity and durability in diverse settings. Moreover, nickel plating enhances the fatigue strength of materials, making them suitable for high-stress applications where endurance is crucial. Additionally, these coatings effectively mitigate corrosion fatigue, further extending the lifespan of the coated surfaces. While nickel coatings exhibit remarkable resistance to various chemicals and environments, they may not withstand harsh conditions involving nitric acid and chloride. However, incorporating a thin layer of microcracked chromium can significantly enhance the longevity of nickel coatings by impeding corrosion propagation. This mechanism redirects corrosion pathways laterally, minimizing direct damage to the underlying nickel layer and thereby increasing the overall effectiveness and lifespan of the coating [4].

Electroless nickel-phosphorus (Ni-P) coatings are well-known for their excellent corrosion resistance [5]. The phosphorus content in Ni-P coatings plays a crucial role. Coatings with high phosphorus content offer superior corrosion protection, while moderate or lower amounts are less preferred for harsh environments. Factors such as pH, temperature, deposition time, chemicals, and residual stress can impact the corrosion resistance of Ni-P coating. Other factors that can impact the corrosion resistance of Ni-P coating include; the incorporation of particles (such as SiC) and proper heat treatment.

2. Electroless Ni-P coatings.

Electroless plating is the selective reduction of metal ions from its aqueous solution onto a substrate immersed in the solution. Electroless deposition results in coatings with fine structure, low porosity proper bonding to substrate, and uniform coverage [6]. Electroless Ni-P coatings have a wide range of industrial applications as a result of their excellent corrosion resistance, wear resistance, and good mechanical and electrical properties [6], [7]. A wide range of substrates such as low-alloyed steels, Al, Mg, and Cu alloys can be used for the deposition. The deposition process involves immersing the substrate into an aqueous bath containing sodium hypophosphate and nickel sulfate [7].

2.1 Phosphorus content

The phosphorus content of Ni-P coatings greatly influences their corrosion resistance. Higher phosphorus contents give excellent corrosion resistance than coatings with lower phosphorus contents [8]. Ni-P coatings are categorized into three grades based on their phosphorus content as per the ASTM 733B-04 standard. The coating grades are low (1-4 wt. % P), average (5-9 wt. % P), and high (10 wt. % P and above) [7],[9]. Research shows that Ni-P coatings with high P content have an amorphous microstructure with a nobler corrosion potential (E_{corr}) resulting in improved corrosion resistance[10]. In a study by [7], electroless deposition was used to apply functionally graded Ni-P coatings with low phosphorus top layers to copper substrates. The results indicated that the wear and corrosion resistance improved as the phosphorus content increased due to the

Advances in Corrosion Science and Surface Engineering   Materials Research Forum LLC
Materials Research Foundations 188 (2026) 139-146   https://doi.org/10.21741/9781644903919-7

formation of a crack-free compact nodular structure. Fig. 1 displays the scanning electron microscopy (SEM) images of the surface morphology of the samples. (a) corresponds to the sample with a low P content, while (b) represents the sample with a medium P/low P content, and (c) depicts the sample with a high P/medium P/low P layers.

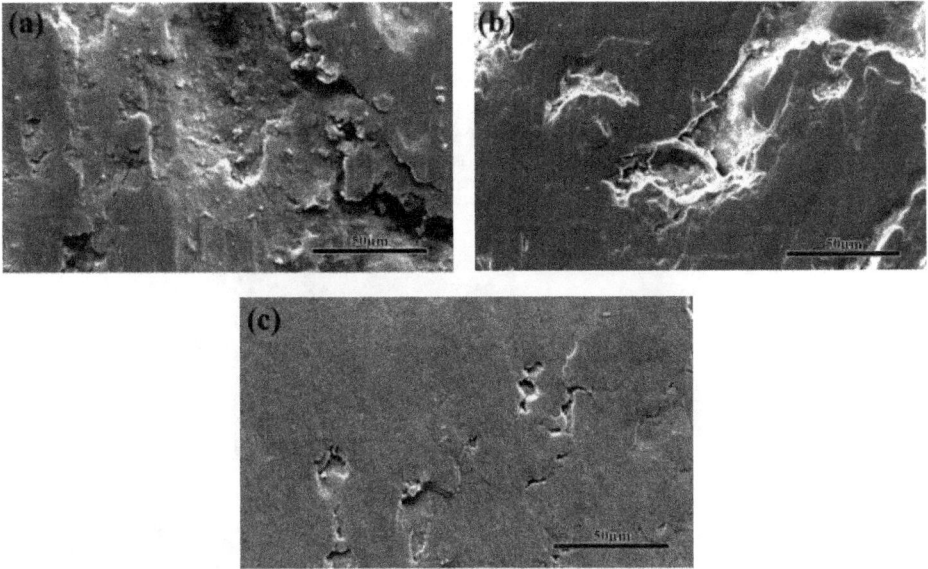

Figure 1. SEM images illustrating the wear surface morphology of the samples with different phosphorus (P) contents: (a) low P, (b) medium P/low P, and (c) high P/medium P/low P layers[7].

2.2   Effect of pH

The rate of deposition in Ni-P electroless coatings is greatly affected by the pH of the electroless bath. The Ni content in the coating deposit increases with increasing pH of the electroless bath. This results in an increase in the rate of deposition. On the other hand, as the pH increases, the phosphorus content reduces due to a decline in the reduction reaction of phosphorus. The reduction reactions for nickel and phosphorus are given by Eq. 1 and Eq. 2 respectively.

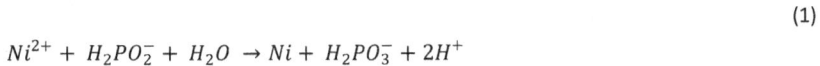

$$Ni^{2+} + H_2PO_2^- + H_2O \rightarrow Ni + H_2PO_3^- + 2H^+ \tag{1}$$

Advances in Corrosion Science and Surface Engineering   Materials Research Forum LLC
Materials Research Foundations 188 (2026) 139-146   https://doi.org/10.21741/9781644903919-7

$$3H_2PO_2^- \rightarrow H_2PO_3^- + H_2O + 2OH^- + 2P \hspace{2cm} (2)$$

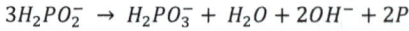

A decrease in the pH of the electroless bath shifts Eq. 2 in the right direction, resulting in a decrease in OH⁻ ions concentration with a corresponding increase in the phosphorus content in the coating. On the other hand, an increase in the pH shifts Eq. 1 to the right causing a decrease in the H⁺ ions concentration and an increase in the Ni content in the coating. This variation in Ni and P content with pH is shown in Fig. 2. As discussed earlier, phosphorus content greatly influences the corrosion resistance of Ni-P coatings. Decreasing the pH of the electroless bath solution increases the phosphorus content, thereby increasing the corrosion resistance.

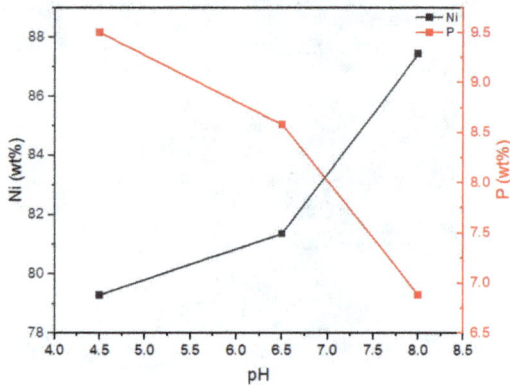

Figure 2. Variation in Ni and P content with electroless bath solution pH. Data adapted from [11]

2.2   Temperature

Temperature plays a significant role in the deposition rate of electroless Ni-P coatings. The reactions that occur during the deposition process are endothermic. Thus, the deposition rate increases with increasing temperature. The operating temperature for acidic baths is around 80-90°C while that of alkaline baths is around a minimum of 40°C. The bath temperature's effect on the deposition rate is illustrated in Fig. 3[12].

Advances in Corrosion Science and Surface Engineering
Materials Research Foundations 188 (2026) 139-146

Materials Research Forum LLC
https://doi.org/10.21741/9781644903919-7

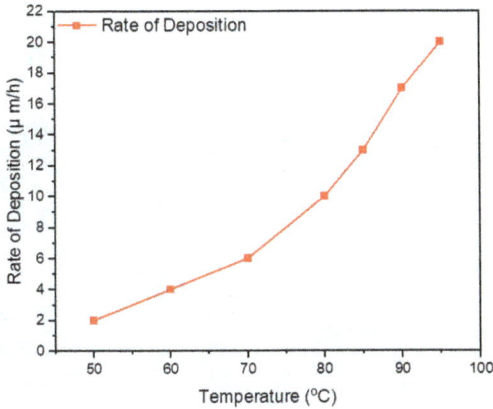

Figure 3. Effect of bath temperature on deposition rate [12]

2.3 Time

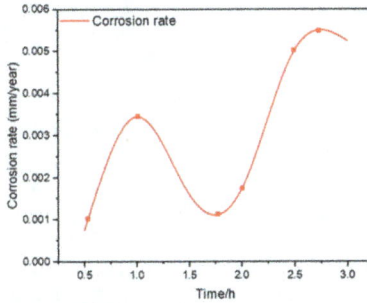

Figure 4. Effect of deposition time on corrosion rate [13]

S.H. Rafizadeh studied the effect of the deposition time on the corrosion current density of as-deposited Ni-P coatings with various phosphorus content. The corrosion current density increased with increasing coating deposition time. The corrosion rate increased due to decreased phosphorus content as shown in Fig. 4.

2.4 Impact of corrosive gases and chemicals

[14] studied the stability of Ni-P coatings and examined the electrolyte penetration through the coating through immersion test in CO_2 and CO_2/H_2S. The experiment was carried out for 10 days. The SEM images show damage in the top layer of the coating along the nodule boundaries under

Advances in Corrosion Science and Surface Engineering   Materials Research Forum LLC
Materials Research Foundations 188 (2026) 139-146   https://doi.org/10.21741/9781644903919-7

both conditions. EDS analysis shows the presence of Ni, P, Fe, C, and O elements in the corrosion product from both environments. Fe in the corrosion products indicates that the corrosion electrolyte has penetrated the coating/substrate interface. In the case of the CO_2/H_2S environment, the presence of the S element in the EDS result confirms the participation of H_2S in the corrosion process as shown in Fig. 5.

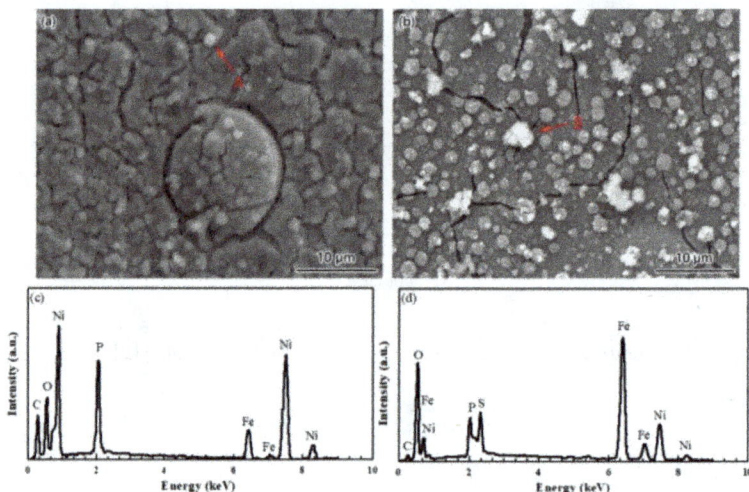

Figure 5. SEM surface morphologies of Ni-P coating after immersion in (a) CO_2 and (b) CO_2/H_2S; (c) (d) EDS results of corrosion products from (a) and (b), respectively [14].

2.5   Residual Stress

Ni-P-based coatings, including single, duplex, multi-layer, and multi-component coatings, have gained significant interest in both academic and industrial circles. Incorporating W or Mo in the Ni-P matrix can enhance the thermal stability and hardness of these coatings. The Ni-P/Ni-W-P and Ni-P/Ni-B duplex coatings are particularly noteworthy for their exceptional wear and corrosion resistance. However, residual stresses may be introduced during coating processing due to thermal mismatch or mechanical and thermal processing [15]. [16] studied the effect of residual stress on the corrosion and wear behavior of electrodeposited nanocrystalline Co-P coatings. The corrosion and wear resistance increased with increased compressive residual stresses of the coating.

Conclusion

This review comprehensively examined the various factors influencing the corrosion resistance of electroless nickel-phosphorus (Ni-P) coatings. Ni-P coatings offer excellent protection against corrosion due to their inherent properties and the influence of controllable deposition parameters. Phosphorus content plays a particularly critical role, with higher phosphorus content leading to enhanced corrosion resistance. Other significant factors affecting corrosion resistance include bath pH, temperature, deposition time, exposure to corrosive environments, and residual stress. Optimizing these parameters is essential for tailoring Ni-P coatings for specific industrial applications. Future research avenues include exploring the effects of particle incorporation and heat treatment processes on further improving the corrosion resistance of Ni-P coatings.

References

[1] A. Popoola, O. Olorunniwo, and O. Ige, "Corrosion Resistance Through the Application of Anti- Corrosion Coatings," *Dev. Corros. Prot.*, no. July, 2014. https://doi.org/10.5772/57420

[2] B. Fotovvati, N. Namdari, and A. Dehghanghadikolaei, "On coating techniques for surface protection: A review," *J. Manuf. Mater. Process.*, vol. 3, no. 1, 2019. https://doi.org/10.3390/jmmp3010028

[3] A. Bahgat Radwan *et al.*, "Properties enhancement of Ni-P electrodeposited coatings by the incorporation of nanoscale Y 2 O 3 particles," *Appl. Surf. Sci.*, vol. 457, pp. 956–967, 2018. https://doi.org/10.1016/j.apsusc.2018.06.241

[4] Z. Ahmad, "CHAPTER 7 - COATINGS," in *Principles of Corrosion Engineering and Corrosion Control*, Z. Ahmad, Ed., Oxford: Butterworth-Heinemann, 2006, pp. 382–437. doi: https://doi.org/10.1016/B978-075065924-6/50008-8

[5] S. J. John, PV George and Sahayaraj, M Edwin and Jappes, JT Winowlin and Leon, "Corrosion characteristics of microwave and furnace annealed electroless Ni-P-TiO2 coatings." SAGE Publications Sage UK: London, England, pp. 1149--1156, 2024.

[6] M. Yaghoobi, B. Bostani, P. Asl farshbaf, and N. P. Ahmadi, "An investigation on Preparation and Effects of Post Heat Treatment on Electroless Nanocrystalline Ni–Sn–P Coatings," *Trans. Indian Inst. Met.*, vol. 71, no. 2, pp. 393–402, 2018. https://doi.org/10.1007/s12666-017-1169-8

[7] A. Hadipour, M. Rahsepar, and H. Hayatdavoudi, "Fabrication and characterisation of functionally graded Ni-P coatings with improved wear and corrosion resistance," *Surf. Eng.*, vol. 35, no. 10, pp. 883–890, 2019. https://doi.org/10.1080/02670844.2018.1539295

[8] R. Elansezhian, B. Ramamoorthy, and P. Kesavan Nair, "Effect of surfactants on the mechanical properties of electroless (Ni-P) coating," *Surf. Coatings Technol.*, vol. 203, no. 5–7, pp. 709–712, 2008. https://doi.org/10.1016/j.surfcoat.2008.08.021

[9] I. A. Shozib, A. Ahmad, A. M. Abdul-Rani, M. Beheshti, and A. A. Aliyu, "A review on the corrosion resistance of electroless Ni-P based composite coatings and electrochemical corrosion testing methods," *Corros. Rev.*, vol. 40, no. 1, pp. 1–37, 2022. https://doi.org/10.1515/corrrev-2020-0091

[10] A. Varmazyar, S. R. Allahkaram, and S. Mahdavi, "Deposition, Characterization and Evaluation of Monolayer and Multilayer Ni, Ni–P and Ni–P–Nano ZnOp Coatings," *Trans. Indian Inst. Met.*, vol. 71, no. 6, pp. 1301–1309, 2018. https://doi.org/10.1007/s12666-018-1279-y

[11] E. M. Fayyad, A. M. Abdullah, A. M. A. Mohamed, G. Jarjoura, Z. Farhat, and M. K. Hassan, "Effect of electroless bath composition on the mechanical, chemical, and

electrochemical properties of new NiP–C 3 N 4 nanocomposite coatings," *Surf. Coatings Technol.*, vol. 362, no. November 2018, pp. 239–251, 2019. https://doi.org/10.1016/j.surfcoat.2019.01.087

[12]   T. Submitted and P. Fulfillment, "Evaluation of Electroless Nickel-Phosphorus (EN) Coatings Permission to Use," no. August 2002, 2003.

[13]   S. H. Rafizadeh, "Effect of coating time and heat treatment on structures and corrosion characteristics of electroless Ni – P alloy deposits," vol. 176, pp. 318–326, 2004. https://doi.org/10.1016/S0257-8972

[14]   J. Li, H. Zeng, C. Sun, and J. L. Luo, "Corrosion behavior of electroless Ni-P coating in the NaCl solution containing CO2 and H2S," *NACE - Int. Corros. Conf. Ser.*, vol. 2019-March, no. March, 2019.

[15]   H. LIU and D. shu QIAN, "Evaluation of residual stress and corrosion behaviour of electroless plated Ni−P/Ni−Mo−P coatings," *Trans. Nonferrous Met. Soc. China (English Ed.*, vol. 28, no. 12, pp. 2499–2510, 2018. https://doi.org/10.1016/S1003-6326(18)64896-4

[16]   N. M. Alanazi, A. M. El-Sherik, S. H. Alamar, and S. Shen, "Influence of residual stresses on corrosion and wear behavior of electrodeposited nanocrystalline cobalt-phosphorus coatings," *Int. J. Electrochem. Sci.*, vol. 8, no. 8, pp. 10350–10358, 2013. https://doi.org/10.1016/s1452-3981(23)13115-4

Advances in Corrosion Science and Surface Engineering　　　　　Materials Research Forum LLC
Materials Research Foundations 188 (2026) 147-160　　　　https://doi.org/10.21741/9781644903919-8

Chapter 8

Materials Selection for Corrosion Resistance

V. Nijarubini[1*], M. Jannathul Firdhouse[2], P. Jeevanantham[3], R. Menaka[1], P. Kavitha[1],
K. Sakthivel[4], G.Kausalya Sasikumar[5], Ramyakrishna Pothu[5,6]

[1]Center for Innovation and Inclusive Research, Sharda University, Greater Noida – 201310,
India

[2]Department of Chemistry, Hajee Karutha Rowther Howdia College, Theni - 625533,
Tamil Nadu, India

[3]Department of Chemistry, National College, Trichy - 620001, Tamil Nadu, India

[4]Department of Biochemistry, SSM College of Arts and Science, Dindigul -624002, India

[5]Centre for Research and Development, KPR Institute of Engineering and Technology,
Coimbatore-641407, India

[5]Centre for Research and Development Cell, Sharda University, Greater Noida-201310, India

[6]School of Physics and Electronics, College of Chemistry and Chemical Engineering, Hunan
University, Changsha 410082, P.R. China.

nijarubinibalaji@gmail.com

Abstract

Corrosion is the gradual degradation of materials due to chemical or electrochemical reactions
with their environment. It occurs when refined metals are converted into stable oxides, influenced
by factors like air, moisture, water, and temperature. These environmental conditions accelerate
material deterioration over time. Corrosion-resistant materials should ideally possess low
reactivity with the surrounding environment, resistance to aggressive chemicals, and the ability to
withstand temperature variations and mechanical stresses. Materials commonly selected for
corrosion resistance include stainless steel, which offers excellent resistance due to the formation
of a protective oxide layer, titanium, known for its resistance to a wide range of corrosive agents,
and high-performance alloys, which are engineered to endure extreme conditions. In recent days,
biodegradable coatings, and green inhibitors have also been used for corrosion resistance, which
produces a sustainable environment. Appropriate material is chosen for corrosion resistance
depending on the environment nature, like moisture, humidity, etc. We apply polymer composites,
nanomaterials, biodegradable coatings, ceramics, green inhibitors, natural fibre reinforced
composites, copper-based alloys and aluminium-based alloys as corrosion inhibitors.

Keywords

Corrosion- Resistance, Humidity, Moisture, Coatings, Nanomaterials

Contents

1.   Introduction

Corrosion, particularly in metals, results from intricate interactions with their environment under specific exposure conditions. Common forms of corrosion include the rusting of iron and its alloys at ambient temperatures, the scaling of steel at high heat, graphitization of cast iron, tarnishing of silver, pitting of stainless steel in chloride-laden environments, and the weakening of concrete when exposed to sulphates. However, some corrosion processes can be harnessed for beneficial purposes, such as creating protective passivation layers on metals, pickling steel, anodizing aluminium, storing energy in dry cells, performing chemical machining, and even blueing steel in alkaline conditions.

With the rapid growth in the development of new materials and ongoing advancements in traditional ones, selecting the ideal material for a specific purpose has become more challenging. To simplify this process, advanced techniques for material selection have emerged, ensuring that materials perform at their best in diverse environments. Material selection plays a pivotal role in engineering design, and by making thoughtful design choices, the negative impact of corrosion can be minimized. The material chosen must meet essential criteria such as mechanical strength, corrosion resistance, and resilience to erosion under the specific conditions. In harsh environments like refineries and chemical plants, high-performance corrosion-resistant alloys are often selected for valves and piping systems, ensuring long-lasting reliability and optimal performance.

1.1   Material selection process

With the rapid advancement in material technology and the availability of over 40,000 metal alloys, along with countless non-metallic options, choosing the right material has become more complex and challenging than ever before. To navigate this complexity, material selection starts

Advances in Corrosion Science and Surface Engineering   Materials Research Forum LLC
Materials Research Foundations 188 (2026) 147-160   https://doi.org/10.21741/9781644903919-8

early, right from the drawing board stage. At this point, crucial factors like engineering design choices and processing methods are thoughtfully considered to ensure that the most suitable material is selected for the product's success.

1.1.1 Factors influencing material selection

(i) Physical and Mechanical Factors

When considering materials, physical factors such as size, shape, and weight play a critical role. Weight is a crucial factor in industries like aerospace, where lightweight, high-performance alloys are essential for improving energy efficiency and boosting operational performance. The size and shape of material also influence the heat treatment processes it undergoes, ensuring it's optimized for its intended use.

Mechanical factors are equally vital, as they determine a material's ability to withstand the stresses it will face in service. Key mechanical properties commonly used in design include density, modulus of elasticity, strength, ductility, fracture toughness, fatigue resistance, corrosion fatigue, creep, impact resistance, and hardness. These properties are fundamental for ensuring materials perform reliably and effectively under varying conditions.

(ii) Functional requirements

The functional requirement is the foundational step in the material selection process. It focuses on optimizing a material's performance based on the product's intended function, geometry, and required properties. Essentially, the functional requirement defines the key characteristics the product must possess. For example, if a product is subjected to a uniaxial load, the material's load-bearing capacity is directly tied to its yield strength. In the case of heat exchanger tubes, the material must not only be an excellent heat conductor but also withstand temperatures higher than the maximum operating conditions, while offering strong resistance to corrosion. By aligning the material choice with these functional requirements, performance and durability are maximized.

(iii) Processing requirements

The processability of a material refers to its ability to be shaped, whether through hot working, cold working, or other techniques, to achieve the desired final product. This includes key aspects like castability, machinability, weldability, and its ability to undergo deformation. Complex shapes are often best created through casting, while smaller shapes are typically made using investment casting. Materials like thermoplastics and ductile metals are best shaped through deformation processes. Fabricability also comes into play, with processes like welding, brazing, soldering, and machining enabling materials to be assembled and refined. Before finishing processes such as surface coating or polishing, materials may undergo thermal and heat treatment to enhance their properties. By carefully selecting the right materials and processing methods, optimal performance and quality can be achieved.

(iv) Cost

The selection of materials is often guided by both cost and availability. Titanium is an outstanding choice for heat exchanger applications in brine, seawater, and other challenging aqueous environments, but its high cost is likely to be of significant consideration. However, the cost of titanium can be justified in certain situations, such as by opting for thinner-walled titanium tubes

(19 mm OD x 0.5 mm) instead of the thicker ones. This approach allows for the use of titanium's superior corrosion resistance while maintaining a cost-effective balance.

(v) Reliability

The material chosen must be capable of performing its intended function reliably over a specified period, without failure. Premature material failure can have disastrous consequences, resulting in catastrophic breakdowns and costly plant shutdowns. For example, the corrosion of steel structures, decks, and bridges can lead to severe structural failures, posing risks to both public safety and infrastructure. In aviation, the failure of critical components like horizontal stabilizers or landing gears can lead to tragic accidents with potentially fatal outcomes. Therefore, selecting the right material is not only about performance but also about ensuring safety, durability, and long-term reliability.

(vi) Resistance to service conditions

Corrosive environments can significantly reduce the lifespan of materials, making it essential to choose materials that are compatible with the conditions they will face. For example, austenitic steels like 304, 304L, 316, and 316L offer superior resistance to the corrosive effects of polluted seawater, outperforming copper alloys in these harsh conditions. Selecting the right material for the environment is crucial to ensuring durability and long-lasting performance, especially when exposed to corrosive elements.

(vii) Codes and other factors

Codes are essential sets of standards established by customers or technical organizations like ASTM (American Society for Testing and Materials) and SAE (Society of Automotive Engineers). These guidelines ensure that materials meet specific quality and performance criteria. Additionally, statutory factors, including regulations related to health, safety, waste disposal, environmental impact, and recycling, must also be carefully considered. These regulations help ensure that materials and processes not only meet technical requirements but also align with broader societal and environmental responsibilities.

2.   Corrosion resistant materials

The need for corrosion-resistant materials has driven the development of various alloys, coatings, and surface treatments designed to resist degradation in different environments. These materials range from metals with inherent corrosion resistance to advanced polymers and composites.

2.1   Polymer composites

Polymer composites are increasingly used for corrosion resistance across various industries due to their enhanced protective properties. Some key applications include:

Marine Industry: Polymer composites are used to coat ships, offshore platforms, and marine structures, protecting them from saltwater corrosion and harsh environmental conditions.

Construction: In buildings and infrastructure, these composites protect steel reinforcement from corrosion, extending the lifespan of structures exposed to moisture and chemicals.

Automotive Industry: Polymer composites are applied to vehicle components to protect them from road salts and environmental factors that cause rust and corrosion.

Advances in Corrosion Science and Surface Engineering
Materials Research Foundations 188 (2026) 147-160

Materials Research Forum LLC
https://doi.org/10.21741/9781644903919-8

Oil and Gas: Corrosion-resistant coatings are applied to pipelines, storage tanks, and rigs to safeguard against aggressive chemicals and moisture in oil and gas operations.

Aerospace: Polymer composites help protect aircraft structures and components from corrosion caused by atmospheric moisture and temperature fluctuations.

Electronics: Coatings with polymer composites protect electronic devices from environmental factors, ensuring durability and performance.

2.1.1 Ceramic Composite Materials (CMCs)

Ceramic Matrix Composites (CMCs) are reinforced with ceramics, offering benefits such as high strength, thermal stability, and wear resistance due to their low electrical and thermal conductivity, hardness, and low thermal expansion. However, they suffer from poor impact energy absorption, limited fatigue resistance, and brittleness, with voids and porosity negatively affecting their mechanical properties. CMCs are ideal for high-temperature applications like turbine blades and aircraft brakes, providing strength, durability, and heat resistance while being lightweight. Their production process, still developing and costly, affects performance and limits widespread use. The fiber-matrix interface is critical for fracture toughness, where coatings can enhance strength. A controlled interface allows debonding and sliding under stress, absorbing energy, and preventing catastrophic failures, ultimately improving CMCs' damage tolerance and durability.

2.1.2 Metal Matrix Composites (MMCs)

Metal Matrix Composites (MMCs) feature a metallic matrix combined with various reinforcements like fibers, whiskers, or particles. The matrix often includes ceramic reinforcements to improve mechanical stability at high temperatures. Fibers used in MMCs, such as steel, carbon, glass, aluminum, and tungsten, each offer unique benefits like high strength, low weight, and thermal stability. Particulate reinforcements, such as alumina and silicon carbide, enhance properties like hardness and density. The particle size impacts the interface strength, larger particles typically improve load transfer and strength, while smaller particles can increase porosity and agglomeration, weakening the bond. For example, coarse alumina particles in Cu-Al_2O_3 composites result in stronger interfaces compared to fine particles, improving overall mechanical performance and reducing porosity.

2.1.3 Epoxy coating

SiO_2 particles were synthesized by hydrolyzing tetra-ethyl orthosilicate (TEOS) in ethanol using aqueous ammonia as a catalyst. A solution of ethanol and water was stirred, and ammonia was added. TEOS was then slowly introduced with continuous stirring, causing turbidity and SiO_2 formation. The particles were collected via centrifugation at 15,000 rpm for 5 minutes, then calcined at 600°C. The resulting silica was crushed and stored in a desiccator for further use. Polypyrrole/SiO_2 composites (PCs) were synthesized by oxidative polymerization of pyrrole using ferric chloride as an oxidant. SiO_2 particles were dispersed in deionized water, and sodium lauryl sulfate (SLS) was added. Pyrrole and ferric chloride solutions were added, initiating polymerization, indicated by a black colour. The reaction occurred at room temperature for 4-5 hours. The composites were separated using filtration, washed with deionized water to remove impurities, and dried in a vacuum oven at 60°C. The corrosion resistance of mild steel coated with conventional epoxy and epoxy with various polypyrrole/SiO_2 and chitosan/polypyrrole/SiO_2

composites was evaluated using Tafel extrapolation, electrochemical impedance spectroscopy, and salt spray tests. Results showed that epoxy coatings with polypyrrole/SiO_2 composites significantly improved corrosion resistance, with a maximum protection efficiency (P.E.) of 99.95% at 3.0 wt% loading. Incorporating chitosan further enhanced the protection, achieving 99.97% P.E. at 2 wt%. Salt spray tests revealed minimal corrosion for coated specimens, indicating the composites' ability to form highly adherent, effective anti-corrosive coatings on mild steel surfaces.

3.   Nanomaterials as Corrosion Resistance

A 70% increase in corrosion resistance is a significant improvement, meaning that structures coated with nanomaterial-based coatings can endure corrosion much longer than those coated with traditional materials. This enhanced protection translates to extended durability and a longer lifespan for the structures, reducing maintenance costs and improving performance over time. By creating a denser, more uniform barrier on the metal surface, we reduce water and corrosive ion penetration. The improved adhesion of the coating to the metal minimizes the risk of delamination, which can cause corrosion. Additionally, some nanomaterials possess active properties that can neutralize or inhibit corrosion at the molecular level, further enhancing the protective effect. Traditional coatings usually rely on organic polymers or epoxy paints. A 70% increase shows that nanomaterials offer far superior protection compared to these conventional methods.

Nanomaterials protect against corrosion through various mechanisms:

- **Physical Barrier:** Nanoparticles create a dense protective layer on the metal surface, effectively blocking water and corrosive ions. Their small size allows them to fill microscopic surface gaps better than traditional materials.

- **Active Inhibition:** Certain nanomaterials like zinc oxide (ZnO) or cerium oxide (CeO_2) release ions that form a passive layer on the metal or neutralize corrosive elements in the environment.

- **Coating Property Enhancement:** Nanomaterials can enhance the mechanical properties and adhesion of coatings, making them more resistant to physical wear and delamination.

Advantages of Nanomaterials in Corrosion prevention

- **High Effectiveness:** Nanomaterials can boost corrosion resistance by up to 70% compared to traditional coatings, offering significantly better protection.

- **Self-Healing Properties:** Certain nanocomposites can "self-repair" minor damage to the coating, enhancing its longevity and maintaining performance over time.

- **Multifunctionality:** Beyond corrosion resistance, nanomaterials can offer additional benefits, such as anti-fouling properties or super-hydrophobicity, enhancing overall performance.

- **Environmentally Friendly:** Using nanomaterials reduces the need for conventional, often toxic, corrosion inhibitors, making the coating more sustainable and eco-friendlier.

Advances in Corrosion Science and Surface Engineering   Materials Research Forum LLC
Materials Research Foundations 188 (2026) 147-160   https://doi.org/10.21741/9781644903919-8

4.   Alloys

An alloy is a homogeneous substance formed by mixing two or more elements, at least one of them being a metal. Alloying helps to improve hardness, lowers the melting point, improves casting property, modifies the electrical conductivity, changes chemical properties, and resists the corrosion of metal.

Purpose of alloying

- Pure metals are very soft and weak and therefore cannot be used for making strong articles. Therefore, alloying them increases their strength.

- Two or more metals can be mixed in different proportions to produce alloys of desirable character for definite applications.

- Alloying reduces the chemical reactivity of a metal with acids and so it is less liable to corrosion.

- The use of alloys instead of expensive pure metals promotes economy in the manufacture and use of components.

- Alloying can enhance or suppress strength and electrical conductivity rendering an alloy more suited than pure metal for specific application.

The most important and useful property of alloying is its ability to resist corrosion. Generally, alloys are more resistant to corrosion than pure metals. Eg: stainless steel, SS (an alloy of Fe, C, Ni, and Cr), which is not corroded by the atmospheric conditions, though pure Fe easily corrodes (rusts) in moist air. Alloys are classified based on the principal metal in the alloy. Stainless steel resists corrosion due to the atmosphere and also by chemicals. They contain essentially chromium together with other elements such as nickel, molybdenum, etc. Chromium is especially effective, if its content is 16% or more. The protection against corrosion is due to the formation of a dense tough film of chromium oxide at the surface of the metal. Alloying metals with different elements enhance their corrosion resistance by improving their ability to form protective surface layers, resist galvanic effects, and withstand specific environmental conditions. Each alloying combination tailors the metal to specific corrosion challenges, enhancing durability in diverse environments. Given below are some alloying combinations and their importance in corrosion resistance,

(i) Fe-Cr Stainless Steel alloys were fabricated by alloying with 0, 0.10 and 0.20 wt.% Sn to investigate the effect of Sn addition on the microstructure, passive characteristics and localized corrosion behaviour in a chloride medium at room temperature. Fe-16CrxSn (x= 0, 1, 2 wt.%) buttons were prepared using vacuum arc melting. The cast alloys were homogenized for approximately 90 min at 1150 °C and hot rolled to 5 mm thick plates. The alloys exhibited identical corrosion resistant behaviour in NaCl medium. Lower corrosion current densities and higher impedance values at high anodic potentials were observed for alloy SS3 (Fe-14Cr-9Ni-0.2Sn), demonstrating improved localized corrosion resistance in NaCl solution. Scanning electrochemical microscopic results confirmed that alloy SS3 remained stable even after 24 h of exposure and displayed no noticeable surface heterogeneity. With the addition of Sn, the pitting potential increased, with the highest pitting potential observed for SS3 alloy. This is because the alloying elements (Ni, Cr, Sn) work synergistically and produce a more compact and dense passive film on the surface of SS alloy. Reduced anodic current density observed even after 24 h of exposure in

Advances in Corrosion Science and Surface Engineering   Materials Research Forum LLC
Materials Research Foundations 188 (2026) 147-160   https://doi.org/10.21741/9781644903919-8

alloy SS3 compared to other samples, indicating that the anodic metal dissolution was successfully inhibited by the formation of a compact and dense passive film on this alloy.

(ii) The T20Nb13Zr (TNZ) alloy was fabricated from Ti, Nb, and Zr powders, synthesized by ball milling and sintered using spark plasma sintering. Samples with a 20 mm diameter and 3 mm thickness were subjected to heat treatment. The biomedical TiNbZr alloy was cooled at different rates, and the solution-treated samples were aged at various temperatures to study the influence of cooling rate and aging temperature on microstructure, mechanical properties, and corrosion resistance in stimulated body fluid (SBF). Corrosion tests were performed in 30 wt% sulfuric acid. The heat-treated TNZ alloy exhibited an enhanced combination of mechanical properties and corrosion resistance compared to the commercial Ti6Al4V alloy. Improved corrosion resistance was achieved by adding small amounts of Ta to ASTM Gr13 titanium alloy. The ingots (Gr13-0.3% Ta, Gr13-0.6% Ta, and Gr13-1.0% Ta) were fabricated by vacuum arc melting, then subjected to hot rolling and reduced to plates with a 4 mm thickness. The presence of Ta in the Ti matrix enhances the stability of TiO_2 by forming highly corrosion-resistant Ta_2O_5. Open circuit potential (OCP) results showed that the transition from the passive state to the corroding state was delayed with the increase of Ta content. This behaviour was supported by Electrochemical impedance spectroscopy (EIS) results, where the shrinkage of the loop, indicating passivity loss, was retarded and/or inhibited by the addition of Ta. Microstructural analysis revealed that Ta promotes the growth of precipitates such as the β-Ti phase and the formation of a complex-structured passive film. The high content of Ta oxide in the TiO_2 passive film for Gr13-1.0% offered exceptional stability, maintaining passivity for up to 30 hours in 30 wt% sulfuric acid solution.

(iii) Alloy compositions, TC4-xNi-yNb (x, y = 0, 0.5) were prepared using titanium sponge, electrolytic nickel, Al-55V alloy, Ti-Nb intermediate alloy, TiO_2, and aluminium beans. The ingots, sized 210 mm × 560 mm, were created by vacuum self-consuming melting, then forged into blocks, hot-rolled at 900°C with 35% deformation, and air-cooled. Annealing at 500°C for 1 hour was followed to relieve stress. The samples (30 × 20 × 2 mm) were immersed in a 1 mol/L HCl solution in a constant temperature water bath at 25 ± 1°C for 2, 4, 6, 8, and 10 days. The addition of Ni elements resulted in high values for self-corrosion potential, polarization resistance and passivation film thickness, along with low self-corrosion current density, showcasing excellent corrosion resistance. The addition of Nb elements increased the alloy's passivation current density, but had a minor impact on self-corrosion potential and self-corrosion current density. TC4 and TC4-0.5Nb alloys showed significant corrosion weight loss in the initial corrosion stage, while the corrosion rate of TC4-0.5Nb alloy gradually decreased and tended to stabilize with prolonged immersion time. X-ray photoelectron spectroscopy (XPS) results suggested that the passivation film on the alloy surface was primarily composed of TiO_2, accompanied by the presence of oxides of various elements. The addition of Ni and Nb elements had different degrees of influence on the composition and structure of the alloy passivation film. The thermodynamic stability and passivation film thickness of TC4 titanium alloy increased and the oxide formed by Nb stabilized the passivation film, enhancing the corrosion resistance of TC4 titanium alloy.

(iv) The Copper-Nickel 90/10 Alloy produced by WAAM (Wire Arc Additive Manufacturing) 3D printing technology. The alloy 3D walls produced are characterized by average hardness in the range of 138–160 HV10, ultimate tensile strength in the range of 495–520 MPa, conventional yield strength of 342–358 MPa, and elongation of 16.6–17.9%. The electrochemical measurements in a 1M NaCl environment are characterized by an average value of corrosion potential of −689 mV, a corrosion current density of 3.6×10^{-9} A/cm^2, and a resistance of 2.6×107 Ω. The average corrosion rate is approximately 7.4×10^{-5} mm/year and this result is below 0.025 mm/year, which indicates that the material is practically non-corrosive or corrosion occurs very slowly. The work shows that it is possible to obtain higher mechanical properties of Cu-Ni 90/10 alloy 3D objects produced using the WAAM method compared to cast materials, which opens up the possibility of using this alloy to produce objects with more complex shapes and for use in corrosive working conditions.

(v) Aluminium-silicon (Al-Si) containing 7 wt.%, 12 wt.%, 14 wt.% of silicon and two different aluminium-magnesium (Al-Mg) alloys containing 2.5 wt.%, 4.5 wt.% of magnesium were prepared by the casting route method. The microstructural analysis was done by using both optical and scanning electron microscopic techniques, hardness values were measured with a Vickers hardness tester, tensile properties were determined by using an Instron universal testing machine, and the wear properties were studied by using a ball-on-plate wear tester. The chemical composition analysis of the prepared alloys revealed increased hardness, yield strength and ultimate tensile strength, with increasing Si content in case of Al-Si alloy as well as with increasing Mg content in case of Al-Mg alloy. The increment in strength is associated with reduced total percentage of elongation. However, the wear rate decreases with increasing Si content and Mg content of the respective alloys.

(vi) Cobalt–chromium (Co-Cr) alloy was fabricated via conventional powder metallurgy method followed by sintering in an argon atmosphere. The microstructures of Co$_{70}$Cr$_{30}$ alloys before and after thermal treatment at 1200 °C for 2 hours in an argon atmosphere with heating/cooling rates of 10 °C/min are revealed by Scanning Electron Microscopy. The metal ion (Co and Cr) release from the pre- and post- thermal treated alloys is studied by immersing in simulated body fluid (SBF) solution, containing an ion concentration close to that of human blood plasma, The release of Co ion continuously increased with increasing immersion time for both pre- and post- thermal treatment. On the other hand, Cr ion releasing seems to decrease with time. For both ions, the post-thermal treated samples show higher releasing amounts, because the thermal treatment process allows Co-Cr alloy to relieve the internal stress, minimize porosity, exhibit higher ductility and lower hardness, therefore, all metal ions easily release.

5. Green inhibitors

Green inhibitors are eco-friendly substances derived from natural sources that prevent or reduce metallic corrosion without harming the environment. They are non-toxic, biodegradable, and renewable. Organic green inhibitors, such as surfactants, plant extracts, and biopolymers, are commonly used in acidic conditions, while inorganic inhibitors, like rare earth elements, are used in neutral media. These inhibitors form a protective layer on the metal surface, reducing corrosion. Research into green inhibitors has grown due to increased environmental awareness and legislation. They are applied in various industries, including lubricants, concrete, and oil fields, with varying extraction costs depending on the material.

(i) Plant Extract as Green inhibitors

Plant extracts offer a promising alternative to harmful chemical corrosion inhibitors, as they are environmentally friendly and contain bioactive phytochemicals that can effectively prevent corrosion. Various plant parts, including leaves, roots, stems, and bark, are used to extract natural inhibitors, with common examples being neem, Aloe vera, and teas. *Azadirachta indica* (neem), for instance, has demonstrated corrosion inhibition effectiveness, with neem leaf powder in water showing up to 53.09% inhibition efficiency at optimal concentrations and exposure times. Aloe vera, both in powder and gel form, also serves as a corrosion inhibitor, with 10% extract in H_2SO_4 achieving up to 99.1% inhibition on stainless steel. Additionally, green and white tea extracts, rich in phenolic compounds, have been shown to reduce corrosion rates significantly, with green tea reducing corrosion by 71.43%. Natural essential oils, obtained via steam distillation, also exhibit corrosion resistance. For example, *Elettaria* cardamomum oil achieved 88% inhibition, and rosemary oil showed 92.54% efficiency. These plant-derived inhibitors provide sustainable, cost-effective solutions for corrosion prevention across various industries.

(ii) Fatty acids and esters as Green inhibitors

Fatty acids and esters derived from natural oils, such as oleic acid from olive oil and lauric acid from coconut oil, exhibit corrosion inhibition properties by forming protective films on metal surfaces. Esters like methyl esters of fatty acids are also effective inhibitors. For example, Vicia faba peel extract contains compounds like palmitic acid and ethyl palmitate, which form a protective layer on mild steel in seawater. Jatropha curcas seed oil, composed of fatty acids like palmitic and oleic acids, demonstrates up to 97% inhibition efficiency in acidic rainwater. Similarly, sunflower seed oils and Aleurites moluccana seed oils have shown significant corrosion resistance, achieving high inhibition efficiencies under varying conditions.

(iii) Amino acids as corrosion inhibitors

Amino acids such as glutamine (Gln), glutamic acid (Glu), asparagine (Asn), and aspartic acid (Asp) have been shown to inhibit copper corrosion in 0.5 M HCl solution. The effectiveness of these amino acids as inhibitors follows the order: Gln > Asn > Glu > Asp. These amino acids act as mixed inhibitors and adhere to the copper surface, following the Langmuir adsorption isotherm. Strong physical adsorption is observed, supported by adsorption free energy data. For Cu-Ni alloys, particularly Cu-5Ni, amino acids like cysteine demonstrate significant inhibition, achieving up to 85% corrosion inhibition efficiency in chloride solutions. Cysteine, with two adsorption centers, is highly effective in neutral chloride solutions. Additionally, cysteine and phenylalanine have been tested as corrosion inhibitors for bronze in acidic conditions, with cysteine showing superior performance. These amino acids offer an eco-friendly alternative to traditional hazardous corrosion inhibitors, providing effective protection for metals in harsh environments.

(iv) Biopolymer as Corrosion inhibitors

Eco-friendly biopolymers like dextrin have shown significant inhibitory effects on mild steel corrosion in acidic solutions, such as 15% HCl. Dextrin and poly(vinyl acetate) function as mixed-type inhibitors, adhering to the Langmuir adsorption isotherm, with an inhibitory efficiency of 87% for copper at 0.1 mg/L. The effectiveness of biopolymers and plant extracts as corrosion inhibitors has been widely studied using techniques such as gravimetric, potentiodynamic polarization, and electrochemical impedance spectroscopy (EIS). For example, plant extracts from Piper longum, Strychnos nuxvomica, and Mucuna pruriens showed inhibition efficiencies of 91.6%, 80%, and 71.6%, respectively, in 3 M HNO_3, with inhibition increasing as extract concentration rose. These plant extracts follow the Langmuir adsorption isotherm and predominantly exhibit cathodic inhibition behavior. Similarly, the natural extract of *Hyoscyamus muticus* was tested as a copper corrosion inhibitor in 1 M HNO_3 solution. The inhibitor showed a decrease in efficacy with increased temperature and extract concentration. The adsorption of the extract also adhered to the Langmuir adsorption isotherm, with cathodic-type behavior dominating, as shown in polarization curves. These findings highlight the potential of biopolymers and plant extracts as effective, eco-friendly alternatives to traditional corrosive inhibitors.

Applications of Green Inhibitors

- Green corrosion inhibitors provide exceptional benefits in preventing metal degradation, offering an eco-conscious alternative to conventional methods.

- These inhibitors form an impenetrable shield on metal surfaces, blocking harmful elements like oxygen and moisture from causing corrosion, thus dramatically reducing susceptibility to erosion.

- They effectively decelerate corrosion rates, ensuring that metals and alloys maintain their integrity over prolonged exposure to corrosive environments.

- The most remarkable advantage of green inhibitors is their environmental friendliness. Derived from natural, renewable sources, they have a significantly lower ecological footprint compared to toxic, non-biodegradable traditional inhibitors.

- Green inhibitors represent a groundbreaking shift toward sustainable protection for metals while minimizing harmful environmental impact.

Conclusion

Corrosion is a pervasive and significant problem across many industries, leading to the degradation of materials and failure of components, which causes serious economic and safety implications. The chapter delves into a wide array of materials engineered to combat corrosion, including metals, alloys, polymers, ceramics, and composites, each of which brings unique strengths and limitations. A deep understanding of how these materials behave in corrosive environments is vital for making informed decisions that ensure both performance and longevity. In a world where harsh conditions such as exposure to chemicals, moisture, extreme temperatures, and mechanical stress are increasingly common, protecting materials from corrosion is more important than ever. As a result, material selection for corrosion resistance has become a pivotal area of innovation in engineering and materials science. However, metals are not immune to all forms of corrosion, and the fight against corrosion is multifaceted and requires a thorough understanding of materials and their behaviour in different environments. Future innovations, including improved coatings, advanced alloys, and new composite materials, will continue to push the boundaries of corrosion resistance, ensuring the longevity and safety of critical infrastructure and components in an ever-changing world.

References

[1] Oliveira, T.L.L., Hadded, M., Mimouni, S. and Schaan, R.B., 2025. The Role of Non-Destructive Testing of Composite Materials for Aerospace Applications. NDT, 3(1), p.3. https://doi.org/10.3390/ndt3010003

[2] Ruhi, G. and Dhawan, S.K., 2013. Conducting polymer nano composite epoxy coatings for anticorrosive applications. Modern electrochemical methods in nano, surface and corrosion science. https://doi.org/10.5772/58388

[3] Sade, J., 2024. Nano Material Innovation in Enhancing Corrosion Resistance of Offshore Structures. Maritime Park: Journal of Maritime Technology and Society, pp.139-144. https://doi.org/10.62012/mp.v3i3.39554

[4] Jia, X., Yuan, S., Li, B., Miu, H., Yuan, J., Wang, C., Zhu, Z. and Zhang, Y., 2020. Carbon Nanomaterials: Application and prospects of urban and industrial wastewater pollution treatment based on abrasion and corrosion resistance. Frontiers in Chemistry, 8, p.600594. https://doi.org/10.3389/fchem.2020.600594

[5] Rakesh, N.L., Selvakumar, K. and Mohanavel, V., 2024. Physical, Mechanical and Thermal Characterization of Areca, Pineapple and Glass Fiber Reinforced Polymer Composites for Aerospace Applications. J. Environ. Nanotechnol, 13(3), pp.345-352. https://doi.org/10.13074/jent.2024.09.243792

[6] Xiao, K., Ge, J., Zhang, Y., Wang, J., Feng, W., Ou-Yang, X., Yu, Y., Ye, W. and Hui, S., 2025. Effect of Ni and Nb Elements on Corrosion Resistance and Behavior of TC4 Alloy in Hydrochloric Acid. Materials, 18(2), p.246. https://doi.org/10.3390/ma18020246

[7] Zhao, Y., Ma, H., Gao, Z., Huang, Z., Wu, Y. and Lv, K., 2025. Enhancement of the Corrosion and Wear Resistance of an Epoxy Coating Using a Combination of Mullite Powder and PVB. Coatings, 15(1), p.41. https://doi.org/10.3390/coatings15010041

[8] Gong, K., Zheng, C., Zhang, D., Lv, T. and Ju, X., 2024, June. Study on seawater corrosion resistance of copper alloy. In Journal of Physics: Conference Series (Vol. 2783, No. 1, p. 012060). IOP Publishing. https://doi.org/10.1088/1742-6596/2783/1/012060

[9] Ningrum, M.H., Laksmita, A.N., Salam, D.M., Saputra, S.H., Saefudin, M.F., Devi, A.P., Maharani, R. and Fernandes, A., 2024. Oleochemicals as green corrosion inhibitors: a review. Int J Corros Scale Inhibit, 3(13), pp.1375-1393. https://doi.org/10.17675/2305-6894-2024-13-3-2

[10] Galleguillos Madrid, F.M., Soliz, A., Cáceres, L., Bergendahl, M., Leiva-Guajardo, S., Portillo, C., Olivares, D., Toro, N., Jimenez-Arevalo, V. and Páez, M., 2024. Green Corrosion Inhibitors for Metal and Alloys Protection in Contact with Aqueous Saline. Materials, 17(16), p.3996. https://doi.org/10.3390/ma17163996

[11] Verma, C., Ebenso, E.E., Bahadur, I. and Quraishi, M.A., 2018. An overview on plant extracts as environmental sustainable and green corrosion inhibitors for metals and alloys in aggressive corrosive media. Journal of molecular liquids, 266, pp.577-590. https://doi.org/10.1016/j.molliq.2018.06.110

[12] Meena, O., Kaushal, S., Kumar, S. and Dalal, J., 2024. Euphorbia neriifolia extracts as green corrosion inhibitors for aluminium in hydrochloric and nitric acid media. Discover Materials, 4(1), p.102. https://doi.org/10.1007/s43939-024-00157-8

[13] Górnik, M., Lachowicz, M. and Łatka, L., Corrosion resistance of PPTA Ni-based hardfacing layers. Materials Science-Poland, 42(4), pp.66-78. https://doi.org/10.2478/msp-2024-0040

[14] Y. Fu, C. Liu, H. Hao, Y. Dong Xu, and X. rong Zhu, 2021, Effect of ageing treatment on microstructures, mechanical properties and corrosion behavior of Mg-Zn-RE-Zr alloy micro-alloyed with Ca and Sr, China Foundry, 18(2), pp. 131-140. https://doi.org/10.1007/s41230-021-0146-3

[15] S. Pugal Mani, M. Kalaiarasan, K. Ravichandran, N. Rajendran, and Y. Meng, 2021, Corrosion resistant and conductive TiN/TiAlN multilayer coating on 316L SS: a promising metallic bipolar plate for proton exchange membrane fuel cell, J Mater Sci, 56(17), pp.10575-10596. https://doi.org/10.1007/s10853-020-05682-4

[16] R. Li et al., 2022, Effect of annealing treatment on microstructure, mechanical property and anti-corrosion behavior of X2CrNi12 ferritic stainless steel, Journal of Materials Research and Technology, 18, pp. 448-460. https://doi.org/10.1016/j.jmrt.2022.02.117

[17] P. Mishra, P. Akerfeldt, F. Svahn, E. Nilsson, F. Forouzan, and M. L. Antti, 2023, Microstructural characterization and mechanical properties of additively manufactured 21-6-9 stainless steel for aerospace applications, Journal of Materials Research and Technology, 25, pp. 1483-1494. https://doi.org/10.1016/j.jmrt.2023.06.047

[18] Y. Xu, Y. Li, T. Chen, C. Dong, K. Zhang, and X. Bao, 2024, A short review of medical-grade stainless steel: Corrosion resistance and novel techniques," Journal of Materials Research and Technology, 29, pp. 2788-2798. https://doi.org/10.1016/j.jmrt.2024.01.240

[19] M. C. Reboul and B. Baroux, 2011, Metallurgical aspects of corrosion resistance of aluminium alloys, Materials and Corrosion, 62(3), pp. 215-233. https://doi.org/10.1002/maco.201005650

[20] M. A. Hussein, M. Azeem, A. M. Kumar, N. Al-Aqeeli, N. K. Ankah, and A. A. Sorour, 2019, Influence of Thermal Treatment on the Microstructure, Mechanical Properties, and Corrosion Resistance of Newly Developed Ti20Nb13Zr Biomedical Alloy in a Simulated Body Environment, J Mater Eng Perform, 28(3), pp.1337-1349. https://doi.org/10.1007/s11665-019-03908-4

[21] A. M. Kumar and I. ul H. Toor, 2023, Localized corrosion evaluation of newly developed stainless-steel alloys in chloride medium through dynamic and localized micro electrochemical techniques, Journal of Materials Research and Technology, 26, pp. 5668-5682. https://doi.org/10.1016/j.jmrt.2023.08.228

[22] B. Seo, H. K. Park, C. S. Park, and K. Park, 2023, Role of Ta in improving corrosion resistance of titanium alloys under highly reducing condition, Journal of Materials Research and Technology, 23, pp. 4955-4964. https://doi.org/10.1016/j.jmrt.2023.02.158

[23] J. Smalc, M. Voncina, P. Mrvar, T. Balasko, V. Krutis, and M. Petris, 2023, The Influence of Foundry Scrap Returns on Chemical Composition and Microstructure Development of AlSi9Cu3 Alloy, Crystals (Basel), 13(5). https://doi.org/10.3390/cryst13050757

[24] P. Zhou and K. Ogle, 2018, The corrosion of copper and copper alloys, Encyclopedia of Interfacial Chemistry: Surface Science and Electrochemistry, pp. 478-489. https://doi.org/10.1016/B978-0-12-409547-2.13429-8

[25] M. K. Abbass, K. S. Hassan, and A. S. Alwan, 2015, Study of Corrosion Resistance of Aluminum Alloy 6061/SiC Composites in 3.5% NaCl Solution," International Journal of

Materials, Mechanics and Manufacturing, 3(1), pp. 31-35, 2015.
https://doi.org/10.7763/IJMMM.2015.V3.161

[26] K. Xiao et al., 2025, Effect of Ni and Nb Elements on Corrosion Resistance and Behavior of TC4 Alloy in Hydrochloric Acid," Materials, 18(2). https://doi.org/10.3390/ma18020246

[27] V. Kain, 2011, Corrosion-Resistant Materials, Functional Materials: Preparation, Processing and Applications, pp. 507-547. https://doi.org/10.1016/B978-0-12-385142-0.00012-X

[28] M. Phasani, J. Abe, P. Popoola, O. Aramide, and M. Dada, 2025, Synthesis and characterization of Cobalt-chromium based alloys via spark plasma sintering for biomedical applications, Next Materials, 6, pp.100291. https://doi.org/10.1016/j.nxmate.2024.100291

[29] H. S. Akkera, N. N. K. Reddy, M. Poloju, M. C. Sekhar, C. Yuvaraj, and G. S. Prasad, 2016, Fabrication of cast aluminium-silicon (Al-Si) and aluminium-magnesium (Al-Mg) alloys and their properties, Acta Metallurgica Slovaca, 22(4), pp. 212-221. https://doi.org/10.12776/ams.v22i4.760

[30] H. S. Akkera, N. N. K. Reddy, M. Poloju, M. C. Sekhar, C. Yuvaraj, and G. S. Prasad, 2016, Fabrication of cast aluminium-silicon (Al-Si) and aluminium-magnesium (Al-Mg) alloys and their properties," Acta Metallurgica Slovaca, 22(4), pp. 212-221. https://doi.org/10.12776/ams.v22i4.760

[31] M. Maleta, J. Kulasa, A. Kowalski, P. Kwaśniewski, S. Boczkal, and M. Nowak, 2024, Microstructure, Mechanical and Corrosion Properties of Copper-Nickel 90/10 Alloy Produced by CMT-WAAM Method, Materials, 17(1). https://doi.org/10.3390/ma17010050

Advances in Corrosion Science and Surface Engineering   Materials Research Forum LLC
Materials Research Foundations 188 (2026) 161-192   https://doi.org/10.21741/9781644903919-9

Chapter 9

Above Durability: Special Advantages of PTFE-Reinforced Ni-P Coatings for Corrosion Protection

Onur Güler[1,*]

[1]Karadeniz Technical University, Metallurgical and Materials Engineering, 61080, Trabzon, Turkey

onurguler@ktu.edu.tr

Abstract

The integration of PTFE (polytetrafluoroethylene) within nickel-phosphorus (Ni-P) coatings has sparked significant interest in the realm of corrosion protection. This compelling synthesis not only fortifies corrosion resistance but also introduces a novel interplay between PTFE particles and the Ni-P matrix. The study systematically explores the microstructural intricacies, shedding light on how this composite coating mitigates corrosion challenges. Impressively, the resulting coating demonstrates outstanding resilience against corrosion and concurrently exhibits reduced friction coefficients. This review champions the strides made in the development of PTFE-infused Ni-P coatings, advocating for their pivotal role in addressing corrosion concerns across diverse industrial applications.

Keywords

Ni-P-PTFE Composite Coatings, Corrosion Resistance, Microstructural Development, PTFE Distribution, Surface Enhancement

Contents

1.   Introduction

Ni-P (nickel-phosphorus) coatings have a rich historical trajectory, evolving as a pivotal facet of surface engineering. Originating mid-20th century, these coatings have undergone continuous refinement, propelled by an inherent drive to enhance material properties [1,2]. Notably, Ni-P coatings have carved their niche in corrosion resistance, standing as a bulwark against the deleterious effects of environmental factors. Extensive research has illuminated the nuanced interplay between the Ni-P matrix and corrosive environments. The inherent corrosion resistance of nickel, coupled with phosphorus's ability to mitigate surface imperfections, forms the bedrock of this formidable defense mechanism [3,4]. Microstructural analyses delve into the intricate dynamics at play, unraveling the synergy that culminates in a robust shield against corrosion. Applications of Ni-P coatings span a diverse spectrum, ranging from aerospace components to industrial machinery. The adaptability of these coatings is underscored by their efficacy in mitigating corrosion-induced material degradation, thereby extending the service life of critical components. This adaptability is not only a testament to the versatility of Ni-P coatings but also a testament to their enduring relevance in addressing contemporary engineering challenges [5–7]. The selection of substrate materials significantly influences the performance of Ni-P coatings. Understanding the synergies and interactions between the substrate and the coating is imperative for optimizing the protective capabilities of Ni-P [8–10]. Scientific endeavors focus on elucidating the intricate bond formation and material compatibility, providing insights that contribute to the judicious selection of substrates for specific applications. In essence, the journey of Ni-P coatings traverses time, science, and application domains. From its historical roots to its contemporary prowess in corrosion resistance, the multifaceted nature of Ni-P coatings continues to captivate researchers and engineers alike, offering a dynamic canvas for innovation and exploration in the ever-evolving landscape of surface enhancement.

Despite the well-established significance of Ni-P coatings in corrosion resistance, the advent of particle-reinforced Ni-P composite coatings has arisen from a nuanced need for enhanced and tailored material properties. This evolution stems from the inherent limitations of conventional Ni-P coatings in addressing specific challenges. Incorporating different particles into the Ni-P matrix improves the corrosion resistance of coatings [11–14]. For instance, the addition of SiC (Silicon Carbide) [15] or WC (Tungsten Carbide) [16] contributes to increased surface hardness and wear resistance, ultimately protecting the coating against corrosive environments. Al_2O_3 (Aluminum Oxide) [17] particles, known for their chemical inertness, enhance corrosion resistance by stabilizing the metallic surface of the coating in aggressive conditions. Similarly, Si_3N_4 (Silicon Nitride) [18] ceramic particles provide durability to the coating and resist corrosion, particularly in high-temperature applications. These strategic additions of particles are designed to fortify the coatings' ability to withstand corrosion without compromising their mechanical and protective properties.

On the other hand, the incorporation of PTFE (polytetrafluoroethylene) into Ni-P (nickel-phosphorus) coatings introduces unique advantages in enhancing corrosion resistance through

Advances in Corrosion Science and Surface Engineering   Materials Research Forum LLC
Materials Research Foundations 188 (2026) 161-192   https://doi.org/10.21741/9781644903919-9

intricate mechanisms. Ni-P-PTFE coatings began to be applied in the early 1980s [19,20]. PTFE, renowned for its chemical inertness, forms a protective barrier on the coating's surface, acting as a shield against corrosive agents. The non-reactive nature of fluorine in PTFE prevents the initiation of corrosive processes, thereby impeding the penetration of corrosive species into the coating. Additionally, the low surface energy of PTFE minimizes the adhesion of water molecules and corrosive ions, reducing the likelihood of corrosion initiation [21]. Furthermore, the lubricious properties of PTFE play a pivotal role in preventing abrasion-induced corrosion. The reduced friction coefficients attributed to PTFE contribute to a smoother surface, minimizing wear and abrasion on the coating. This, in turn, mitigates the likelihood of corrosion initiation at sites prone to mechanical stress [22]. Moreover, PTFE's ability to fill microstructural voids and irregularities in the Ni-P matrix creates a more uniform and denser protective layer. This uniformity serves as an additional barrier against corrosive agents, hindering their access to vulnerable regions within the coating. The synergistic effect of these mechanisms results in a robust corrosion-resistant Ni-P-PTFE composite coating [23,24]. In conclusion, the incorporation of PTFE into Ni-P coatings orchestrates a multifaceted defense against corrosion. The chemical inertness, reduced friction, and capacity to fill microstructural gaps collectively contribute to a formidable protective barrier, exemplifying the tailored and effective nature of Ni-P-PTFE composite coatings in combatting corrosion. The evolution towards PTFE particle-reinforced Ni-P composite coatings is driven by a quest for heightened performance and versatility [25–27]. These unique properties make Ni-P-PTFE composite coatings well-suited for a multitude of applications. In the aerospace industry, where components are exposed to extreme conditions, these coatings provide effective corrosion protection [28–30]. Similarly, in automotive applications, where exposure to road salts and environmental elements poses a challenge, Ni-P-PTFE coatings prove invaluable [22,31]. Moreover, their use extends to marine applications, protecting surfaces against the corrosive effects of saltwater. By strategically incorporating PTFE particles into the matrix, researchers and engineers aim to transcend the limitations of traditional Ni-P coatings, ushering in a new era of tailored surface engineering that meets the evolving demands of modern industries [32].

Despite numerous reviews focusing on the corrosion properties of Ni-P composite coatings in the literature [33–35], there is a notable gap in research specifically addressing the contributions of PTFE particles in enhancing these properties. When the keyword 'Ni-P-PTFE' is entered 'search documents' option and 'all fields' selected in the 'Search within' field and activated on Scopus [36], it becomes evident that the number of studies conducted on Ni-P-PTFE in recent years has been steadily increasing. This observation indicates a growing interest and focus within the academic and research community on the exploration of Ni-P-PTFE composite materials. The surge in the number of publications suggests an expanding body of knowledge and an enhanced understanding of the properties, applications, and potential advancements related to Ni-P-PTFE in various fields.

Recognizing the oversight in neglecting the impact of PTFE, this study aims to bridge this gap by investigating the production methods of PTFE-enhanced Ni-P composite coatings. It will delve into the microstructural developments that underlie corrosion improvements, emphasizing the need to consider additional composite-forming elements beyond PTFE. Through a scientific exploration of these aspects, the study aims to provide valuable insights into the corrosion resistance enhancements achieved by PTFE-inclusive Ni-P composite coatings, shedding light on their potential in diverse industrial applications.

2.   Common Challenges in Ni-P-PTFE Coatings

The use of Ni-P-PTFE coatings is commonly preferred in industries to provide properties such as friction reduction, corrosion resistance, and wear resistance. However, these coatings also have some potential issues. Here are some of the most common problems encountered with Ni-P-PTFE coatings:

Adhesion Issues: Improper or insufficient adhesion to the surface of the coating can adversely affect its durability and performance. This can lead to peeling or cracking of the coating over time. Adhesion problems can often be mitigated by optimizing surface preparation techniques before coating application. Studies have shown that surface roughness and cleanliness significantly affect adhesion strength. Techniques such as mechanical abrasion, chemical etching, or surface activation using plasma treatment have been proposed to enhance adhesion by promoting surface roughness and creating active sites for bonding [37]. Poor adhesion between the coating and substrate can create pathways for corrosive agents to penetrate beneath the coating, leading to localized corrosion such as blistering or delamination. Furthermore, voids at the interface may facilitate galvanic corrosion by fostering the creation of micro-galvanic cells between unlike materials, hastening substrate corrosion.

Porosity: Formation of porosity or pores on the coating surface can reduce corrosion resistance and jeopardize the integrity of the coating. Porosity can weaken the protection against corrosion of the underlying metal. To address porosity issues, researchers have explored various approaches such as modifying coating composition, adjusting deposition parameters, and post-treatment methods [9,38]. Incorporating additives or adjusting the P content in the Ni-P matrix has shown promise in reducing porosity. Additionally, optimizing deposition parameters such as bath composition, temperature, and current density can help control porosity levels [39]. Post-treatment methods such as sealing processes or impregnation with sealants can be investigated to improve corrosion resistance by reducing porosity [40]. Porosity within the coating provides sites for corrosive species to accumulate and initiate corrosion. In a corrosive environment, electrolytes can penetrate through the pores, leading to localized corrosion attacks on the substrate material. This can result in accelerated corrosion rates and reduced overall coating performance.

Surface Roughness: Excessive surface roughness of the coating can result in unwanted increases in friction. Additionally, rough surfaces may be more susceptible to corrosion. Controlling surface roughness is crucial for optimizing the performance of Ni-P-PTFE coatings. Studies have investigated the effect of coating deposition parameters, substrate preparation methods, and post-treatment techniques on surface roughness. Utilizing polishing or grinding processes before coating application can help achieve the desired surface finish. Furthermore, adjusting parameters such as bath agitation, deposition time, and current density during electroplating can influence surface roughness. Post-treatment methods like burnishing or chemical-mechanical polishing have been explored to further refine surface roughness and enhance coating properties [41–43]. Increased surface roughness can enhance the adhesion of corrosive species and create crevices where moisture and contaminants can accumulate, promoting corrosion initiation and propagation. Rough surfaces can also hinder the formation of protective passive films, further exacerbating corrosion susceptibility.

Advances in Corrosion Science and Surface Engineering   Materials Research Forum LLC
Materials Research Foundations 188 (2026) 161-192   https://doi.org/10.21741/9781644903919-9

Surface Defects: Cracks, bubbles, or other defects formed on the surface during coating can adversely affect its performance. These defects can weaken the structure of the coating, leading to cracking or peeling. Strategies for minimizing surface defects involve improving coating deposition techniques, optimizing process parameters, and implementing quality control measures. Researchers have proposed advanced deposition methods such as pulse plating or electroless deposition to mitigate defects like cracks or voids. Additionally, precise control of bath composition, temperature, and agitation during electroplating can minimize defect formation. Quality control measures such as inspection protocols, defect detection techniques, and adherence to industry standards are essential for ensuring defect-free coatings [44]. Surface defects such as cracks or voids act as preferential sites for corrosion initiation due to their increased surface area and reduced protective barrier properties. Corrosive agents can infiltrate these defects, leading to localized corrosion attacks and compromising the integrity of the coating.

PTFE Distribution Issues: Non-uniform distribution of PTFE particles can imbalance the friction-reducing properties of the coating. This can result in undesired increases in friction or inadequate wear resistance in some areas of the coating. Achieving a uniform distribution of PTFE particles within the Ni-P matrix is crucial for optimizing friction-reducing properties. Studies have focused on enhancing dispersion methods and modifying coating formulations to improve PTFE distribution. Techniques such as ultrasonic agitation, mechanical stirring, or surfactant-assisted dispersion have been explored to achieve homogenous PTFE distribution. Furthermore, adjusting deposition parameters such as current density or bath composition can influence particle incorporation and distribution. Characterization techniques like scanning electron microscopy (SEM) or energy-dispersive X-ray spectroscopy (EDS) are employed to assess PTFE dispersion and its impact on coating performance [45,46]. Uneven dispersion of PTFE particles may lead to coating sections exhibiting inconsistent friction-reducing characteristics. In regions lacking adequate PTFE presence, friction-related abrasion can degrade the protective layer, uncovering the substrate and making it vulnerable to corrosion. Non-uniform distribution can likewise disrupt the establishment of a continuous passive film, heightening susceptibility to corrosion.

Chemical Resistance Issues: Lack of resistance to certain chemicals can limit the use of the coating in specific industrial environments. Coatings may degrade or dissolve when exposed to aggressive chemicals. Improving chemical resistance involves optimizing coating composition, incorporating corrosion inhibitors, and implementing barrier protection strategies. Studies have investigated the effect of alloying elements, PTFE content, and post-treatment methods on chemical resistance [47–49].

Alloying nickel-phosphorus with elements like boron or tungsten has been shown to enhance chemical stability. Incorporating corrosion inhibitors such as cerium salts or organic compounds into the coating matrix can provide additional protection against aggressive environments. Barrier protection techniques such as multi-layer coatings or hybrid systems combining Ni-P-PTFE with other materials have also been explored to improve chemical resistance [50]. Poor chemical resistance can result in the degradation or dissolution of the coating when exposed to corrosive chemicals. Chemical attack can lead to the breakdown of the protective barrier, allowing corrosive agents to directly contact the substrate material and initiate corrosion processes.

Temperature Resistance: The coating's resistance to high temperatures may be limited. Issues such as mismatch of coating materials or thermal degradation of PTFE may occur at elevated temperatures. Enhancing temperature resistance requires careful selection of coating materials, optimization of deposition parameters, and consideration of operating conditions. Research efforts have focused on developing novel alloy compositions and composite structures to withstand high temperatures. Alloying elements such as cobalt or molybdenum have been investigated to improve thermal stability and mechanical properties [51–53]. Furthermore, optimizing deposition parameters such as bath temperature, pH, and current density can influence coating microstructure and performance at elevated temperatures. Evaluation techniques like thermal analysis or mechanical testing under thermal cycling conditions are employed to assess the temperature resistance of Ni-P-PTFE coatings [54]. Inadequate temperature resistance can cause degradation or phase transformations within the coating, compromising its protective properties. Elevated temperatures can accelerate chemical reactions, leading to increased corrosion rates and reduced coating stability. Thermal expansion mismatches between the coating and substrate can also induce mechanical stresses, promoting cracking or delamination and facilitating corrosion initiation.

Application Techniques and Quality: The application method and quality of the coating affect the performance of the final product. The quality of equipment used during application can determine the coating's quality and consistency. Ensuring high-quality coatings involves optimizing application techniques, adhering to industry standards, and implementing rigorous quality control measures. Studies have explored advanced coating deposition methods such as electroplating automation [55] or robotic spraying to improve efficiency and consistency [56]. Adhering to standardized procedures and specifications such as ASTM or ISO guidelines is essential for maintaining quality and reliability. Quality control measures such as process monitoring, inspection protocols, and performance testing are integral parts of ensuring consistent coating quality [57]. Collaboration between researchers, coating manufacturers, and end-users is crucial for advancing application techniques and quality standards in Ni-P-PTFE coating technology. Inconsistent or poor-quality coatings may exhibit variations in composition, microstructure, or thickness, resulting in localized corrosion susceptibility. Inadequate coverage or insufficient adhesion can create discontinuities in the protective barrier, allowing corrosive agents to reach the substrate and initiate corrosion. Proper application techniques and quality control measures are essential for ensuring uniform coating coverage and adherence, thereby maximizing corrosion resistance.

To enhance the corrosion resistance of Ni-P-PTFE coatings and ensure long-term performance in corrosive environments, it is essential to address issues related to coating parameters, material selection, and quality control measures. Proper material selection, careful quality control during the coating process, and regular maintenance can minimize these issues. In industrial applications, monitoring and maintenance of coating processes are particularly important. Researchers and industry experts employ a multifaceted approach incorporating scientific principles, engineering innovations, and practical applications to enhance performance and reliability. In conclusion, optimizing coating parameters, selecting appropriate materials, and implementing quality control measures are crucial for enhancing the corrosion resistance of Ni-P-PTFE coatings and ensuring their long-term performance in corrosive environments.

3.   Effective Solutions and Strategies for Ni-P-PTFE Coatings

Ger and Hwang investigated the effect of surfactants on the incorporation of PTFE particles into electroless Ni-P coating [58,59]. The researchers presented their findings on the connection between the deposition rate of the Ni-P matrix and PTFE particles, with the concentration of

Advances in Corrosion Science and Surface Engineering   Materials Research Forum LLC
Materials Research Foundations 188 (2026) 161-192   https://doi.org/10.21741/9781644903919-9

surfactants and PTFE loading. The research indicates that the way in which the Ni-P matrix and PTFE particles combine is significantly influenced by the weak adsorption of PTFE particles on the substrate, the strong adsorption of surfactants, and the extent of surface coverage of the surfactant on the substrate [58]. The researchers have proposed a theoretical model that can help us understand the co-deposition behaviors better. According to the model, the surface coating material inhibits the deposition of the accumulated Ni-P layer on the substrate. The research has also found that an increase in the amount of surfactant generated on the surface of PTFE particles decreases the deposition rate of PTFE particles within the Ni-P layer. As a result, a reduction in competition during deposition among PTFE-surfactant-Ni/P has been reported. The researchers have identified the optimum amount of PTFE, approximately 9%, along with the distribution of PTFE, as shown in Figure 1. The recommended model has facilitated a better understanding of the various factors and mechanisms at play in the electrodeposition of PTFE particles with Ni-P [58].

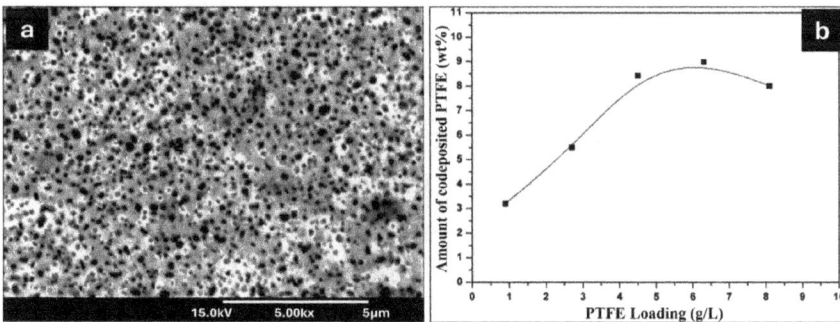

Figure 1. Analyzing of PTFE reinforced Ni-P matrix coatings; (a) SEM image of PTFE particles and (b) optimum PTFE ratio in Ni-P matrix [58], with permission from Elsevier.

In the study of Liu and Zhao [46], the surfactant with a positive charge showed better effectiveness in suspending PTFE particles in plating solutions compared to its neutral counterparts. When non-ionic surfactants were used, the PTFE particles with a diameter of 1.5 µm were absorbed into the Ni-P-PTFE coatings due to particle aggregation in the plating solutions. However, smaller PTFE particles ranging from 0.05 to 0.2 µm were uniformly dispersed throughout the Ni-P-PTFE coating when the cationic surfactant was used. Blending cationic and non-ionic surfactants did not offer any discernible advantage. The maximum PTFE content in the coatings primarily depended on the ratio between the surfactant concentration and the concentration of PTFE particles in the plating solution. When using the cationic surfactant, the deposition rates were comparatively lower as opposed to other surfactants. When the pH and concentrations of surfactant and PTFE were constant, the deposition rate of Ni-P-PTFE increased proportionately with the increase in plating solution temperature. Similarly, when the plating solution temperature and concentrations of surfactant and PTFE were constant, the deposition rate showed a marked escalation with increasing pH. The maximum volumetric PTFE content in the coating reached approximately 28% when using the cationic surfactant.

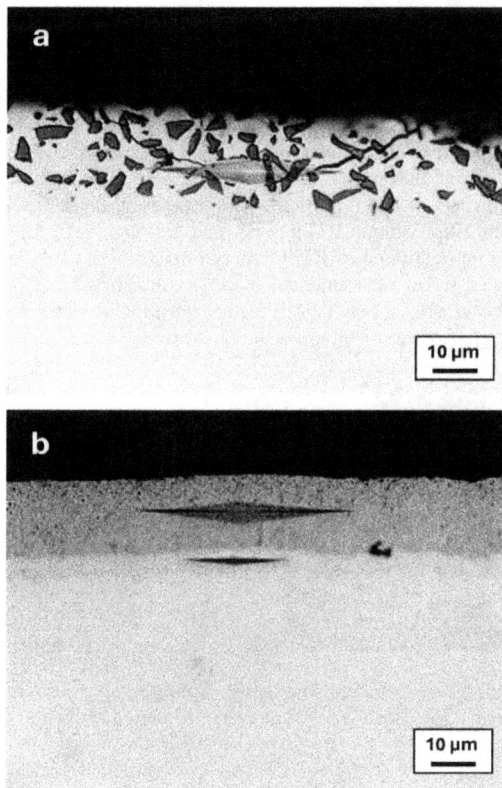

Figure 2. Knoop hardness trace; (a) Ni-P-SiC layer with fracture and (b) Ni-P and Ni-P-PTFE layer [48], with permission from Elsevier.

Increasing the proportion of PTFE within the Ni-P matrix using different surfactants results in heightened hydrophobicity due to its exceptionally low surface energy. Consequently, the surface exhibits water-repellent properties, facilitating self-cleaning characteristics [60,61]. However, excessive PTFE content can lead to adverse effects: Excessive PTFE content may diminish the coating's mechanical resilience. The inherently low friction coefficient of PTFE could compromise the coating's resistance to abrasion and scratching, impede strong adhesion to underlying surfaces, increase the risk of delamination or detachment, and may pose challenges to uniform application, leading to inconsistencies in coating quality and performance. Thus, careful control of PTFE content is essential to strike a balance between reaping the benefits of its low surface energy and mitigating the adverse effects of over-addition.

Advances in Corrosion Science and Surface Engineering   Materials Research Forum LLC
Materials Research Foundations 188 (2026) 161-192   https://doi.org/10.21741/9781644903919-9

To further enhance the superior properties of the electroless Ni-P-PTFE coatings, the addition of harder particles can compensate for the slight decrease in hardness resulting from the PTFE addition. Accordingly, Straffelini and his team [48] introduced SiC particles into the electroless Ni-P-PTFE layer to enhance hardness values on the steel substrates. The hardness of Ni-P coatings was determined to be 820 HK, Ni-P-PTFE coatings exhibited a hardness of 800 HK, Ni-P-SiC coatings had 810 HK and Ni-P-PTFE-SiC coatings showed a hardness of 880 HK. The brittleness of the Ni-P-SiC layer (Figure 2) was reduced both by the addition of PTFE and compensated by the hardness-increasing effect of SiC.

On the other hand, copper has a strong affinity for both nickel and PTFE, facilitating improved interfacial bonding between the coating layers. This enhanced bonding helps to anchor the PTFE particles more securely within the Ni-P matrix, thereby improving adhesion. Copper is known to act as a grain refiner in metal matrices. By promoting the formation of finer grains within the Ni-P-PTFE coating, the surface area available for bonding between the coating and substrate increases. This finer microstructure can enhance mechanical interlocking and adhesion [62]. Copper can enhance the ductility of the coating matrix. Increased ductility can mitigate the formation of microcracks or delamination at the coating-substrate interface, leading to improved adhesion and resistance to mechanical stresses. The presence of copper can also improve the wetting and spreading characteristics of the coating during deposition [63]. This can lead to better coverage and more uniform distribution of PTFE particles within the Ni-P matrix, promoting stronger interfacial bonding and adhesion. In the research conducted by Zhao and Liu [62], Ni-Cu-P-PTFE coatings were applied onto copper and steel surfaces using the electroless plating technique. Copper was obtained from $CuSO4$. The proportion of cationic surfactant concentration to PTFE concentration in the plating solution had a significant effect on the amount of PTFE present in the final coating. The dispersion of PTFE particles throughout the Ni-Cu-P matrix was achieved uniformly, without any additional mechanical or ultrasonic agitation methods. The ratio of accumulated elements was directly influenced by pH and temperature, while the introduction of copper resulted in improved interfacial adhesion, thus delaying corrosion and related issues, Figure 3.

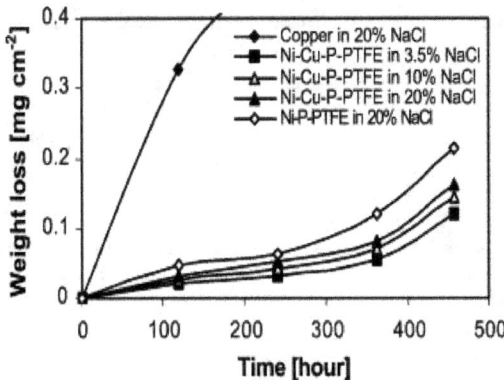

Figure 3. Corrosion loss enhancement because of Cu reinforcement [60], with permission from Elsevier.

Mafi and Dehghanian [45] This study aimed to investigate the impact of incorporating various surfactants, namely cationic, anionic, and non-ionic, at different concentrations within the plating solution on the deposition rate, PTFE content, and surface morphology of electroless Ni-P/PTFE composite coatings. The objective of this research was to identify the best-suited surfactant type and concentration for creating a flawlessly dispersed PTFE particle arrangement within the Ni-P matrix, free of any imperfections. By integrating PTFE and selecting the appropriate surfactant type and concentration, the researchers were able to achieve this outcome successfully. Two surfactants, namely CTAB and PVP, facilitated the uniform dispersion of PTFE particles within the coating. Conversely, SDS failed to effectively incorporate the PTFE particles into the deposit, as shown in Figure 4. The results demonstrate that surfactant selection is a crucial factor in determining the uniformity of PTFE particle dispersion within the Ni-P matrix, thereby affecting the overall quality of the electroless composite coating. In conclusion, this study provides valuable insights into the surfactant-mediated preparation of electroless Ni-P/PTFE coatings. It emphasizes the importance of selecting the appropriate surfactant type and concentration to achieve the desired level of particle dispersion, free of any imperfections. The results of this research could benefit industries that rely on electroless Ni-P/PTFE coatings, such as automotive, aerospace, and electronics, by improving the quality and durability of the coatings.

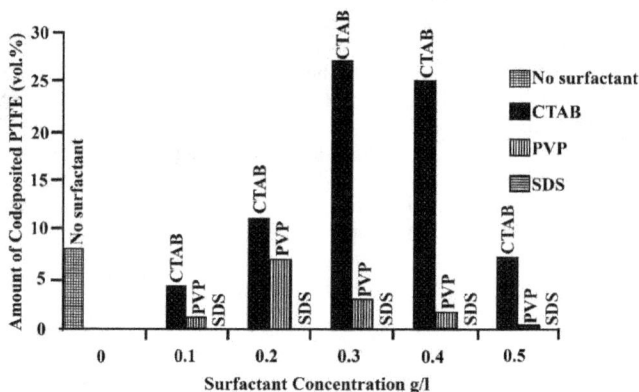

Figure 4. The percentage of PTFE particles codeposited with varying concentrations of surfactants in the plating bath [45], with permission from Elsevier.

In conclusion, it is evident that various steps in the application process, such as surfactant application procedures, surface preparation techniques, the addition of different elements to the Ni-P-PTFE layer, and the modification of coating process parameters, directly influence the properties of the Ni-P-PTFE coating layer. These steps play crucial roles in the development of interfacial properties and can lead to enhanced corrosion resistance in the final coatings. By carefully adjusting these parameters, it is possible to optimize the interfacial adhesion between the coating and the substrate, thereby improving the overall durability and corrosion resistance of the coating. For instance, the choice of surfactant and its concentration can affect the distribution of

PTFE particles within the coating, while variations in surface preparation methods can influence the cleanliness and roughness of the substrate surface, ultimately impacting adhesion. Furthermore, the incorporation of different elements into the Ni-P-PTFE matrix can alter its chemical composition and microstructure, thereby affecting its corrosion resistance properties. Adjusting coating process parameters such as temperature, pH, and deposition time can further fine-tune the coating's structure and properties to meet specific corrosion protection requirements. In essence, by carefully controlling these factors throughout the coating process, it is possible to tailor Ni-P-PTFE coatings with superior interfacial development and enhanced corrosion properties, thus extending the lifespan and reliability of coated components in various applications.

4. Enhancing Corrosion Resistance: Recent Ni-P-PTFE Coating Studies

This section will discuss recent studies focusing on the improvement of corrosion properties based on the interfacial development between the substrate and Ni-P-PTFE-based coatings, considering the influence of microstructural evolution of the coatings, the effects of various added particles, application to different substrate surfaces, and coating methods. These studies highlight advancements in technology, emphasizing the importance of understanding the microstructural development of Ni-P-PTFE-based coatings and its correlation with corrosion resistance. By considering factors such as the incorporation of different particles, application to diverse substrate surfaces, and variations in coating methods, researchers aim to enhance the corrosion resistance properties of these coatings. The discussion will provide insights into how recent technological developments have contributed to the optimization of corrosion protection through tailored microstructural development, thereby addressing the growing demand for durable and reliable protective Ni-P-PTFE-based coatings in various industrial applications.

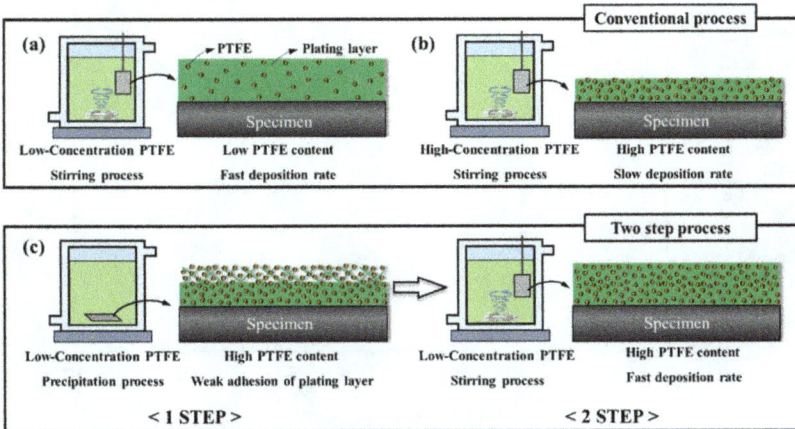

Figure 5. Comparison of conventional and new electrodeposition methods of Ni-P-PTFE [64], licensed by CC-BY-4.0.

Advances in Corrosion Science and Surface Engineering   Materials Research Forum LLC
Materials Research Foundations 188 (2026) 161-192   https://doi.org/10.21741/9781644903919-9

Lee and his colleagues [64] propose a new technique for electroless plating. The technique aims to improve the content of PTFE within the coating layer. This new method prevents the clumping of PTFE and increases the co-deposition rate, even when PTFE concentrations are low. However, unlike traditional plating methods, this approach involves an extra step. The specimen is initially positioned on the floor to allow the solid particles to settle without any agitation. The plating process is then carried out using the conventional stirring method. This innovative plating (Figure 5) technique is expected to provide outstanding wear resistance and low surface energy, particularly when PTFE concentrations are low. Additionally, even though various investigations were conducted to examine plating layers exceeding 10 μm in thickness [65], the use of thinner plating layers enhances product precision. In this study, the researchers analyzed a less than 5 μm thick electroless composite plated layer on high-carbon steel that consisted of a Ni-P-PTFE thin film.

It is evident that the impact of higher PTFE content achieved with less PTFE addition as seen in Figure 6 [64], compared to traditional coating methods, on the surface coating properties should not be overlooked. Therefore, it is imperative to conduct comprehensive studies focusing on the effect of these final coatings on corrosion resistance. This statement emphasizes the importance of thoroughly investigating how the innovative electroless plating technique, which allows for higher PTFE content with reduced PTFE addition, affects the corrosion resistance of the coatings. By conducting such studies, researchers can gain insights into the potential benefits of these coatings in mitigating corrosion and enhancing the durability of coated materials in various applications.

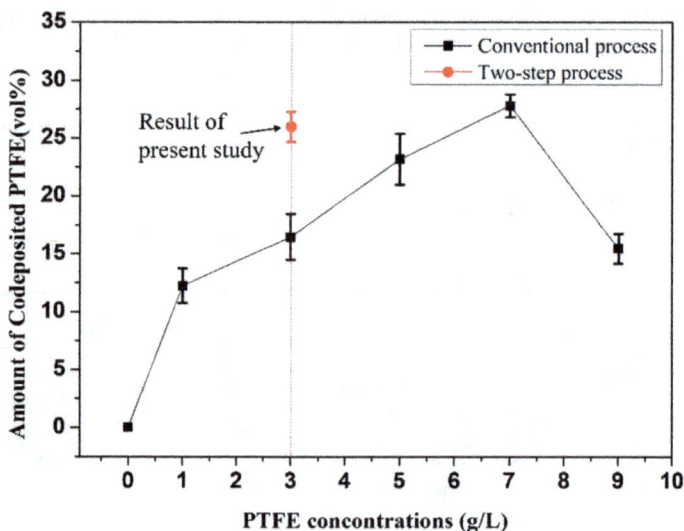

Figure 6. Comparison of amount of codeposited PTFE content in conventional and new electrodeposition methods of Ni-P-PTFE [64], licensed by CC-BY-4.0

Advances in Corrosion Science and Surface Engineering   Materials Research Forum LLC
Materials Research Foundations 188 (2026) 161-192   https://doi.org/10.21741/9781644903919-9

Gao and colleagues [66] aimed to improve the mechanical adhesion of subsequent coating layers by initially electropolishing steel surfaces to create nano-pores through anodizing, followed by anodic oxidation. Subsequently, they applied Ni-P-SiC coatings using graphite anode and steel cathode electrodes in an electrolyte containing Ni source, reducing agent, and SiC particles. Afterward, they immersed the surfaces in a PTFE solution to impregnate PTFE into the surface roughnesses. Finally, they subjected the surfaces to heat treatment at 350 °C to smoothen them. This multi-coating development process resulted in coatings with superior corrosion resistance. While Ni-P-SiC coatings exhibited a corrosion current density (i_{corr}) value of 1.6 μA·cm^{-2}, Ni-P-SiC-PTFE coatings achieved a value of 0.09 μA·cm^{-2}. These coatings exhibited superior properties, demonstrating significantly low current density even in potentiodynamic tests conducted in a strong HF solution, attributed to both SiC and PTFE contributions. This study highlights the effectiveness of the multi-coating approach in enhancing the corrosion resistance of steel surfaces. By incorporating nano-pores, SiC particles, and PTFE impregnation, the coatings achieved remarkable corrosion resistance, surpassing traditional Ni-P-SiC coatings. Additionally, the low corrosion current densities observed in strong HF solution tests indicate the potential of these coatings in harsh environments, suggesting their suitability for various industrial applications requiring robust corrosion protection.

Boakye and colleagues [67] fabricated electroless Ni-P-PTFE coatings by introducing PTFE particle additives into the electroless Ni-P plating solution. They varied the PTFE additive ratio at 5, 10, and 15 g/L, and found that the coating surface lacked homogeneity due to non-uniform PTFE distribution when the PTFE ratio was 15 g/L as seen with black holes in SEM images of coatings in Figure 7. They determined that a PTFE additive ratio of 10 g/L provided optimal homogeneity in the Ni-P-PTFE coatings.

It can be inferred from this that non-uniform coating layers affect corrosion outcomes. Non-uniform coating layers can have significant implications on corrosion resistance. When the distribution of PTFE particles in the coating is uneven, certain areas of the coated surface may have higher concentrations of PTFE than others. This non-uniform distribution can create localized variations in properties such as surface roughness, chemical composition, and barrier properties against corrosive agents. As a result, areas with lower PTFE content may be more susceptible to corrosion compared to regions with higher PTFE content, leading to accelerated corrosion rates and reduced coating durability. Non-uniform coating layers can significantly affect corrosion resistance. Uneven distribution of PTFE particles can lead to localized variations in surface properties such as roughness, chemical composition, and barrier effectiveness against corrosive agents. Consequently, regions with lower PTFE content may experience accelerated corrosion rates compared to areas with higher PTFE content, compromising coating durability.

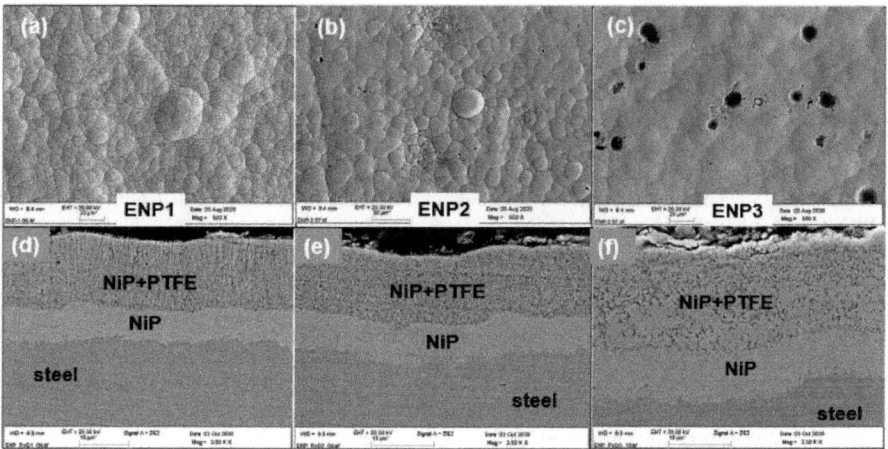

Figure 7. Surface (a-c) and cross-sectional (d-f) images of Ni-P-PTFE coatings containing different ratios of PTFE; EP1 (5% PTFE), EP2 (10% PTFE), and EP3 (15% PTFE) [67], licensed by CC-BY-4.0.

Huang and colleagues [68] explored the influence of varying PTFE concentrations (1, 3, 5, and 7 ml/L) in electroless Ni-P-PTFE coatings on the corrosion resistance of steel in a CO_2 environment. Adding PTFE to a material has been noticed to greatly improve its ability to prevent hydrogen from passing through and reduce its tendency to crack due to hydrogen exposure. This is because PTFE helps create a stronger coating on the material's surface, which acts as a better barrier against hydrogen. By doing so, PTFE helps to limit the amount of hydrogen atoms that can enter the material, thus reducing the likelihood of hydrogen-induced cracking. It's interesting to note that the corrosion resistance of composite coatings initially increases and then decreases with increasing PTFE content. This unusual trend is due to the stable molecular bonds of PTFE and its ability to optimize the microstructure of the coating. The interplay between PTFE and the coating matrix is complex.

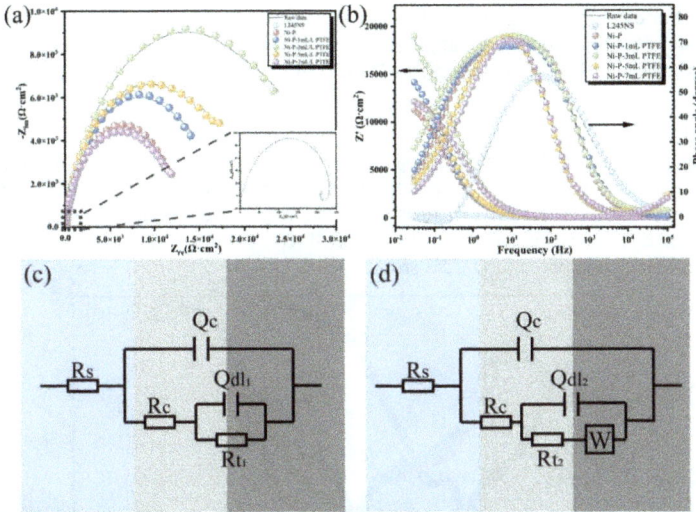

Figure 8. Electrochemical impedance spectroscopy (EIS) measurements of Ni-P/PTFE coatings in CO_2 environment; The curves of (a) Nyquist and (b) Bode; (c) equivalent circuits of Ni-P, Ni-P-1, and 3 mL/L PTFE) and (d) Ni-P-5 and 7 mL/L PTFE [68], with permission from Elsevier.

Notably, the addition of PTFE (3 mL/L) in Ni-P composite coatings resulted in significant improvements in corrosion resistance as demonstrated by the EIS results in Figure 8. It was found that the electrical properties of the coating varied depending on the amount of PTFE additive used. When 1 mL/L of PTFE was added, the coating adhered to a commonly accepted model circuit, which is illustrated in Figure 8c. The enhanced porosity of the coating, thanks to the addition of PTFE, has established a direct pathway for diffusion between the substrate and solution. This has led to a transfer process that not only occurs through the coating but also via the voids within it. As a result, the equivalent circuit illustrates a combination of a high-frequency capacitive semicircle, a low-frequency capacitive semicircle, and a Warburg (W) impedance, as shown in Figure 8d. The combination of PTFE and Ni-P gives rise to a unique "maze effect" that effectively strengthens the coating's ability to withstand corrosive environments. It is important to keep in mind that excessive amounts of PTFE can actually hinder the coating's resistance to corrosion. This is because an overabundance of PTFE can create more porous coatings, which can create a direct path for diffusion between the substrate and the solution, ultimately compromising the coating's overall effectiveness against corrosion.

It can be understood that examining the effects of different PTFE addition ratios over a wider range, such as adding PTFE at lower and higher concentrations. This can help determine the optimal PTFE addition ratio to optimize the coating's corrosion resistance. Using alternative solution compositions that better reflect real-field conditions beyond CO_2-saturated NaCl solution. For instance, using solutions containing other chemicals found in the oil and natural gas industry can be used to simulate corrosion behavior more realistically. Employing different mixing and application methods such as electroless coating can be used to improve the coating's homogeneity and quality. For example, using ultrasonic mixing or different spray techniques can be carried out to fabrication of coatings too. These approaches can provide valuable insights into enhancing the performance of electroless composite Ni-P-PTFE coatings for applications in the oil and natural gas industry.

Liang et al. [69] conducted research on the corrosion properties of Ni-P-Cu coatings with different PTFE ratios applied to Q235B steel using the electroless plating method. Copper and CTAB addition were used as a cationic surfactant, while the incorporation of PTFE aimed to provide low surface energy and improve corrosion resistance. The results from both weight loss experiments and electrochemical tests indicated that the Ni-Cu-P-PTFE composite coating offered superior corrosion resistance compared to Q235B steel.

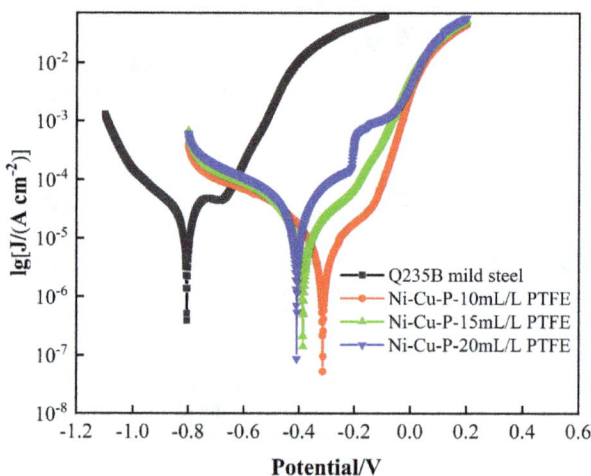

Figure 9. Tafel curves of the samples [69], licensed by CC-BY-4.0.

Figure 9 shows the polarization curves obtained from experiments conducted on Q235B mild steel and Ni-Cu-P-PTFE composite coatings with different PTFE reinforcement ratios in a 3.5 wt% NaCl solution. The composite coating with a PTFE concentration of 10 mL/L exhibited the highest corrosion resistance, with a weight loss rate of 0.24 mg·cm^{-2}, corrosion current density of 7.255×10^{-6} A·cm^{-2}, and corrosion voltage of -0.314 V. However, increasing the concentration of PTFE led to a deterioration in the corrosion resistance of the Ni-Cu-P-PTFE composite plating due to the rise in porosity within the Ni-P matrix coating layer. PTFE's high hydrophobicity and chemical stability enhance the corrosion resistance of the Ni-Cu-P-PTFE composite coating. However, a high concentration of PTFE weakens efficacy due to limited dispersion of CTAB, resulting in larger spacing between PTFE particles and increased defects.

While the inclusion of original research is commendable, the implementation of additional processes such as pore formation on the substrate surface or the application of graded coatings before the plating processes could indeed facilitate the advancement of such coatings and the

Advances in Corrosion Science and Surface Engineering   Materials Research Forum LLC
Materials Research Foundations 188 (2026) 161-192   https://doi.org/10.21741/9781644903919-9

investigation of their impact on corrosion resistance. By introducing these pre-treatment steps, it becomes possible to enhance the adhesion between the coating and the substrate, optimize the microstructure of the coating, and tailor its properties to specific applications. Additionally, graded coatings can provide gradual transitions in composition or properties, which may further improve the overall performance and durability of the coating system. Therefore, integrating such pre-treatment techniques can offer valuable insights into the development and corrosion resistance enhancement of these coatings.

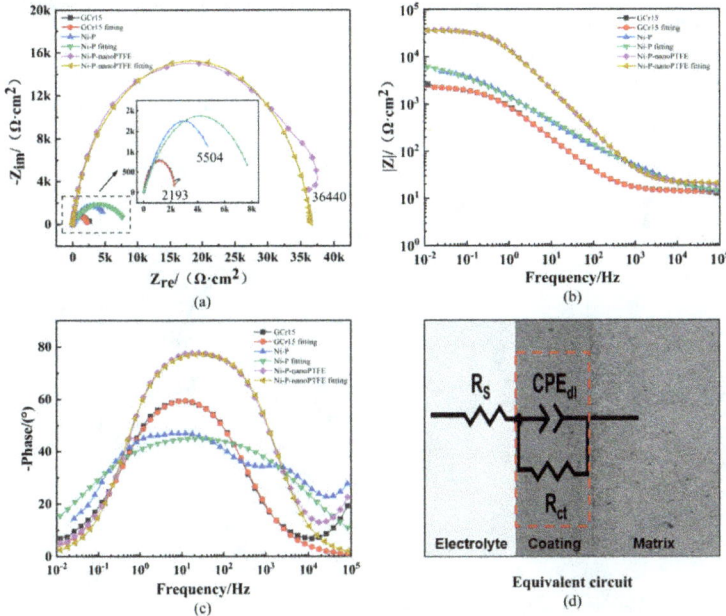

Figure 10. The Nyquist and Bode diagrams depict the characteristics of the samples, alongside their equivalent circuit representation. Panel (a) illustrates the Nyquist plot, (b) displays the Bode diagram, (c) delineates the phase response, and (d) presents the equivalent circuit schematic [70], licensed by CC-BY-4.0.

Mei and colleagues [70] fabricated a Ni-P/nano-PTFE composite coating on GCr15 steel spinning rings through an electroless coating process. The composite coatings were studied, and it was found that the highest PTFE content was obtained at a PTFE emulsion concentration of 8 mL/L, which resulted in a weight percentage of 2.16%. The analysis showed that the corrosion resistance of the composite coating was significantly improved compared to the Ni-P coating. The corrosion potential shifted to a more positive value of -421 mV, which was a 7.6% increase from the Ni-P coating. Moreover, the corrosion current was substantially reduced from 6.71 μA to 1.54 μA, representing a reduction of 77%. At the same time, the impedance increased remarkably from 5504 $\Omega \cdot cm^2$ to 36,440 $\Omega \cdot cm^2$, indicating a substantial enhancement of 562%. The size of the Ni-P-nano

Advances in Corrosion Science and Surface Engineering
Materials Research Foundations 188 (2026) 161-192

Materials Research Forum LLC
https://doi.org/10.21741/9781644903919-9

PTFE composite coating sample also had an impact on the protective capability, as larger sizes correlated positively with better protective capability. Higher impedance modulus values in the low-frequency region were found to correlate with improved corrosion resistance. The equivalent circuit diagram illustrated different components representing various electrical properties, such as resistance (R), electric double-layer capacitance constant phase angle element (CPE), and solution resistance (Rs) (Figure 10).

Ma et al. [71] focused on the effect of changing the phosphorus content in Ni-P-PTFE coatings. They varied the phosphoric acid content, which served as the phosphorus source. Additionally, they applied these coatings for the first time using the electroforming method. An increase in phosphoric acid content increased the reduced PTFE content. However, with the increase, there was a tendency for agglomeration, especially in coatings obtained with phosphoric acid concentrations higher than 8 g/L, leading to increased current density values. This is clearly evident when examining the Nyquist curves provided in Figure 11.

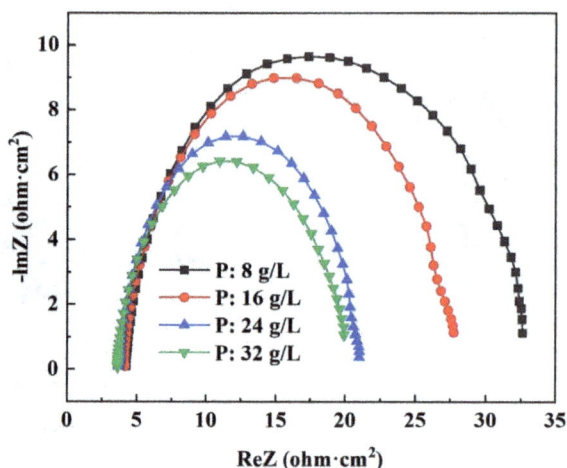

Figure 11. Nyquist curves of the Ni-P-PTFE coatings with different P ratios [71], with permission from Elsevier.

A wide range of ReZ (resistance with a real part) indicates that the resistance of the electrode varies over a wide range of electrode potentials. Corrosion resistance refers to a material's resistance to environmental effects. Furthermore, in this study, the calculated Charge Transfer Resistance (ohm) values for coatings obtained with 8 g/L, 16 g/L, 24 g/L, and 32 g/L of phosphoric acid were 28.10, 23.02, 17.26, and 15.74, respectively. This indicates that an increase in charge transfer resistance generally contributes to the formation of protective layers, such as oxidation or passivation, on the electrode surface. These layers protect the material's surface and provide resistance to environmental effects. Therefore, an increase in charge transfer resistance typically enhances the material's durability against corrosive environments. In addition to the study's focus on the effect of phosphorus content variation in Ni-P-PTFE coatings, the impact of using novel coating methods involving different compositions, such as PTFE and Cu additives, could be explored. By investigating the influence of these additives on the developed coatings, further advancements can be achieved, leading to more enhanced coatings. These improvements could be facilitated by conducting additional studies that focus on optimizing the composition of the coatings through systematic experimentation. For example, researchers could explore different ratios of PTFE and Cu additives in the coating solution to determine the most effective combination for improving corrosion resistance and other desired properties. Moreover, techniques such as surface analysis, electrochemical testing, and mechanical characterization can be employed to comprehensively evaluate the performance of the modified coatings. Overall, by systematically exploring various additives and their effects on coating properties, researchers can pave the way for the development of more advanced and effective coatings with superior performance characteristics.

Li and colleagues [72] applied Ni–P-Al_2O_3-PTFE coatings on Q235 low-carbon steel using the novel jet electrodeposition method, a relatively new approach for this type of coating. They aimed to increase coating hardness with the addition of hard Al_2O_3 particles, while also obtaining lubricating properties with PTFE. Furthermore, this production technique allows for the modification of coating properties through adjustments in current density, resulting in varied microstructural developments. The HV hardness value increased from around 310 HV to approximately 560 HV as the current density increased from 5 A/dm^2 to 10 A/dm^2. This enhancement can be attributed to the higher nanoparticle content present in the composite coating. The uniform distribution of hard Al_2O_3 particles within the composite coating reduces the occurrence of defects like dislocations and twins, thereby providing dispersion strengthening. Additionally, nanoparticles inhibit grain growth during deposition, leading to a denser structure of the composite coating and consequent microhardness enhancement. The most significant limitation of this study is the lack of investigation into corrosion resistance. This omission is crucial because the resistance of materials to environmental degradation, such as corrosion, is a vital aspect of their performance, especially in real-world applications. This scenario serves as a poignant reminder that despite significant progress, there remains a plethora of unexplored avenues and untapped potential within Ni-P-PTFE coatings. The current state of knowledge suggests that numerous aspects of these coatings are yet to be fully understood and optimized. From exploring novel structural developments to refining coating compositions and embracing innovative methodologies, there exists a wealth of opportunities for further advancement. This underscores the dynamic and evolving nature of materials science and the continuous quest for improvement and innovation.

Boakye and colleagues [73] investigated the corrosion resistance of two different coatings in simulated high-temperature geothermal conditions. The first coating was a polymer coating modified with graphene oxide (GO), while the second coating was a duplex electroless Ni-P coating supplemented with PTFE. The study tested both coatings with and without H2S and CO2 gases, which are typical in geothermal applications.

The researchers found that adding GO nanosheets to the polymer coating resulted in reduced wetting ability and helped to mitigate corrosion effects on the substrate at 120°C. They also observed that the Ni-P/PTFE duplex coating with the lowest phosphorus content performed well in a pure water environment at 120°C, while higher phosphorus content showed promise in the more aggressive two-phase CO2/H2S environment. The coating method used in this study has a wide variety of compositions and parameters available. By investigating the impact of altering these components on corrosion properties, researchers can identify enhanced coating compositions and parameters.

Vasconcelos et al. [74] highlighted the importance of investigating the properties of Ni-P-PTFE coatings on natural butadiene rubber and different substrates. The study emphasized the need to ensure the uniform and effective incorporation of PTFE into the Ni-P coating, and thus a cationic surfactant CTAB was introduced into the plating bath. The optimized level of PTFE incorporation reached 6.8%, which resulted in composite coating adhesion to NBR being 20% higher compared to Ni-P coatings alone. The researchers assessed the surface characteristics of Ni-P-PTFE coatings in terms of PTFE particle dispersion and quantity. SEM micrographs in Figure 12 illustrated the distribution of PTFE particles, with arrows indicating their presence. It was observed that a uniform dispersion of particles was achieved in all samples, except for Ni-P-PTFE (10-2), which corresponded to the highest PTFE concentration with the lowest CTAB concentration as given in Figure 12.

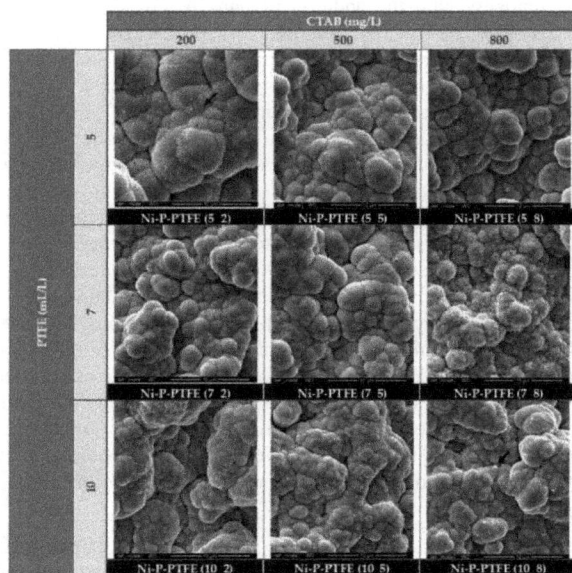

Figure 12. SEM micrographs were taken of Ni-P-PTFE samples on NBR with varying concentrations of CTAB and PTFE [74], licensed by CC-BY-4.0

The study conducted by Vasconcelos and colleagues, as mentioned above, underscores a significant gap in research regarding the investigation of corrosion resistance. The absence of exploration of corrosion properties in their study highlights a notable deficiency in the literature, particularly in the realm of Ni-P-PTFE-based coatings applied on polymer substrates. Therefore, there exists a compelling need to delve into the corrosion characteristics of such coatings and to explore the application of advanced coatings on these substrates. This area of study holds promise for addressing this gap and advancing our understanding of corrosion protection mechanisms, especially in the context of polymer substrates. Further research in this direction would not only enhance our knowledge of corrosion-resistant coatings but also contribute to the development of more effective protective measures for polymer-based materials in various industrial applications.

Li and his team [75] applied Ni-P-PTFE-Al$_2$O$_3$ coatings on Q235 mild steel surfaces using the pulse electrodeposition technique, while varying the duty cycle percentage. Coating hardness exceeding 670 HV was achieved at a 30% duty cycle, thanks to the homogeneous distribution of PTFE and Al$_2$O$_3$ particles within the Ni-P matrix structure. This resulted in both microstructural development and hardness enhancement. At higher duty cycle ratios, pore formation occurred due to agglomeration. However, corrosion properties were not examined in the study, leaving a gap in the research. Additionally, as indicated by the literature provided earlier, it is evident that corrosion properties could be improved by altering the ratios, morphology, and particle size of PTFE and Al$_2$O$_3$ alongside this novel production technique.

5.   The application areas of Ni-P-PTFE-based corrosion protective coatings

Applying Ni-P-PTFE-based corrosion protective coatings serves a variety of industries and applications due to their unique properties and capabilities. These coatings are known for their corrosion resistance, wear resistance, low friction, and non-stick properties, making them suitable for a wide range of environments and conditions. In this detailed exploration, we will delve into the application areas of Ni-P-PTFE-based coatings, the reasons behind their usage, and the specific benefits they offer.

Aerospace Industry: In the aerospace sector, where components are exposed to harsh environmental conditions [76], Ni-P-PTFE coatings provide excellent corrosion protection. Aircraft parts such as landing gear, hydraulic systems, and engine components benefit from these coatings due to their ability to withstand extreme temperatures, moisture, and chemical exposure. Additionally, the low friction properties of PTFE contribute to reduced wear and improved performance of moving parts, leading to increased efficiency and longevity. For materials, components such as landing gear, hydraulic systems, and engine parts are often made of aluminum alloys [77], stainless steels [78], or titanium alloys [79]. Ni-P-PTFE coatings act as a barrier between the substrate material and the corrosive environment encountered during flight. The Ni-P matrix provides excellent corrosion resistance, while PTFE imparts low friction and non-stick properties, reducing wear and preventing the buildup of corrosive substances.

Automotive Industry: The automotive industry utilizes Ni-P-PTFE coatings for various applications, including piston rings, engine valves, and transmission components. These coatings offer superior corrosion resistance, which is crucial for automotive parts exposed to corrosive elements like salt, moisture, and road chemicals. Furthermore, the low friction properties of PTFE reduce frictional losses, resulting in improved fuel efficiency and reduced wear on engine components. For materials, engine components like piston rings and valves are typically made of steel [80] or aluminum alloys [81]. Ni-P-PTFE coatings provide a protective layer against corrosive elements such as moisture, road salt, and engine fluids. The nickel-phosphorus matrix forms a barrier against corrosion, while the PTFE reduces friction between moving parts, minimizing wear and enhancing durability.

Oil and Gas Industry: In the oil and gas sector, where equipment is subjected to corrosive environments, Ni-P-PTFE coatings play a vital role in protecting critical components such as valves, pipelines, and offshore structures. These coatings provide a barrier against corrosive substances present in oil, gas, and drilling fluids, thereby extending the service life of equipment and reducing maintenance costs. Additionally, the non-stick properties of PTFE help prevent the buildup of deposits and fouling on surfaces, ensuring smooth operation and minimizing downtime. For materials, valves, pipelines, and offshore structures are often made of carbon steel or stainless steel [82,83]. Ni-P-PTFE coatings protect equipment from corrosion caused by exposure to corrosive substances like oil, gas, and drilling fluids. The Ni-P matrix offers corrosion resistance, while PTFE's non-stick properties prevent the adhesion of corrosive materials, reducing the likelihood of corrosion initiation and propagation.

Chemical Processing Industry: Equipment used in chemical processing plants is often exposed to corrosive chemicals and high temperatures. Ni-P-PTFE coatings offer excellent chemical resistance and thermal stability, making them ideal for protecting valves, pumps, and reactors in this industry. The non-stick properties of PTFE also help prevent the adhesion of sticky or viscous substances to equipment surfaces, facilitating easier cleaning and maintenance. For materials, equipment such as valves, pumps, and reactors may be constructed from stainless steel, Hastelloy, or other corrosion-resistant alloys [84–86]. Ni-P-PTFE coatings provide a protective barrier against chemical corrosion and high temperatures encountered in chemical processing environments. The nickel-phosphorus matrix offers resistance to chemical attack, while PTFE's low friction properties prevent the buildup of corrosive substances and facilitate easy cleaning.

Marine Industry: In the marine sector, where structures are constantly exposed to saltwater and atmospheric corrosion, Ni-P-PTFE coatings provide effective protection for components such as ship propellers, shafts, and offshore platforms. These coatings offer superior resistance to saltwater corrosion and fouling, helping to maintain the structural integrity of marine equipment and extend its service life. Ship propellers, shafts, and offshore platforms are typically made of stainless steel or corrosion-resistant alloys [87–90]. Ni-P-PTFE coatings protect marine equipment from corrosion caused by exposure to saltwater and atmospheric conditions. The Ni-P matrix provides corrosion resistance, while PTFE's non-stick properties prevent fouling by marine organisms and reduce the accumulation of corrosive salts, extending the service life of marine structures.

Medical and Pharmaceutical Industry: In medical and pharmaceutical applications, where cleanliness and sterility are paramount, Ni-P-PTFE coatings are used to protect equipment such as surgical instruments, medical devices, and pharmaceutical processing equipment. These coatings provide a smooth, non-reactive surface that is easy to clean and sterilize, helping to prevent contamination and ensure product purity. Surgical instruments, medical devices, and pharmaceutical equipment may be made of stainless steel, titanium, or other biocompatible alloys [91–93]. Ni-P-PTFE coatings offer corrosion protection for medical and pharmaceutical equipment, ensuring cleanliness and sterility. The nickel-phosphorus matrix provides corrosion resistance, while PTFE's non-reactive surface prevents the adhesion of contaminants and facilitates easy sterilization.

Food Processing Industry: In the food processing industry, where hygiene and cleanliness are critical, Ni-P-PTFE coatings are applied to equipment such as mixing blades, conveyor belts, and food packaging machinery. These coatings offer excellent non-stick properties, preventing food from sticking to surfaces and reducing the risk of contamination. Additionally, the corrosion resistance of these coatings ensures the longevity of equipment in aggressive food processing environments. Mixing blades, conveyor belts, and packaging machinery are often made of stainless steel or food-grade plastics [94–96]. Ni-P-PTFE coatings protect food processing equipment from corrosion and contamination. The Ni-P matrix provides corrosion resistance, while PTFE's non-stick properties prevent food particles from adhering to surfaces, reducing the risk of contamination, and ensuring product purity.

Overall, the application areas of Ni-P-PTFE-based corrosion protective coatings are diverse and extensive, spanning various industries where protection against corrosion, wear, and friction is essential. By providing superior performance in harsh environments and offering unique properties such as corrosion resistance, low friction, and non-stick characteristics, these coatings play a crucial role in enhancing the durability, reliability, and efficiency of critical components and equipment.

Conclusions

Investigating the effective factors on the corrosion resistance of Ni-P-PTFE-based coatings through the analysis of literature examples highlights the significance of advanced coating methods and coating parameters. Advanced coating methods can enhance corrosion resistance by facilitating more effective and homogeneous coating deposition, as well as allowing control over coating properties. These methods include optimizing electrolytic coating parameters, utilizing coating materials with nano-properties, and employing post-coating thermal treatments. Firstly, the importance of electrolytic coating parameters should be emphasized. Factors such as current density, deposition time, electrolyte composition, and temperature determine the microstructure, composition, and corrosion resistance of the coating. For instance, low current density typically results in a more homogeneous coating, whereas high current density may offer faster coating rates but could compromise quality. Similarly, appropriate deposition time and temperature are necessary to achieve the desired coating quality. Furthermore, the use of nanomaterials represents a significant advancement in enhancing corrosion resistance. Incorporating nanoscale materials can improve the mechanical and chemical properties of the coating material. For example, integrating nano-sized PTFE particles into the Ni-P matrix can reduce the coefficient of friction and enhance wear resistance, thereby increasing corrosion resistance. Post-coating thermal treatments are also commonly used to improve coating properties. Thermal treatments can alter the composition and structural properties of the coating, enhancing mechanical strength and corrosion resistance. This process can rearrange the coating's crystal structures and promote stronger bonding with the substrate. In conclusion, careful consideration of advanced coating methods and coating parameters is essential to enhance the corrosion resistance of Ni-P-PTFE-based coatings. Optimizing electrolytic coating parameters, utilizing coating materials with nano-properties, and employing post-coating thermal treatments enable the development of more durable and long-lasting coatings for industrial applications. Effective implementation of these methods can result in materials that are more resistant to corrosion, thus reducing long-term costs and maintenance requirements.

Future Perspectives

Changing the morphology and size of PTFE particles, including both nano and micro-sized particles in the Ni-P matrix, could potentially mitigate agglomeration issues. Additionally, employing different morphologies of nano-sized PTFE particles may offer enhanced dispersion within the matrix. Applying metal coatings onto the surface of PTFE can promote interfacial

Advances in Corrosion Science and Surface Engineering   Materials Research Forum LLC
Materials Research Foundations 188 (2026) 161-192   https://doi.org/10.21741/9781644903919-9

development, enhancing the compatibility between PTFE and the metal matrix. The addition of metal additives such as Cu in powdered form to Ni-P-PTFE coatings in various ratios could offer tailored enhancements to the coating properties. After mixing PTFE-metal powder blends through milling, introducing them into Ni-P plating solutions could provide a novel approach for incorporating metal-PTFE mixtures into the coatings. Investigating the effects of anodic oxidation treatments on substrate surfaces with different morphologies before coating processes can provide insights into the influence of porous substrate surfaces on coating properties. Research on applying Ni-P-PTFE-based coatings to polymer substrate surfaces is relatively limited. However, applying such coatings to suitable polymer surfaces could lead to unique and valuable studies, particularly concerning interfacial compatibility. While most studies focus on electroless plating methods, exploring electrodeposition techniques offers a novel avenue for research, as parameters such as electrode distance, current density, temperature, agitation rate, and anode type directly influence coating properties. Innovative coating methods beyond electroless plating, such as electrodeposition or hybrid approaches combining multiple deposition techniques, could offer enhanced control over coating properties and interfacial interactions. Investigating the influence of specific parameters in electrodeposition, such as electrode distance, current density, and temperature, on the properties of Ni-P-PTFE coatings could yield valuable insights. Additionally, exploring advanced characterization techniques, such as in situ monitoring of deposition processes or nanoscale analysis of coating structures, can provide a deeper understanding of the coating mechanisms and performance. Integrating these approaches into future research endeavors can pave the way for the development of high-performance Ni-P-PTFE coatings tailored for various industrial applications

References

[1]A. Brenner, D.E. Couch, E.K. Williams, Electrodeposition of alloys of phosphorus with nickel or cobalt, J. Res. Nat. Bur. Stand 44 (1950) 109.

[2]A. Brenner, G.E. Riddell, Nickel plating by chemical reduction, US Patent US2532282 (1950).

[3]T. Mimani, S.M. Mayanna, The effect of microstructure on the corrosion behaviour of electroless Ni☐P alloys in acidic media, Surf Coat Technol 79 (1996) 246–251. https://doi.org/10.1016/0257-8972(95)02446-8

[4]W. Sha, J.S. Pan, Electroplating Ni☐P films and their corrosion property, J Alloys Compd 182 (1992) L1–L3. https://doi.org/10.1016/0925-8388(92)90568-T

[5] C. Sun, S. Shuang, H. Zeng, V. Fattahpour, M. Mahmoudi, J.-L. Luo, Investigation of Corrosion Properties of a High-Phosphorus Ni-P Coating and Corrosion Resistant Alloys in 3.5 wt.% NaCl Solution, (2018) 1-10. https://doi.org/10.5006/C2018-11526

[6]A. Sosa Domínguez, J.J. Pérez Bueno, I. Zamudio Torres, M.L. Mendoza López, Characterization and corrosion resistance of electroless black Ni-P coatings of double black layer on carbon steel, Surf Coat Technol 326 (2017) 192–199. https://doi.org/10.1016/J.SURFCOAT.2017.07.044

[7]B. Qin, L. Li, J. Wang, G. Chen, Z. Huang, Y. Liu, J. Dou, Erosion-Corrosion Behavior of Electroless Ni–P Coating on M2052 Alloy in Artificial Seawater, ISIJ International 62 (2022) 550–560. https://doi.org/10.2355/ISIJINTERNATIONAL.ISIJINT-2021-359

[8]B. Jiang, S.L. Jiang, A.L. Ma, Y.G. Zheng, Erosion-corrosion behavior of electroless Ni-P coating on copper-nickel alloy in 3.5 wt.% sodium chloride solution, J Mater Eng Perform 23 (2014) 230–237. https://doi.org/10.1007/S11665-013-0763-0/FIGURES/11

[9]H. Ashassi-Sorkhabi, S.H. Rafizadeh, Effect of coating time and heat treatment on structures and corrosion characteristics of electroless Ni–P alloy deposits, Surf Coat Technol 176 (2004) 318–326. https://doi.org/10.1016/S0257-8972(03)00746-1

[10]   Y. Su, B. Zhou, L. Liu, J. Lian, G. Li, Electromagnetic shielding and corrosion resistance of electroless Ni-P and Ni-P-Cu coatings on polymer/carbon fiber composites, Polym Compos 36 (2015) 923–930. https://doi.org/10.1002/PC.23012

[11]   D. Ahmadkhaniha, F. Eriksson, P. Leisner, C. Zanella, Effect of SiC particle size and heat-treatment on microhardness and corrosion resistance of NiP electrodeposited coatings, J Alloys Compd 769 (2018) 1080–1087. https://doi.org/10.1016/J.JALLCOM.2018.08.013

[12]   S.H.M. Anijdan, M. Sabzi, M.R. Zadeh, M. Farzam, The influence of pH, rotating speed and Cu content reinforcement nano-particles on wear/corrosion response of Ni-P-Cu nano-composite coatings, Tribol Int 127 (2018) 108–121. https://doi.org/10.1016/J.TRIBOINT.2018.05.040

[13]   J.N. Balaraju, Kalavati, K.S. Rajam, Influence of particle size on the microstructure, hardness and corrosion resistance of electroless Ni–P–Al2O3 composite coatings, Surf Coat Technol 200 (2006) 3933–3941. https://doi.org/10.1016/J.SURFCOAT.2005.03.007

[14]   G. Pedrizzetti, L. Paglia, V. Genova, S. Cinotti, M. Bellacci, F. Marra, G. Pulci, Microstructural, mechanical and corrosion characterization of electroless Ni-P composite coatings modified with ZrO2 reinforcing nanoparticles, Surf Coat Technol 473 (2023) 129981. https://doi.org/10.1016/J.SURFCOAT.2023.129981

[15]   F. Bigdeli, S.R. Allahkaram, An investigation on corrosion resistance of as-applied and heat treated Ni–P/nanoSiC coatings, Mater Des 30 (2009) 4450–4453. https://doi.org/10.1016/J.MATDES.2009.04.020

[16]   H. Luo, M. Leitch, Y. Behnamian, Y. Ma, H. Zeng, J.L. Luo, Development of electroless Ni–P/nano-WC composite coatings and investigation on its properties, Surf Coat Technol 277 (2015) 99–106. https://doi.org/10.1016/J.SURFCOAT.2015.07.011

[17]   J.N. Balaraju, Kalavati, K.S. Rajam, Influence of particle size on the microstructure, hardness and corrosion resistance of electroless Ni–P–Al2O3 composite coatings, Surf Coat Technol 200 (2006) 3933–3941. https://doi.org/10.1016/J.SURFCOAT.2005.03.007

[18]   D.R. Dhakal, Y.K. Kshetri, B. Chaudhary, T.H. Kim, S.W. Lee, B.S. Kim, Y. Song, H.S. Kim, H.H. Kim, Particle-size-dependent anticorrosion performance of the Si3 N4-nanoparticle-incorporated electroless Ni-P coating, Coatings 12 (2022) 9. https://doi.org/10.3390/COATINGS12010009/S1

[19]   S.S. Tulsi, Composite PTFE-Nickel coatings for low friction applications, Mater Des 4 (1983) 919–923. https://doi.org/10.1016/0261-3069(84)90004-9

[20]   S.S. Tulsi, Electroless Nickel-PTFE Composite Coatings, Transactions of the IMF 61 (1983) 142–149. https://doi.org/10.1080/00202967.1983.11870654

[21] Y. Liu, Q.Z.-P. and surface finishing, undefined 2004, Study of PTFE content & anti-corrosion properties of electroless Ni-P-PTFE coatings, Sterc.Org (2004). https://sterc.org/pdf/psf2004/040448.pdf (accessed February 10, 2024).

[22] A. Sharma, A.K. Singh, Electroless Ni-P-PTFE-Al2O3 dispersion nanocomposite coating for corrosion and wear resistance, J Mater Eng Perform 23 (2014) 142–151. https://doi.org/10.1007/S11665-013-0710-0/FIGURES/9

[23] Z. Chen, L. Zhu, L. Ren, J. Liu, Electroless Plating of Ni-P and Ni-P-PTFE on Micro-Arc Oxidation Coatings for Improved Tribological Performance, Materials Research 25 (2022) e20220096. https://doi.org/10.1590/1980-5373-MR-2022-0096

[24] Y. Li, L. Zheng, B. Sun, C. Zhang, H. Zhao, Z. Qu, X. Xu, Preparation and characterization of Ni-P-Al2O3-PTFE nanocomposite coatings by unidirectional jet electrodeposition, Mater Today Commun 35 (2023) 105647. https://doi.org/10.1016/J.MTCOMM.2023.105647

[25] H. Zhang, J. Zou, N. Lin, B. Tang, REVIEW ON ELECTROLESS PLATING Ni–P COATINGS FOR IMPROVING SURFACE PERFORMANCE OF STEEL, Https://Doi.Org/10.1142/S0218625X14300020 21 (2014). https://doi.org/10.1142/S0218625X14300020

[26] S. Cao, M. Zou, B. Zhao, H. Gao, G. Wang, Investigation of corrosion and fouling resistance of Ni–P-nanoparticles composite coating using online monitoring technology, International Journal of Thermal Sciences 184 (2023) 107953. https://doi.org/10.1016/J.IJTHERMALSCI.2022.107953

[27] K.W. Liew, H.J. Kong, K.O. Low, C.K. Kok, D. Lee, The effect of heat treatment duration on mechanical and tribological characteristics of Ni–P–PTFE coating on low carbon high tensile steel, Materials & Design (1980-2015) 62 (2014) 430–442. https://doi.org/10.1016/J.MATDES.2014.05.047

[28] R. Asmatulu, Nanocoatings for corrosion protection of aerospace alloys, Corrosion Protection and Control Using Nanomaterials (2012) 357–374. https://doi.org/10.1533/9780857095800.2.357

[29] R. Kannan, M. Selvambikai, S. Jyothi, E. Selvakumar, S. Venkateswaran, E. Shobhana, P. Devaki, An investigations on structural, mechanical and magnetic properties of electroplated NiP nano crystalline thin films for aerospace and automotive applications, Mater Res Express 6 (2019) 116435. https://doi.org/10.1088/2053-1591/AB4AF7

[30] C.K. Lee, Structure, electrochemical and wear-corrosion properties of electroless nickel–phosphorus deposition on CFRP composites, Mater Chem Phys 114 (2009) 125–133. https://doi.org/10.1016/J.MATCHEMPHYS.2008.08.088

[31] A. Sharma, A.K. Singh, Corrosion and wear study of Ni-P-PTFE-Al2O3 coating: The effect of heat treatment, Central European Journal of Engineering 4 (2014) 80–89. https://doi.org/10.2478/S13531-013-0137-2/MACHINEREADABLECITATION/RIS

[32] J. Tian, X. Liu, J. Wang, X. Wang, Y. Yin, Electrochemical anticorrosion behaviors of the electroless deposited Ni–P and Ni–P–PTFE coatings in sterilized and unsterilized

seawater, Mater Chem Phys 124 (2010) 751–759.
https://doi.org/10.1016/J.MATCHEMPHYS.2010.07.053

[33]   Z. Li, C. Bian, L. Hu, Exploration of the Corrosion Behavior of Electroless Plated Ni-P
Amorphous Alloys via X-ray Photoelectron Spectroscopy, Molecules 2023, Vol. 28, Page 377
28 (2023) 377. https://doi.org/10.3390/MOLECULES28010377

[34]   M. Farhan, O. Fayyaz, M.G. Qamar, R.A. Shakoor, J. Bhadra, N.J. Al-Thani, Mechanical
and corrosion characteristics of TiC reinforced Ni-P based nanocomposite coatings, Mater
Today Commun 36 (2023) 106901. https://doi.org/10.1016/J.MTCOMM.2023.106901

[35]   Y. Wu, Z. Zhang, Z. Leng, J. Zhang, S. Yang, W. Shen, K. Xu, H. Zhu, Y. Liu,
Improvement of the corrosion resistance of amorphous Ni-P coatings modified by a laser–
electrodeposition hybrid process: Effect of morphology evolution on the electrochemical
corrosion behavior, Appl Surf Sci 624 (2023) 157016.
https://doi.org/10.1016/J.APSUSC.2023.157016

[36]   Scopus - Document search | Signed in, (n.d.).
https://www.scopus.com/search/form.uri?display=authorLookup#basic (accessed February
10, 2024).

[37]   Q. Zhao, Y. Liu, H. Müller-Steinhagen, G. Liu, Graded Ni–P–PTFE coatings and their
potential applications, Surf Coat Technol 155 (2002) 279–284. https://doi.org/10.1016/S0257-
8972(02)00116-0

[38]   J. Hou, S. Wang, Z. Zhou, The Effect of Ni-P Alloy Pre-Plating on the Performance of
Ni-P/Ni-P-PTFE Composite Coatings, Key Eng Mater 561 (2013) 537–541.
https://doi.org/10.4028/WWW.SCIENTIFIC.NET/KEM.561.537

[39]   A. Zarebidaki, S.R. Allahkaram, Porosity measurement of electroless Ni-P coatings
reinforced by CNT or SiC particles, Surface Engineering 28 (2012) 400–405.
https://doi.org/10.1179/1743294411Y.0000000087/ASSET/IMAGES/LARGE/10.1179_1743
294411Y.0000000087-FIG6.JPEG

[40]   H.Y.- Bin, W.C.- Sheng, L. De-Gang, L.Z.- Jie, S.X.- Jun, Z.Q.- Yong, L.Y.- Hui, Study
on Corrosion Resistance of Electroless Plating Ni-P Complex Coating, (2005).
https://dx.doi.org/ (accessed February 10, 2024).

[41]   P. Sahoo, Optimization of electroless Ni?P coatings
based on multiple roughness characteristics, Surface and Interface Analysis 40 (2008) 1552–
1561. https://doi.org/10.1002/SIA.2945

[42]   R. Elansezhian, B. Ramamoorthy, P. Kesavan Nair, Effect of surfactants on the
mechanical properties of electroless (Ni–P) coating, Surf Coat Technol 203 (2008) 709–712.
https://doi.org/10.1016/J.SURFCOAT.2008.08.021

[43]   C.S. Chang, K.H. Hou, M. Der Ger, C.K. Chung, J.F. Lin, Effects of annealing
temperature on microstructure, surface roughness, mechanical and tribological properties of
Ni–P and Ni–P/SiC films, Surf Coat Technol 288 (2016) 135–143.
https://doi.org/10.1016/J.SURFCOAT.2016.01.020

[44]   M. Nishira, K. Yamagishi, H. Matsuda, M. Suzuki, O. Takano, Uniform Dispersibility of PTFE Particles in Electroless Composite Plating, Transactions of the IMF 74 (1996) 62–64. https://doi.org/10.1080/00202967.1996.11871095

[45]   I.R. Mafi, C. Dehghanian, Comparison of the coating properties and corrosion rates in electroless Ni–P/PTFE composites prepared by different types of surfactants, Appl Surf Sci 257 (2011) 8653–8658. https://doi.org/10.1016/J.APSUSC.2011.05.043

[46]   Y. Liu, Q. Zhao, Effects of surfactants on the PTFE particle sizes in electroless plating Ni-P-PTFE coatings, Transactions of the IMF 81 (2003) 168–171. https://doi.org/10.1080/00202967.2003.11871529

[47]   Q. Zhao, Y. Liu, Investigation of graded Ni–Cu–P–PTFE composite coatings with antiscaling properties, Appl Surf Sci 229 (2004) 56–62. https://doi.org/10.1016/J.APSUSC.2004.01.044

[48]   G. Straffelini, D. Colombo, A. Molinari, Surface durability of electroless Ni–P composite deposits, Wear 236 (1999) 179–188. https://doi.org/10.1016/S0043-1648(99)00273-2

[49]   A. Sharma, A.K. Singh, Corrosion and wear resistance study of Ni-P and Ni-P-PTFE nanocomposite coatings, Central European Journal of Engineering 1 (2011) 234–243. https://doi.org/10.2478/S13531-011-0023-8/MACHINEREADABLECITATION/RIS

[50]   H. Omidvar, M. Sajjadnejad, G. Stremsdoerfer, Y. Meas, A. Mozafari, Manufacturing Ternary Alloy NiBP-PTFE Composite Coatings by Dynamic Chemical Plating Process, Materials and Manufacturing Processes 31 (2016) 31–36. https://doi.org/10.1080/10426914.2014.994753

[51]   A. Kumar, A. Singh, M. Kumar, D. Kumar, S. Barthwal, Study on thermal stability of electroless deposited Ni-Co-P alloy thin film, Journal of Materials Science: Materials in Electronics 22 (2011) 1495–1500. https://doi.org/10.1007/S10854-011-0336-7/FIGURES/6

[52]   G. Zhao, R. Wang, S. Liu, D. Wu, Y. Zhang, T. Wang, Y. Zou, Study on the role of element Mo in improving thermal stability and corrosion resistance of amorphous Ni-P deposit, J Non Cryst Solids 549 (2020) 120358. https://doi.org/10.1016/J.JNONCRYSOL.2020.120358

[53]   P. Sahoo, S.K. Das, Tribology of electroless nickel coatings – A review, Mater Des 32 (2011) 1760–1775. https://doi.org/10.1016/J.MATDES.2010.11.013

[54]   G.O. Boakye, A.M. Ormsdóttir, B.G. Gunnarsson, S. Irukuvarghula, R. Khan, S.N. Karlsdóttir, The Effect of Polytetrafluoroethylene (PTFE) Particles on Microstructural and Tribological Properties of Electroless Ni-P+PTFE Duplex Coatings Developed for Geothermal Applications, Coatings 2021, Vol. 11, Page 670 11 (2021) 670. https://doi.org/10.3390/COATINGS11060670

[55]   P. Peelers, G. V.D. Hoorn, T. Daenen, A. Kurowski, G. Staikov, Properties of electroless and electroplated Ni–P and its application in microgalvanics, Electrochim Acta 47 (2001) 161–169. https://doi.org/10.1016/S0013-4686(01)00546-1

[56]   P.G. Engleman, N.B. Dahotre, C.A. Blue, D.C. Harper, R. Ott, HIGH DENSITY INFRARED PROCESSING OF WC/Ni–11P COMPOSITE COATINGS,

Https://Doi.Org/10.1179/026708401225002811 18 (2002) 113–119.
https://doi.org/10.1179/026708401225002811

[57]   S. Papavinasam, R.W. Revie, Review of standards for evaluating coatings to control external corrosion of pipelines, Corrosion Reviews 26 (2008) 295–371. https://doi.org/10.1515/CORRREV.2008.295/MACHINEREADABLECITATION/RIS

[58]   M. Der Ger, B.J. Hwang, Effect of surfactants on codeposition of PTFE particles with electroless Ni-P coating, Mater Chem Phys 76 (2002) 38–45. https://doi.org/10.1016/S0254-0584(01)00513-2

[59]   M.-D. Ger, B.J. Hwang, Role of Surfactants in Codeposition of PTFE Particles with Electroless Ni-P Coating, Journal of the Chinese Institute of Chemical Engineers 32 (2001) 503–509. https://doi.org/10.6967/JCICE.200111.0503

[60]   Q. Zhao, Y. Liu, Electroless Ni-Cu-P-PTFE composite coatings and their anticorrosion properties, Surf Coat Technol 200 (2005) 2510–2514. https://doi.org/10.1016/j.surfcoat.2004.06.011

[61]   M. Tajbakhsh, O. Yaghobizadeh, M. Farhadi Nia, Investigation of the physical and mechanical properties of Ni–P and Ni–P–PTFE nanocomposite coatings deposited on aluminum alloy 7023, Proceedings of the Institution of Mechanical Engineers, Part E: Journal of Process Mechanical Engineering 233 (2019) 94–103. https://doi.org/10.1177/0954408917744159

[62]   Q. Zhao, Y. Liu, E.W. Abel, Effect of Cu content in electroless Ni–Cu–P–PTFE composite coatings on their anti-corrosion properties, Mater Chem Phys 87 (2004) 332–335. https://doi.org/10.1016/J.MATCHEMPHYS.2004.05.028

[63]   S. Armyanov, J. Georgieva, D. Tachev, E. Valova, N. Nyagolova, S. Mehta, D. Leibman, A. Ruffini, Electroless deposition of Ni-Cu-P alloys in acidic solutions, Electrochemical and Solid-State Letters 2 (1999) 323–325. https://doi.org/10.1149/1.1390824/XML

[64]   M. Lee, J. Park, K. Son, D. Kim, K. Kim, M. Kang, Electroless Ni-P-PTFE Composite Plating with Rapid Deposition and High PTFE Concentration through a Two-Step Process, Coatings 2022, Vol. 12, Page 1199 12 (2022) 1199. https://doi.org/10.3390/COATINGS12081199

[65]   H.H. Sheu, S.Y. Jian, M.H. Lin, C.I. Hsu, K.H. Hou, M. Der Ger, Electroless Ni-P/PTFE Self-Lubricating Composite Thin Films Applied for Medium-carbon Steel Substrate, Int J Electrochem Sci 12 (2017) 5464–5482. https://doi.org/10.20964/2017.06.30

[66]   P. ping Gao, M. lian Gao, A. ru Wu, X. bo Wu, C. xun Liu, Y. Zhang, H. kun Zhou, X. min Peng, Z. yong Xie, Electrochemical characteristics of electroplating and impregnation Ni-P/SiC/PTFE composite coating on 316L stainless steel, J Cent South Univ 27 (2020) 3615–3624. https://doi.org/10.1007/S11771-020-4508-6/METRICS

[67]   G.O. Boakye, A.M. Ormsdóttir, B.G. Gunnarsson, S. Irukuvarghula, R. Khan, S.N. Karlsdóttir, The Effect of Polytetrafluoroethylene (PTFE) Particles on Microstructural and Tribological Properties of Electroless Ni-P+PTFE Duplex Coatings Developed for Geothermal Applications, Coatings 2021, Vol. 11, Page 670 11 (2021) 670. https://doi.org/10.3390/COATINGS11060670

[68]   C. Huang, Z. Zhang, J. yan, L. Sun, J. Wang, Enhancing wear and corrosion resistance of electroless Ni-P coatings in CO2-saturated NaCl solution through polytetrafluoroethylene incorporation, Corros Sci 226 (2024) 111620. https://doi.org/10.1016/J.CORSCI.2023.111620

[69]   X. Liang, P. Wu, L. Lan, Y. Wang, Y. Ning, Y. Wang, Y. Qin, Effect of Polytetrafluoroethylene (PTFE) Content on the Properties of Ni-Cu-P-PTFE Composite Coatings, Materials 2023, Vol. 16, Page 1966 16 (2023) 1966. https://doi.org/10.3390/MA16051966

[70]   S.; Mei, C.; Zhou, Z.; Hu, Z.; Xiao, Q.; Zheng, X. Chai, S. Mei, C. Zhou, Z. Hu, Z. Xiao, Q. Zheng, X. Chai, Preparation of a Ni-P-nanoPTFE Composite Coating on the Surface of GCr15 Steel for Spinning Rings via a Defoamer and Transition Layer and Its Wear and Corrosion Resistance, Materials 2023, Vol. 16, Page 4427 16 (2023) 4427. https://doi.org/10.3390/MA16124427

[71]   Z. Ma, B. Jiang, D. Drummer, L. Zhang, Influence of phosphorous acid concentration on the self-lubricating properties of electroformed Ni-P-PTFE ternary composites, Surf Coat Technol 477 (2024) 130375. https://doi.org/10.1016/J.SURFCOAT.2024.130375

[72]   Y. Li, L. Zheng, X. Xu, Y. Zhang, M. Zhang, M. Liu, Effect of current density on properties of Ni–P-Al2O3-PTFE nanocomposite coatings by jet electrodeposition, International Journal of Advanced Manufacturing Technology 125 (2023) 5743–5755. https://doi.org/10.1007/S00170-023-11088-8/FIGURES/13

[73]   G. Oppong Boakye, E.O. Straume, D. Kovalov, S.N. Karlsdottir, Wear-reducing nickel-phosphorus and graphene oxide-based composite coatings: Microstructure and corrosion behavior in high temperature geothermal environment, Corros Sci 209 (2022) 110809. https://doi.org/10.1016/J.CORSCI.2022.110809

[74]   B. Vasconcelos, R. Serra, J. Oliveira, C. Fonseca, Characterization and Tribological Behavior of Electroless-Deposited Ni-P-PTFE Films on NBR Substrates for Dynamic Contact Applications, Coatings 2022, Vol. 12, Page 1410 12 (2022) 1410. https://doi.org/10.3390/COATINGS12101410

[75]   Y. Li, L. Zheng, M. Liu, Z. Qu, X. Xu, Y. Zhang, M. Zhang, H. Han, Z. Yang, Effect of duty ratio on the performance of pulsed electrodeposition Ni–P–Al2O3–PTFE nanocomposite coatings, Appl Phys A Mater Sci Process 128 (2022) 1–11. https://doi.org/10.1007/S00339-022-05787-4/FIGURES/12

[76]   P. Kramer, K.S. Williams, K.A. Schultz, F. Friedersdorf, D.A. Jackson, T. Sweitzer, Effect of Mechanical Stress and Environmental Conditions on Degradation of Aerospace Coatings That Guard Against Atmospheric Corrosion, (2018). https://dx.doi.org/ (accessed February 10, 2024).

[77]   J.T. Staley, Corrosion of Aluminium Aerospace Alloys, Materials Science Forum 877 (2017) 485–491. https://doi.org/10.4028/WWW.SCIENTIFIC.NET/MSF.877.485

[78]   A. Kvryan, N.A. Carter, H.K. Trivedi, M.F. Hurley, Accelerated Testing to Investigate Corrosion Mechanisms of Carburized and Carbonitrided Martensitic Stainless Steel for Aerospace Bearings in Harsh Environments, Tribology Transactions 63 (2020) 265–279. https://doi.org/10.1080/10402004.2019.1685726

[79]   R.R. Boyer, Titanium for aerospace: Rationale and applications, Advanced Performance Materials 2 (1995) 349–368. https://doi.org/10.1007/BF00705316/METRICS

[80]   M.A. Maleque, S.Y. Cetin, M. Hassan, M. Hafiz Sulaiman, A.H. Rosli, A systematic review on corrosive-wear of automotive components materials, Jurnal Tribologi 35 (2022) 33–49.

[81]   C. Author, ORIGINAL ARTICLES Magnesium and Aluminum Alloys in Automotive Industry, J Appl Sci Res 8 (2012) 4865–4875. http://www.worldaluminium.org, (accessed February 10, 2024).

[82]   D.M. Aylor, R.J. Ferrara, R.A. Hays, R.M. Kain, Crevice Corrosion Performance of Candidate Naval Ship Seawater Valve Materials in Quiescent and Flowing Natural Seawater, (1999). https://dx.doi.org/ (accessed February 10, 2024).

[83]   M.S. Thomas, A. Okeremi, External Pitting And Crevice Corrosion Of 316L Stainless Steel Instrument Tubing In Marine Environments And Proposed Solution, (2008). https://dx.doi.org/ (accessed February 10, 2024).

[84]   W. Ding, A. Bonk, T. Bauer, Corrosion behavior of metallic alloys in molten chloride salts for thermal energy storage in concentrated solar power plants: A review, Front Chem Sci Eng 12 (2018) 564–576. https://doi.org/10.1007/S11705-018-1720-0/METRICS

[85]   W.S. Tait, Controlling Corrosion of Chemical Processing Equipment, Handbook of Environmental Degradation Of Materials: Third Edition (2018) 583–600. https://doi.org/10.1016/B978-0-323-52472-8.00028-9

[86]   T.J. Glover, Application of stainless steels in chemical plant corrosive environments, Anti-Corrosion Methods and Materials 29 (1982) 11–12. https://doi.org/10.1108/EB007190/FULL/XML

[87]   J. Bhandari, F. Khan, R. Abbassi, V. Garaniya, R. Ojeda, Modelling of pitting corrosion in marine and offshore steel structures – A technical review, J Loss Prev Process Ind 37 (2015) 39–62. https://doi.org/10.1016/J.JLP.2015.06.008

[88]   G.E. Moller, The Successful Use of Austenitic Stainless Steels in Sea Water, Proceedings of the Annual Offshore Technology Conference 1976-May (1976) 959–976. https://doi.org/10.4043/2699-MS

[89]   S.X. Li, R. Akid, Corrosion fatigue life prediction of a steel shaft material in seawater, Eng Fail Anal 34 (2013) 324–334. https://doi.org/10.1016/J.ENGFAILANAL.2013.08.004.

[90]   Phull B, Abdullahi AA, Chapter 09209 - Marine Corrosion, Reference Module in Materials Science and Materials Engineering (2016) 1–39. https://doi.org/10.1016/B978-0-12-803581-8.09209-2

[91]   A.A. Daniyan, T.L. Akpomejero, O.O. Ige, P.A. Olubambi, Corrosion Prevention of Biomedical Implants: Surface Coating Techniques Perspective, ChemistrySelect 8 (2023) e202300223. https://doi.org/10.1002/SLCT.202300223

[92]   M.T. Mohammed, Z.A. Khan, A.N. Siddiquee, Surface Modifications of Titanium Materials for developing Corrosion Behavior in Human Body Environment: A Review,

Procedia Materials Science 6 (2014) 1610–1618.
https://doi.org/10.1016/J.MSPRO.2014.07.144

[93]   N.S. Radhi, Z. Al-Khafaji, INVESTIGATION BIOMEDICAL CORROSION OF
IMPLANT ALLOYS IN PHYSIOLOGICAL ENVIRONMENT, SCOPUS Indexed Journal
Www.Tjprc.Org (n.d.). www.tjprc.org (accessed February 10, 2024).

[94]   B.W. Waters, J.M. Tatum, Y.C. Hung, Effect of chlorine-based sanitizers properties on
corrosion of metals commonly found in food processing environment, J Food Eng 121 (2014)
159–165. https://doi.org/10.1016/J.JFOODENG.2013.08.027

[95]   A. Montanari, Basic Principles of Corrosion of Food Metal Packaging, in(2015) 105–132.
https://doi.org/10.1007/978-3-319-14827-4_6

[96]   G.K. Deshwal, N.R. Panjagari, Review on metal packaging: materials, forms, food
applications, safety and recyclability, J Food Sci Technol 57 (2020) 2377–2392.
https://doi.org/10.1007/S13197-019-04172-Z/FIGURES/5

Advances in Corrosion Science and Surface Engineering   Materials Research Forum LLC
Materials Research Foundations 188 (2026) 193-207   https://doi.org/10.21741/9781644903919-10

Chapter 10

Non-Destructive Testing (NDT) Techniques

Senthilkumar C.[1,3*], Utchimahali Muthuraja P.[2], Senbagaraj R.[3], G. Kausalya Sasikumar[2,] Ramyakrishna Pothu[4,5]

[1]Department of Mechanical Engineering, Shreenivasa Engineering College, Bommidi, Dharmapuri - 635301, India

[2]Centre for Research and Development, KPR Institute of Engineering and Technology, Coimbatore-641402, India

[3]Department of Mechanical Engineering, RVS Technical Campus, Coimbatore-641402, India

[4]Center for Innovation and Inclusive Research, Sharda University, Greater Noida – 201310, India

[5]School of Physics and Electronics, College of Chemistry and Chemical Engineering, Hunan University, Changsha 410082, P.R. China

sndlmech@gmail.com

Abstract

Detection of small-size defects is important to make sure that important components, having high-value use, do not lose their integrity. Detection often allows the parts to be recovered through repair, thus supporting the principles of the circular economy through the extension and reuse of resources. In order to extend the life of new and current systems while ensuring component safety and minimizing downtime to prevent financial losses, corrosion detection is one of the top concerns for the chemical, defense, and transportation sectors. This study aims to evaluate and explain the most widely used non-destructive methods for corrosion problem monitoring, early diagnosis, and repair in the industry. The advantages, disadvantages, and operational processes of certain important non-destructive approaches are thoroughly reviewed.

Keywords

Destructive Testing, Non-Destructive Testing, Corrosion Detection, Eddy Current, Radiography

Contents

1.   Introduction

A material that deteriorates through a process referred to as corrosion as a consequence of interactions with its environment. Alternatively, corrosion is a chemical reaction by which metals transition into their most stable oxidized state, often in the presence of moisture and oxygen. The structural integrity and endurance of substances and structures are primarily subject to threat from corrosion in numerous industries, including construction, oil and gas, aerospace, automotive, and marine. While guaranteeing safety, early corrosion detection can lower maintenance expenses and failures. In-service defects usually have their origins related to creep, thermal cycling, and fatigue, or a combination of the effects of environmental exposure. Industrial machinery's declining lifespan is a serious problem for businesses, resulting in high and sometimes costly expenses. To guarantee optimal machinery operation, the issue necessitates ongoing monitoring, preventative maintenance, and prompt replacement of damaged parts [1].

These expenses go beyond the anticipated direct costs of replacing personnel and parts for maintenance; they also include indirect costs related to unplanned production halts. This not only reduces operational efficiency but also results in additional losses because of manufacturing delays. These hazards also force businesses to keep large inventories of replacement components, which raise costs. Corrosion monitoring is an essential requirement in the industrial environment because it ensures that corrosion prevents possible occurrences related to corrosion and hence is critical input in the practices of full monitoring of corrosion in industries. Proper corrosion monitoring, through early identification of corrosive processes, prolongs infrastructures and prevents major break downs [2]. In this regard, monitoring corrosion is an enabling tool that prevents such an event while ensuring the safeguarding of persons and the viability of vital resources. To identify potential mechanisms of failure, destructive testing pushes components to their limits. Destructive testing, on the other hand, renders the tested goods unusable for regular operations due to the lasting damage caused. In contrast, non-destructive testing examines the components without causing them any irreversible harm. Therefore, without inflicting any harm on the materials, it is the ideal method to assess their endurance. A wide range of techniques are discussed, including radiography techniques, acoustic emissions, eddy current, guided wave

Advances in Corrosion Science and Surface Engineering   Materials Research Forum LLC
Materials Research Foundations 188 (2026) 193-207   https://doi.org/10.21741/9781644903919-10

testing, infrared thermography, and visual and optical testing [3]. The article also includes recent research that explains the equipment needed for these procedures as well as their operating principles.

2.   Principle and types of corrosion

A corrosive process is when materials, especially metals, deteriorate due to reactions that are chemical or electrochemical in their environment. It mostly happens when a material, often a metal, combines with oxygen, water, acids, or other chemicals to generate solid substances that consist of oxides, hydroxides.

2.1   Corrosion reaction

Concurrent anode oxidation and cathode reduction are examples of corrosion. A number of environmental conditions, including temperature, oxygen content, pH levels, moisture content, and the presence of contaminants or salts, affect the rate and degree of corrosion.

Oxidation at Anode:

$$M \rightarrow M^{n+} + ne^- \tag{1}$$

Reduction at Cathode:

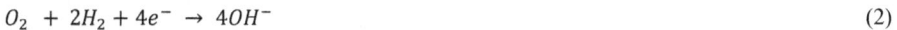

$$O_2 + 2H_2 + 4e^- \rightarrow 4OH^- \tag{2}$$

2.2   Types of corrosion

Certain types of corrosion are distinctive and have special mechanics. Uniform corrosion is the most prevalent kind, in which the surface of the material deteriorates uniformly. Pitting corrosion, on the other hand, happens in small places and can result in pits, which are tiny holes that inflict severe damage even though they only take up a little amount of space. Attacks on narrow openings or gaps, known as crevice corrosion, can restrict access to the environment and cause variations in oxygen content, which speeds up the corrosion process. When two distinct metals come into electrical contact in a corrosive environment, galvanic corrosion occurs. The metal that corrodes more quickly is the more reactive one. Intergranular corrosion, which is mostly caused by contaminants or incorrect heat treatment, attacks a metal's grain boundaries. The combination of tensile stress and a corrosive environment causes stresses to corrode and fracture.

Erosion is when a corrosive fluid and the material surface move relative to one another; corrosion occurs, and material loss rises. When surfaces repeatedly come into touch with one another, the protective layers are removed, increasing the corrosiveness and causing fretting corrosion. Formation of weaker structure resulting from the removal of one element from alloy metal is known as selective leaching. There is also Microbiologically Influenced Corrosion (MIC), which is brought on by bacteria that produce corrosive substances as a result of their activity, speeding up the pace of material degradation. For effective prevention and mitigation, it is necessary to

Advances in Corrosion Science and Surface Engineering   Materials Research Forum LLC
Materials Research Foundations 188 (2026) 193-207   https://doi.org/10.21741/9781644903919-10

comprehend these forms of corrosion. Figure 1 illustrates the different types of corrosion on materials and some examples of corrosions are represented in Figure 2.

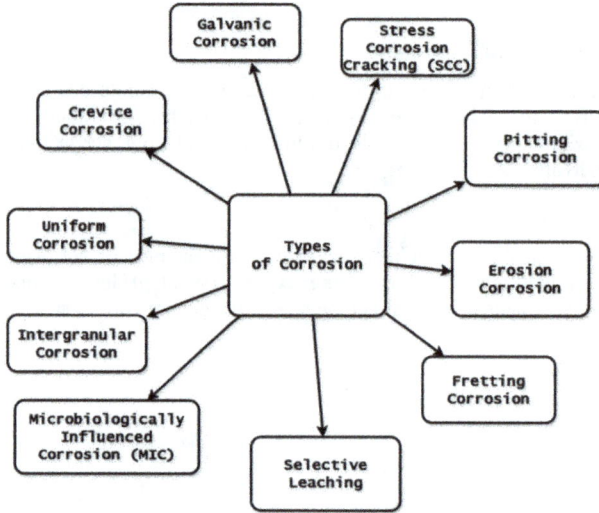

Figure. 1 Illustration of Corrosion Types

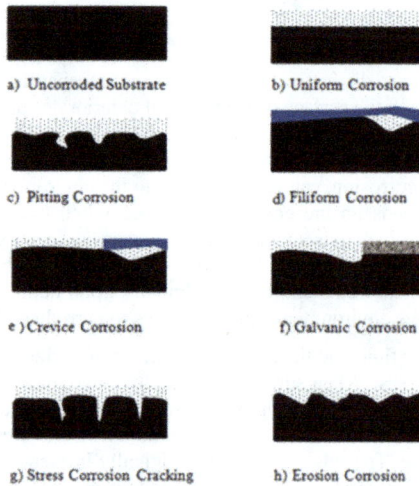

a) Uncorroded Substrate

b) Uniform Corrosion

c) Pitting Corrosion

d) Filiform Corrosion

e) Crevice Corrosion

f) Galvanic Corrosion

g) Stress Corrosion Cracking

h) Erosion Corrosion

Figure. 2 Some Examples of Corrosions [4]

Advances in Corrosion Science and Surface Engineering   Materials Research Forum LLC
Materials Research Foundations 188 (2026) 193-207   https://doi.org/10.21741/9781644903919-10

3.   Detection of corrosion

In terms of identifying, quantifying, and forecasting corrosion damage, corrosion monitoring includes techniques to ascertain the physical and chemical changes of materials. To put it briefly, it discusses methods for gathering data about how quickly materials decay. Despite being used interchangeably, corrosion monitoring and detection are quite different concepts. Both include keeping an eye out for any physical or chemical changes in the material, but monitoring is primarily a predictive role, whereas detection is a protective precaution to guarantee the integrity of the equipment.

3.1   Methods of corrosion detection

As previously stated, there are two types of corrosion monitoring techniques: destructive and non-destructive approaches. Destructive techniques (DT) include linear polarization resistance, electrochemical frequency modulation, electrical resistance, inductive resistance, and electrochemical impedance spectroscopy. Some of the non-destructive methods include: visual inspection, radiography, electromagnetic, guided wave (GW), ultrasonic, and acoustic procedures [5]. Destruction and Non-Destructive Techniques are used to ensure that corrosion is found safely and efficiently and prevent the high cost associated with tear-down inspections.

3.2   Non-Destructive Techniques (NDT)

NDT stands for the inspection of test samples without causing damage. Such operations are carried out periodically to observe rates of deterioration in materials, in-service inspection, and quality control with no performance loss. Data acquired from NDT operations are processed into parameters that can be used to predict component performance and lifespan. Other than conserving labor and energy, NDT operations promote safety and reliability.

Several techniques are employed in non-destructive testing for determining the suitability of materials, parts, or systems without causing any damage. Some of the very basic methods listed as Non-Destructive Eval. Corros. Corros. Crack.," 2019, used are Visual Inspection (VI) and search for apparent surface problems such as corrosion or cracks, and ultrasonic testing, where the examination is made using high-frequency sound waves on voids or cracks and detecting interior defects. Radiographic testing (RT) employs x-rays or gamma rays to create images of interior structures that exhibit variations in density. Magnetic Particle Testing (MT) applies a magnetic field to ferromagnetic materials and utilizes the accumulation of magnetic particles to identify defects. Liquid Penetrant Testing (PT) focuses on surface-breaking defects by using a dye that can penetrate cracks and be clearly visible with developer. Eddy Current Testing (ECT) relies on electromagnetic induction in finding surface or near-surface flaws in materials that are conducting. Thermographic inspection uses infrared cameras to find anomalies in temperature and thus signal possible flaws. The Acoustic Emission Testing, AET is used to locate structural problems; it monitors energy discharge from strained materials. Stereo- and holography are some of the laser testing techniques where lasers are used to detect deformations caused by defects. Lastly, leak testing involves various methods such as mass spectrometry and bubble tests to detect leaks in pressurized parts. Each NDT method is tailored for specific materials and fault types to ensure safety and integrity of buildings without compromising their usability [7].

Advances in Corrosion Science and Surface Engineering   Materials Research Forum LLC
Materials Research Foundations 188 (2026) 193-207   https://doi.org/10.21741/9781644903919-10

3.2.1 Visual inspection technique (VI)

Visual inspection is an NDT method of establishing the state of objects, materials, or buildings that involves the human eye to see the state. It includes a systematic procedure that ensures all areas of the possible examination are covered. In the visual inspection, inspectors apply a methodical approach that includes prior planning and preparation. A lot of dependence on the unaided eye occurs during the visual inspection to observe the state of the surfaces being investigated [8]. This procedure meets the structured criteria as set by industry standards, such as ASNT SNT-TC-1A and ISO 9712, with adequate planning and preparation to meet the needs of a thorough inspection. The use of a variety of inspection tools, such as flashlights, mirrors, magnifying glasses, and borescopes for limited areas, may enhance the identification and evaluation of surface defects in visual inspections.

Visual inspection involves several essential procedures in conducting a detailed assessment. First is planning and preparation as pre-inspection where the inspector reviews the relevant documents, understands the requirements for inspection, and prepares the tools or equipment needed in conducting the examination. After the surface inspection, it is cleaned at the surface, a process applied to removing coatings and dust from the surface. The visual inspection then follows, whereby the inspector will give a close eye to the surface with the intention of looking for flaws such as corrosion or fractures using instruments to magnify. The next phase is evaluation and interpretation, in which the grading of the condition of the item is done based on the comparison made between the observations with the predetermined criteria. The last step would be documentation and reporting in which all the results, conclusions, and recommendations would be documented in a comprehensive report to be used in the future or at a later stage [9]. Figure 3 explains the Visual inspection technique, step-by-step procedure.

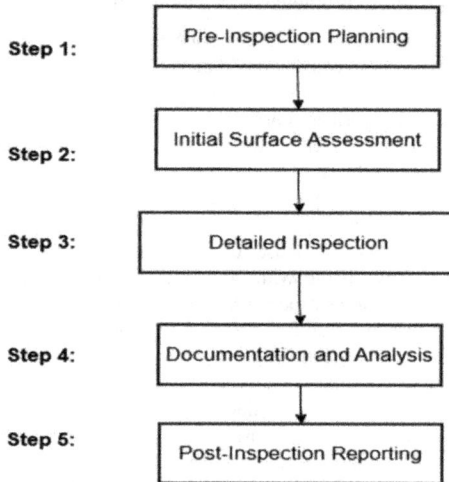

Figure 3. Step by step process of Visual inspection technique

Advances in Corrosion Science and Surface Engineering   Materials Research Forum LLC
Materials Research Foundations 188 (2026) 193-207   https://doi.org/10.21741/9781644903919-10

It is simple, economical, fast to determine surface defects without the use of complicated equipment, non-destructive, preserving the integrity of the object under inspection, applicable to most materials and structures, and very quick in arriving at decisions at the point of inspection [10]. On the other hand, there are several disadvantages associated with the technique. Its effectiveness depends upon the skill, expertise, and eyesight of the inspector and this sometimes brings subjective results. Visual inspection can't find out subsurface or internal defects and can scan only for superficial defects. Furthermore, environmental conditions such as inadequate lighting or insufficient access will render it difficult to scan large and complex objects accurately and scanning may also take time [11].

3.2.2 Acoustic Emissions (AE) technique

AE, or Acoustic Emissions testing, is the procedure of detecting and investigating the liberation of elastic waves in materials or structure. In other words, these waves occur due to an abrupt energy emission from localized sources; ruptures and leaks, which are forms of stress-related events, are considered examples. As such, it is a preferred method for observing the integrity of structures placed under stress-for example, bridge structures, pressure vessels, and pipeline systems. In AE testing, sensors with high sensitivity are mounted on the surface of the test material to detect acoustic emissions generated by damage and alterations in the dynamics of the material. The usually high-frequency vibrations characterizing this phenomenon can be detected by piezoelectric sensors that scan the material. The source and type of emission are then identified by processing these signals using specialist equipment. The procedure begins with meticulous planning, initially choosing the suitable establishing baseline conditions by positioning the sensors and apparatus. The tested object's surface will have sensors installed. After that, the gadget is activated. Although the system analyzes the AE signals that originate, the real structure is either subjected to its regular circumstances or additional stress is imposed. Real-time data analysis identifies anomalous activity or emission sources, and depending on the results, more research or repairs are started. Ultimately, the findings are recorded and analyzed to aid in decision-making. The step by step procedure is illustrated in Figure 4. Acoustic emissions have the benefit of being able to identify damage in real time without affecting the structure's regular operation, which makes them perfect for in-service monitoring. AE has the benefit of identifying damage at the crack propagation stage because of its high sensitivity to both active and passive flaws [12]. This is a reasonably priced way to continuously monitor vital infrastructures, especially when combined with other methods. AE can overlook flaws more intense in the material or structure since it can only detect surface or near-surface events, among other limitations. The positioning of the sensors is also crucial, and in certain situations, a significant number of sensors may be required to provide adequate coverage. Additionally, interpreting data can be complicated, necessitating the use of specialist tools and knowledgeable staff. Moreover, in cases where the ambient noise is too high or when the emission signals have not been well segregated, AE can sometimes result in false positives or misinterpretations [13].

Advances in Corrosion Science and Surface Engineering   Materials Research Forum LLC
Materials Research Foundations 188 (2026) 193-207   https://doi.org/10.21741/9781644903919-10

Process Initiation	Data Acquisition	Data Processing	Data Analysis
Step 1	Step 2	Step 3	Step 4

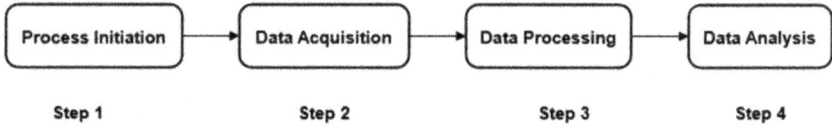

Figure 4. Process of Acoustic Emissions technique

3.2.3 Ultrasonic Inspection (UI) technique

Ultrasonic inspection is a method of non-destructive testing, wherein high-frequency sound waves are applied to identify defects in materials and determine their nature. This is widely used for evaluating the structural fit of a component in various industries such as construction, automobile, and aeronautical. It injects ultrasonic waves into the material via a transducer [14]. When the injected ultrasonic waves hit the discontinuities, such as cracks or voids, they travel back to the transducer for analysis and determine the size and location of the defect. The transducer and coupling medium are chosen first based on their availability in order to transmit the ultrasonic wave properly. A transducer is attached to the surface of the material that sends out ultrasonic pulses. Data processing on the reflected waves is done in order to locate an abnormality. Following this, material is interpreted and finally a report is generated [15]. Detailed procedure is explained in Figure 5. In relation to the benefits, ultrasonic inspection reveals both surface as well as subsurface defects. It shows inaccuracies and also can be applied to almost all types of materials. On the other hand, it is not free of disadvantages that interfere with accuracy: it requires training operators, costlier equipment, and cannot inspect detailed geometries and rough surfaces [16].

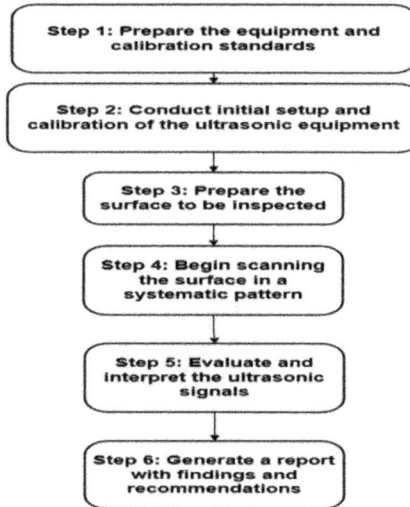

Step 1: Prepare the equipment and calibration standards

Step 2: Conduct initial setup and calibration of the ultrasonic equipment

Step 3: Prepare the surface to be inspected

Step 4: Begin scanning the surface in a systematic pattern

Step 5: Evaluate and interpret the ultrasonic signals

Step 6: Generate a report with findings and recommendations

Figure 5. Ultrasonic Inspection method step-by-step procedure

Advances in Corrosion Science and Surface Engineering   Materials Research Forum LLC
Materials Research Foundations 188 (2026) 193-207   https://doi.org/10.21741/9781644903919-10

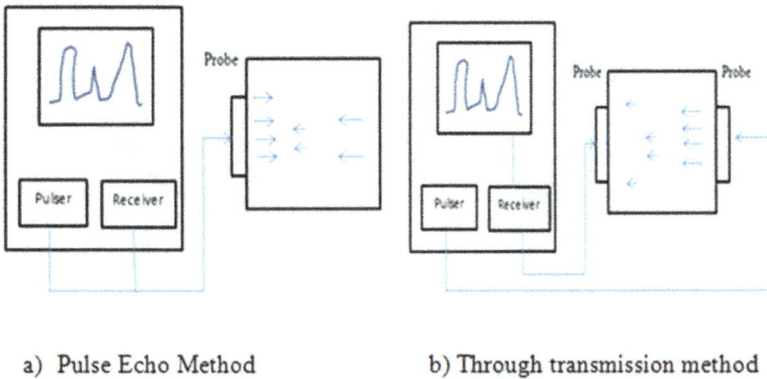

a) Pulse Echo Method   b) Through transmission method

Figure 6. Ultrasonic Inspection method

3.2.4 Eddy Current Testing (ECT)

Eddy Current Testing is stated to be one of the more advanced NDE techniques that find application in carrying out corrosion detection, besides making integrity assessments for structures in conductive materials. The basis for Eddy Current Testing is electromagnetic induction where an alternating magnetic field is made use of so as to create a probe that is capable of inducing eddy currents into the test material. The method is simple. It positions the probe close to the substance's surface. Flaws are determined based on the signals produced due to the generation of a magnetic field [17]. When these currents encounter material discontinuities, such as corrosion or fractures, they affect the local impedance of the probe and thus produce detectable signals that can then be analyzed. That this usually involves a calibration of an ECT device, methodological scanning through a surface of test material, as well as its comparison with standard information or accepted requirements [18]. The mentioned advantages of the ECT lie in its great speed for surface and near-surface scan, high sensibility to in homogeneities of the structure being tested, as well as complete absence of its direct physical contacts. In general, because relatively recent advancements including pulsed eddy current PEC have tended to raise application for detection under insulation and also coatings, its utility is even more specifically needed in coatings situations or for less magnetic layers. However, its own drawback means that it proves difficult to recognize deeply buried defects, especially using highly irregular geometries, and further, effectiveness severely limits when using such non-conducting materials. Advanced software and skilled operators may be required for data interpretation. The new research of ECT has brought in tremendous improvements, such as its analysis with the help of machine learning and multi-frequency approaches, which enhances its accuracy in most practical applications. This ensures that ECT will remain an essential method used during manufacturing, oil & gas, and aerospace to ensure non-occurrence of structural failure and catastrophic failures [19,20].

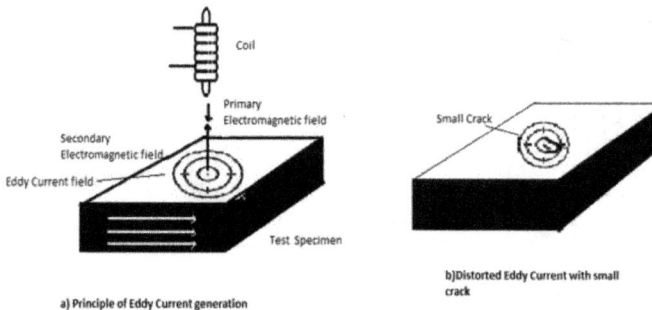

Figure 7. Principle of Eddy Current Generation

3.2.5 Radiographic testing

Radiographic examination is one of the most extensively applied NDE techniques for appraising for fissures, cavities, and other defects of material or sub-materials/contents. Interior corrosion, holes, or some other defects cannot be inspected successfully unless advanced, high-energy types of radiation in the form of gamma or X-rays are adopted. Pipelines and pressure vessel, among a wide range of other materials require high-energy applications. The basis of the radiotherapy (RT) concept is the differential absorption of radiation as it passes through a material. Essentially, if the material thickness is reduced due to corrosion or cracks, then more radiation will flow through and create contrasting pictures on the detector [21]. RT has several advantages. It allows the inspection of the constructed parts and simultaneously see inside and outside corrosion, thus creating a permanent record of the inspections that can be referenced at a later date. RT is very efficient in detecting flaws hidden from the surface and intricate systems without the need to disassemble [22]. However, RT is not without its limitations. This method is relatively unsuccessful in showing minute flaws or low material thinning and requires examination access to the both sides of the test specimen. It also happens to be really expensive and extremely time-consuming since it uses ionizing radiation to inspect materials which has inherent associated safety hazards necessitating stringent control procedures and an expert workforce for operation. With recent developments in RT speed and accuracy, such as digital radiography as well as computed tomography (CT), real-time imaging and enhanced defect characterization are now possible. RT is still very important despite its difficulties because it guarantees safety and adherence in crucial operations in the evaluation of structural integrity and corrosion detection.

Advances in Corrosion Science and Surface Engineering
Materials Research Foundations 188 (2026) 193-207

Materials Research Forum LLC
https://doi.org/10.21741/9781644903919-10

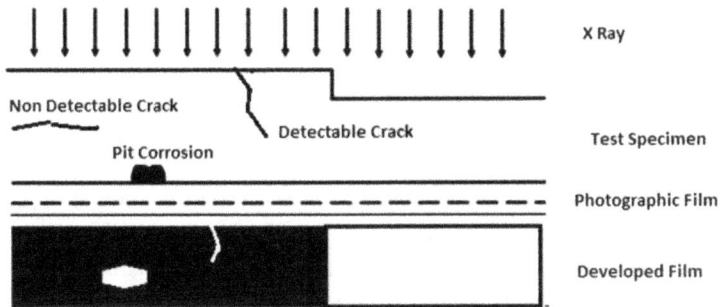

Figure 8. Radiographic method of NDT

3.2.6 Thermography testing

Thermography Testing, or TT is one of those modern NDE techniques widely adopted for corrosion detection that is based upon the principle to measure material thermal properties with the help of infrared radiation. Alterations in temperature at the surface are caused due to defects, such as vacancy, corrosion or delamination inside the material. So, according to this concept corrosion alters the properties or thicknesses of the materials and henceforth affects the transference of heat. Such changes can be considered as temperature changes in infrared images. These can be applied by inspectors to inspect the state of a building or component without touching it physically [23,24].

Whereas passive thermography makes use of operating heat or the naturally occurring heat sources of sunshine, active thermography applies heat from an external source, such as lights, lasers, or hot air. A detailed process: Preparation Procedure This includes getting ready to choose the right equipment, which should work with infrared cameras that have the right resolutions and sensitivities, as well as figuring out the heating technique if active thermography is being used. Calibration: To provide an accurate measurement, the infrared camera has to be adjusted for the substance being tested. Heating: The surface of the test material is heated uniformly by an external heat source in order to perform active thermography. Imaging: The material is heated either during or just after the imaging process to produce the infrared pictures. Analysis: To find thermal patterns linked to corrosion or flaws, images are analyzed using specialist software [25].

3.2.7 Liquid penetrant testing

Liquid Penetrant Testing is the most extensively used and oldest non-destructive testing procedure. It is often referred to as dye penetrant inspection. Liquid penetrant testing makes use of capillary action, or the liquid's ability to run into breaks or holes without the assistance of other forces like as gravity, to detect any type of flaw. After a while, any surplus surface penetrant is removed, and a developer is applied. The developer removes the penetrant from the surface-breaking flaws and exposes it, and the resting time period is known as dwell time. The Liquid Penetrant Testing (LPT) approach is commonly used to evaluate nonferrous and nonmagnetic materials such as aluminum

Advances in Corrosion Science and Surface Engineering   Materials Research Forum LLC
Materials Research Foundations 188 (2026) 193-207   https://doi.org/10.21741/9781644903919-10

and chrome steel, as well as alternative materials that cannot be inspected using the Magnetic Particle technique [26].

The following are the steps involved in performing the Liquid Penetrant test:Surface Preparation: This is one of the most important phases in a liquid penetrant inspection. Water, oil, grease, and other impurities that could stop penetrant from penetrating defects must be removed from the surface. If the sample has undergone mechanical processes like machining, sanding, or grit blasting, etching might also be necessary. Metal can be spread across the defect opening by these and other mechanical processes, blocking the penetrant's entry. Penetrant Application: Additionally, it might contaminate the liquid penetrant and alter its fluorescent qualities. In certain cases, the harmful substances on the material's surface are removed using etchants or pickle liquor. Penetrant materials are classified into three types such as fluorescent penetrant dye, Contrast dye (usually red in color), Dual-function dye (both visible and fluorescent). Penetrant Dwell: The metal surface is subsequently covered with the liquid dye. Therefore, a waiting period known as the dwell time is used to make sure the dye penetrates areas with flaws. Excess Penetrant Removal: To ensure that none of the dye within the flaws is removed, excess dye should be carefully washed (sometimes with an emulsifier). Developer Application: The developer is left on the part surface sufficient time to allow the trapped penetrant to be removed from any surface imperfections. A finely divided powder applied over the surface of a part to help bring out penetrant indications. Indication Development: A minimum of 10 minutes is typically required for the developer to stand on the part surface long enough to allow the trapped penetrant to be extracted from any surface imperfections; for narrow cracks, much longer times may be required. Inspection: Inspection is subsequently conducted under adequate lighting to locate indications from any defects which may be present. Clean Surface: The process ends with a thorough cleaning of the part surface in order to get rid of the developer from the sections that were chosen acceptable [27].

3.2.8 Magnetic particle inspection

It is a standard inspection technique for detecting surface-breaking faults in ferromagnetic materials. The material being examined is magnetic, either by direct magnetization or by employing an electromagnetic yoke. Fine magnetic particles, either dry or suspended in a liquid carrier, are then deposited on the surface. If a flaw exists, the magnetic field is disrupted, causing the particles to cluster and form visual indicators that can be easily identified [28]. In the early 1920s, William Hoke found that magnetic particles might be combined with magnetism to locate faults. Hoke observed that a surface or subsurface imperfection in a magnetized material distorts and extends the magnetic field beyond the component. This discovery was brought to his attention at the machine shop. He found that metallic grindings from hard steel parts (held by a magnetic chuck while being ground. left patterns on the parts' faces that correlated to surface fissures. Applying a fine ferromagnetic powder to the pieces resulted in a powder buildup over faults, creating a visible indicator of MPT. The widespread use of magnetic particle testing (MPT), a non-destructive testing technique, allows for the discovery of surface and near-surface faults in ferromagnetic materials. It has been utilized in a wide range of industries, including manufacturing, mining, oil and gas, power generation, petrochemicals, vehicle production, and infrastructure. The benefits of MPT help to explain why it is so popular. It is a non-destructive inspection approach since it allows the detection of defects without damaging the test [29].

The steps of inspection follows as, In ferritic materials, whether in the dressed or as-welded state, as well as the parent material and related heat-affected zones on both sides of the weld that are at

Advances in Corrosion Science and Surface Engineering   Materials Research Forum LLC
Materials Research Foundations 188 (2026) 193-207   https://doi.org/10.21741/9781644903919-10

least an inch from the weld. Surface Preparation; Any elements that might delay the test or hide undesirable discontinuities must be removed from the area to be inspected and at least one inch on each side prior to the test. Slag, splatter, oil, scale, rough surfaces, and protective coatings are a few examples of these. In situations where surface irregularities might hide signs of undesirable discontinuities, surface preparation through grinding, machining, or other techniques may be required. The test surface temperature cannot be higher than 135°F for magnetic inks and 600°F for dry powders [30]. The test surface should be dry and clean for dry powders. If required, the area to be inspected can be precleaned using a cloth lightly soaked with cleanser before parts are inspected using magnetic inks. Equipment and Consumable Control; The yoke must be able to lift a minimum of 10 pounds at the maximum pole spacing that will be utilized, and its magnetizing force must be inspected at least once a year or following any damage and/or repairs. Magnetic powders and magnetic inks can be used once. Lighting Conditions; Before starting the inspection, the inspector must give the dark at least five minutes to settle down. The inspector must make sure that there is sufficient lighting at the part's surface when doing an examination in white lighting. Direction of Magnetizing Field; The magnetizing field will be applied in two directions, roughly perpendicular to each other, in a sequential manner. The Burmah Castrol magnetic field indicator can be used to determine the direction of the field; it will provide its strongest indicators when positioned across the flux direction.

Conclusion

The solidity of materials and structures is seriously threatened by corrosion failures that include many components. As a result, non-destructive monitoring and inspection techniques have been given priority in recent developments in corrosion science and engineering in order to evaluate structural health while reducing damage and financial effect. Environmental conditions, pollutants, heat treatment, and material choice all have an impact on corrosion. The application of appropriate detection and characterization techniques would depend on the kind of corrosion. No one inspection method could identify every form of corrosion because of variations in corrosion caused by the material, location, and environmental factors. The majority of corrosion control techniques may be classified as either preventative or remedial. In essence, preventive corrosion management involves servicing or repairing components before to failures, while corrective corrosion management involves replacing impacted parts once issues occur. While visual examination can be helpful in detecting surface problems, it is not a reliable method for detecting deeper defects. For greater efficacy, sophisticated techniques are typically coupled with ocular inspections. Many sectors employ the non-destructive methods included here, such as AE, ECT, and ultrasonic testing, to assess the structural and material integrity without causing harm. Every technique has distinct advantages and disadvantages, making them appropriate for certain uses in various industrial contexts.

Reference

[1]B.A. Egbokhaebho, B.I. Olalere, J.O. Gidiagba, J.I. Okparaeke, A.A. F awole, N.N. -Ehiobu, Review On Non-Destructive Techniques For Early Flaw Detection In Inspections, Mater. Corros. Eng. Manag. 4 (2023). https://doi.org/10.26480/macem.02.2023.44.50

[2]V. Marcantonio, D. Monarca, A. Colantoni, Cecchini, an Introduction To Corrosion Monitoring, Int. J. Press. Vessel. Pip. 96 (2016).

[3] C.J. Hellier, Handbook of Nondestructive Evaluation, Second Edition, 2013.

[4] V. Vasagar, M.K. Hassan, A.M. Abdullah, A. V. Karre, B. Chen, K. Kim, N. Al-Qahtani, T. Cai, Non-destructive techniques for corrosion detection: A review, Corros. Eng. Sci. Technol. 59 (2024) 56–85. https://doi.org/10.1177/1478422X241229621

[5] A. SHUBBAR, Z. Al-khafaji, M. Nasr, M. Falah, Using Non-Destructive Tests For Evaluating Flyover Footbridge: Case Study, Knowledge-Based Eng. Sci. 1 (2020). https://doi.org/10.51526/kbes.2020.1.01.23-39

[6] Non-Destructive Evaluation of Corrosion and Corrosion-assisted Cracking, 2019. https://doi.org/10.1002/9781118987735

[7] G. Lalitha, K.B. Showry, Experimental Study on Non Destructive Testing Techniques (NDTT), Int. J. Eng. Res. Gen. Sci. 3 (2015).

[8] P. Pfändler, K. Bodie, G. Crotta, M. Pantic, R. Siegwart, U. Angst, Non-destructive corrosion inspection of reinforced concrete structures using an autonomous flying robot, Autom. Constr. 158 (2024). https://doi.org/10.1016/j.autcon.2023.105241

[9] D. Seo, J. Kim, S. Park, An Experimental Study on Defect Detection of Anchor Bolts Using Non-Destructive Testing Techniques, Materials (Basel). 16 (2023). https://doi.org/10.3390/ma16134861

[10] S.A. Bin Idris, F.A. Jafar, N. Abdullah, Study on corrosion features analysis for visual inspection & monitoring system: An NDT technique, J. Teknol. 77 (2015). https://doi.org/10.11113/jt.v77.6608

[11] F. Akgul, Inspection and evaluation of a network of concrete bridges based on multiple NDT techniques, Struct. Infrastruct. Eng. (2020). https://doi.org/10.1080/15732479.2020.1790016

[12] A.I. Sagaidak, Technologies for acoustic emission monitoring of concrete and reinforced concrete structures, Bull. Sci. Res. Cent. Constr. 38 (2023). https://doi.org/10.37538/2224-9494-2023-3(38)-62-81

[13] E. Verstrynge, C. Van Steen, E. Vandecruys, M. Wevers, Steel corrosion damage monitoring in reinforced concrete structures with the acoustic emission technique: A review, Constr. Build. Mater. 349 (2022). https://doi.org/10.1016/j.conbuildmat.2022.128732

[14] H. Taheri, A.A. Hassen, Nondestructive ultrasonic inspection of composite materials: A comparative advantage of phased array ultrasonic, Appl. Sci. 9 (2019). https://doi.org/10.3390/app9081628

[15] H. Shi, M. Ebrahimi, P. Zhou, K. Shao, J. Li, Ultrasonic and phased-array inspection in titanium-based alloys: A review, Proc. Inst. Mech. Eng. Part E J. Process Mech. Eng. 237 (2023). https://doi.org/10.1177/09544089221114253

[16] J.K. Shah, H.B.F. Braga, A. Mukherjee, B. Uy, Ultrasonic monitoring of corroding bolted joints, Eng. Fail. Anal. 102 (2019). https://doi.org/10.1016/j.engfailanal.2019.04.016

[17] N.P. de Alcantara, F.M. da Silva, M.T. Guimarães, M.D. Pereira, Corrosion assessment of steel bars used in reinforced concrete structures by means of eddy current testing, Sensors. 16 (2016). https://doi.org/10.3390/s16010015

[18]   H. Shaikh, N. Sivaibharasi, B. Sasi, T. Anita, R. Amirthalingam, B.P.C. Rao, T. Jayakumar, H.S. Khatak, B. Raj, Use of eddy current testing method in detection and evaluation of sensitisation and intergranular corrosion in austenitic stainless steels, Corros. Sci. 48 (2006). https://doi.org/10.1016/j.corsci.2005.05.017

[19]   M. Janovec, Černan, Škultéty, Use of non-destructive eddy current technique to detect simulated corrosion of aircraft structures, Koroze a Ochr. Mater. 64 (2020). https://doi.org/10.2478/kom-2020-0008

[20]   R. Ghoni, M. Dollah, A. Sulaiman, F. Mamat Ibrahim, Defect Characterization Based on Eddy Current Technique: Technical Review, Adv. Mech. Eng. 2014 (2014). https://doi.org/10.1155/2014/182496

[21]   A. Movafeghi, N. Mohammadzadeh, E. Yahaghi, J. Nekouei, P. Rostami, G. Moradi, Defect Detection of Industrial Radiography Images of Ammonia Pipes by a Sparse Coding Model, J. Nondestruct. Eval. 37 (2018). https://doi.org/10.1007/s10921-017-0458-9

[22]   C. Rathinasuriyan, V.S. Senthil Kumar, A.G. Shanbhag, Radiography and corrosion analysis of sub-merged Friction Stir Welding of AA6061-T6 alloy, in: Procedia Eng., 2014. https://doi.org/10.1016/j.proeng.2014.12.355

[23]   S. Liu, H. Liu, Z. Liu, Quantification of pitting corrosion from thermography using deep neural networks, Rev. Sci. Instrum. 92 (2021). https://doi.org/10.1063/5.0026653

[24]   Z. Liu, M. Genest, D. Krys, Processing thermography images for pitting corrosion quantification on small diameter ductile iron pipe, NDT E Int. 47 (2012). https://doi.org/10.1016/j.ndteint.2012.01.003

[25]   G. Cadelano, A. Bortolin, G. Ferrarini, B. Molinas, D. Giantin, P. Zonta, P. Bison, Corrosion Detection in Pipelines Using Infrared Thermography: Experiments and Data Processing Methods, J. Nondestruct. Eval. 35 (2016). https://doi.org/10.1007/s10921-016-0365-5

[26]   C. Chris Roshan, H. Vasanth Ram, J. Solomon, Non-destructive testing by liquid penetrant testing and ultrasonic testing-A review, Int. J. Adv. Res. Ideas Innov. Technol. 5 (2019).

[27]   G. Caturano, G. Cavaccini, A. Ciliberto, V. Pianese, R. Fazio, Liquid Penetrant Testing: Industrial Process, SIMAI Congr. 3 (2009).

[28]   A.I. Sacarea, G. Oancea, L. Parv, Magnetic particle inspection optimization solution within the frame of ndt 4.0, Processes. 9 (2021). https://doi.org/10.3390/pr9061067

[29]   Q. Wu, K. Dong, X. Qin, Z. Hu, X. Xiong, Magnetic particle inspection: Status, advances, and challenges — Demands for automatic non-destructive testing, NDT E Int. 143 (2024). https://doi.org/10.1016/j.ndteint.2023.103030

[30]   D.J. Eisenmann, D. Enyart, C. Lo, L. Brasche, Review of progress in magnetic particle inspection, in: AIP Conf. Proc., 2014. https://doi.org/10.1063/1.4865001

Materials Research Forum LLC
https://doi.org/10.21741/9781644903919-11

Chapter 11

Machine Learning (ML)-aided Modeling of Cyanopyran-based Corrosion Inhibitiors

Jagadeesan Saranya[1]*, Yalla Jeevan Nagendra Kumar[2], Abdelkader Zarrouk[3], G. Kausalya Sasikumar[4], Rajender Boddula[5]

[1]Department of Chemistry, PSG College of Arts and Science, Coimbatore 641014, India

[2]Department of Information Technology, Gokaraju Rangaraju Institute of Engineering and Technology, Hyderabad, India

[3]Faculty of Sciences, Laboratory of Materials, Nanotechnology and Environment, Mohammed V University in Rabat, Agdal-Rabat, Morocco

[4]Centre for Research and Development, KPR Institute of Engineering and Technology, Coimbatore, India

[5]School of Sciences, Woxsen University, Hyderabad - 502345, Telangana, India

jcsaranya.chem96@gmail.com

Abstract

Corrosion is a significant challenge across multiple industries, impacting material longevity and safety. This chapter explores interdisciplinary approaches that combine computational methods with experimental validation to ensure practical applicability in real-world scenarios. Computational approaches have revolutionized the study of corrosion mechanisms, prediction, and mitigation strategies. In this study, machine learning methods have been studied to evaluate the corrosion performance of the inhibitor 2-amino-4-(4-hydroxyphenyl)-6-(p-tolyl)-4H-pyran-3-carbonitrile (HCN) on mild steel in 1M H_2SO_4. Experimental studies like the weight loss method were carried out to test the inhibition efficiency of the inhibitor molecule, which revealed that the inhibition efficiency increased with the increase in the concentration of the inhibitor. Quantum chemical studies were also performed to study the interaction of inhibitor molecules with metal. Regarding machine learning studies, it is identified that the Random Forest was the best algorithm that predicted the entire time profile of corrosion rates with the mean squared error ranging from 0.005 to 0.093. The sensitivity of corrosion rates to changes in the environmental variables is well-predicted by the trained Random Forest model.

Keywords

Inhibition Efficiency, Corrosion, Cyanopyran, Algorithm, Random Forest, Artificial Neural Network

Advances in Corrosion Science and Surface Engineering   Materials Research Forum LLC
Materials Research Foundations 188 (2026) 208-227   https://doi.org/10.21741/9781644903919-11

Contents

Machine Learning (ML)-aided Modeling of Cyanopyran-based Corrosion Inhibitiors **208**

1.   Introduction

Metals are the most widely used material in the metallurgical and mechanical engineering sectors. Moreover, in oil and gas industries, metallic materials are used majorly as flow components for transporting fluids, making the metals more liable to corrosion. Corrosion of metals is a serious problem that greatly impacts the environment, human health, and economic loss. Metals undergo corrosion due to unfavorable surroundings creating metal ions and free electrons which change the composition, surface, color, and engineering properties and harm the environment. Mild steel is one of the widely used metals in oil and gas industries which is often prone to corrosion when exposed to an aqueous medium[1,2]. Scientists are finding various techniques to mitigate metal corrosion like coating, cathodic protection, and corrosion inhibitors to improve the properties, quality, and lifetime of the metals and their alloys. The application of inhibitors to protect the metal is a traditional and practical method as this method is also cheap, safe, and efficient. Even a minimal inhibitor reduces the corrosion rate in an aggressive environment[3]. Organic heterocyclic compounds comprised of oxygen, nitrogen, and sulfur are considered potential compounds to prevent the metal from corrosion by blocking the active surface of the metal. They form complexes with metal-free electrons, allowing adsorption to protect the metal in the aqueous medium[4]. The organic inhibitor used in this study is a heterocyclic compound containing an oxygen atom, a cyanopyran derivative called 2-amino-4-(4-hydroxyphenyl)-6-(p-tolyl)-4H-pyran-3-carbonitrile

(HCN). We already reported that the inhibitor HCN acted as an excellent inhibitor for preventing mild steel in 1M H_2SO_4.

The experiments carried out for synthesizing the inhibitors, characterizing the inhibitors, and evaluating the corrosion rates using weight loss, electrochemical, surface studies, or other means of studies, through various studies are time-consuming, tedious, and costly but are however necessary, in the absence of reliable theoretical methods. Density functional theory (DFT) and machine learning (ML) methods have proven to be the best methods that could offer possible solutions for the rational discovery of new and effective corrosion inhibitors[5-10]. ML methods can be applied to various fields as it saves time. Few authors reported that the ML methods provide information about the prediction of the corrosion inhibition process. For instance, Stango and Vijayalakshmi 2018 [11] investigated agricultural waste i.e. orange peel as an inhibitor for mild steel corrosion in an HCl medium using the ML technique. Samide et al., 2019 [12] highlighted the Convolutional Neural Network (CNN) approach of deep learning and artificial intelligence techniques used to study the surface morphology of the metal in the presence and absence of corrosion inhibitors like polyvinyl alcohol and polyvinyl alcohol with silver nanoparticles. Dexamethasone drug was studied as an inhibitor for mild steel in 2M HCl by Anadebe et al. 2020 [13]. The authors used adaptive neural fuzzy inference systems (ANFIS) and artificial neural networks (ANN) as ML optimization tools for predicting inhibition properties. ANFIS was the best method and gave 81.24% compared to ANN (79.74%). Supervised learning using the Random Forest algorithm was used to predict the time profile of a corrosion process dependent on time was studied by Aghaaminiha et al. [14]. During the corrosion process, the study was able to envisage the sensitivity of changes in environmental variables like concentration, temperature, etc. Diao et al. [15]. ML technique was applied to study the corrosion of low-alloy steel using the random forest method. The input parameters, such as the chemical composition of the inhibitor and environmental factors were used by the authors Taylor and Tossey [16]. Corrosion of high-resistant alloys and their temperature using ML techniques were studied using both supervised and unsupervised methods by Coelho, Zhang et al. A detailed review of literature on ML applied for corrosion studies. Aeshah et al. [18] provided the outcome of the work and guidance for designing and synthesizing effective and new pyrimidine derivatives using data-driven ML models. Anadebe et al. [19] studied the corrosion inhibition of salbutamol drug molecule on mild steel in oilfield acidizing fluid using Artificial neural network-genetic algorithm (ANNGA) and adaptive neural fuzzy inference system-genetic algorithm (ANFIS-GA) tools. ANFIS method was found to be the best method in terms of R^2 (coefficient of determination), χ^2 (chi-square), RMSE (root mean square error), and MPE (model predictive error). The effect of concentration and time are the most important parameters according to ML studies.

Computational modeling is one of the methods used for analyzing the properties, and performance of engineering materials which can analyze and improve the interactions between the multiple parameters and the response of the output. Multiple linear regressions are not suitable for such optimization purposes[20]. In this context, the authors aim to extend the previous studies in which experimental studies like weight loss, polarization, impedance, and surface studies (SEM, and EDS) were already reported[21]. Now, to explore the potential of HCN, a combined DFT and machine learning methods are investigated to understand the corrosion inhibition process at the molecular level of the approach.

Advances in Corrosion Science and Surface Engineering   Materials Research Forum LLC
Materials Research Foundations 188 (2026) 208-227   https://doi.org/10.21741/9781644903919-11

2.   Experimental methods

2.1   Inhibitor

Computational approaches streamline the design and optimization of corrosion inhibitors. By implementing DFT, researchers can identify the optimal inhibitor structures that exhibit strong binding affinities to specific metallic surfaces and predict the desorption tendencies of inhibitors under varying environmental conditions. It will reduce the environmental impact of inhibitors by designing biodegradable and non-toxic alternatives. In this context, the inhibitor 2-amino-4-(4-hydroxyphenyl)-6-(p-tolyl)-4H-pyran-3-carbonitrile (HCN) is synthesized which is non-toxic and can be evaluated as corrosion inhibitors. The structure of the synthesized inhibitor HCN is shown in Fig. 1.

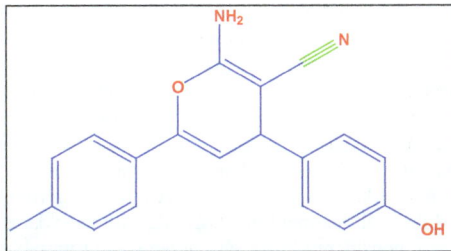

Figure 1. Structure of the inhibitor (HCN)

2.2   Weight loss studies

The parameters like inhibition efficiency (IE %) and surface coverage (θ) were calculated by the formulae used in the literature [21]. The inhibition efficiency (Eq. 1) and corrosion rate (Eq. 2) from WL studies were calculated using the formulae,

$$\text{Inhibition efficiency (IE \%)} = \frac{(W_{bl} - W_{in})}{W_{bl}} \times 100 \tag{1}$$

where W_{bl} = WL without inhibitor; W_{in} = WL with inhibitor,

$$\text{Corrosion rate, CR} = \frac{534 \times \text{Weight loss } (g)}{\text{Density } (g/cm^3) \times \text{Area } (cm) \times \text{Time } (h)} \tag{2}$$

2.3   Quantum chemical method

Quantum chemical methodologies, such as density functional theory (DFT), have transformed corrosion research by providing atomic-level insights into material properties. These methods allow researchers to investigate surface interactions by simulating the relationships between metallic substrates and corrosive agents like oxygen, chloride ions, and water molecules, thus pinpointing vulnerable locations for corrosion starts. This will enhance the examination of adsorption mechanisms. DFT studies are crucial for ascertaining adsorption energies, charge transfer, and electronic characteristics, which are vital for comprehending the bisulfate ions, the binding of inhibitors to surfaces, and their function in mitigating corrosive processes. The enhancement of corrosion inhibitors will be accomplished by the simulation of molecular structures and their electrical characteristics, allowing researchers to develop inhibitors that are both highly effective and environmentally compatible. This chapter discusses quantum chemical investigations employing the inhibitor HCN for mild steel in a sulfuric acid environment, utilizing the G09W software package, which markedly enhances its resistance to acid-induced corrosion.

2.4   ML technique process

The pictorial representation of the ML process is shown in Fig. 2. ML techniques were performed from the data extracted from the experimental results. In the following process of visualizing and analyzing the patterns generated in recording the outputs of experimental data a dataset of 160 records with six attributes has been taken into consideration for building a machine-learning model[22,23]. The attributes are temperatures, concentration, weight, surface area, inhibition efficiency, and corrosion rate. Strictly speaking in the way of constructing models to identify the patterns that are generated in the data set, multiple ML algorithms such as a K-Nearest neighbor (KNN), support vector machine (SVM), kernel support vector machine (kSVM), random forests (RF), and artificial neural networks (ANN) were used. All these algorithms are implemented in R using R studios. In each of these above-mentioned ML algorithms, a dataset containing the above-mentioned features has been considered in which the dependent variable is inhibition efficiency whereas despite corrosion rate also being a dependent variable which is not taken into consideration while building the models due to lack of its presence and identifying the patterns of generating inhibition efficiency. All the rest of the attributes mentioned in the dataset are considered independent variables which are highly important in identifying the patterns. This dataset is a result of sequential experiments done on mild steel in a medium of artificially generated $1M$ H_2SO_4 reacting with HCN where HCN is used as a corrosion inhibitor on mild steel as a whole. It is interesting to study experimental methods like weight loss and compare the results with machine learning and molecular modeling techniques[24,25]. The combination of all these studies is to identify the best atmosphere in terms of temperature, the relative amount of HCN required for the effective prevention of mild steel corrosion in an acid medium, and the adsorption sites of the inhibitor.

Advances in Corrosion Science and Surface Engineering   Materials Research Forum LLC
Materials Research Foundations 188 (2026) 208-227   https://doi.org/10.21741/9781644903919-11

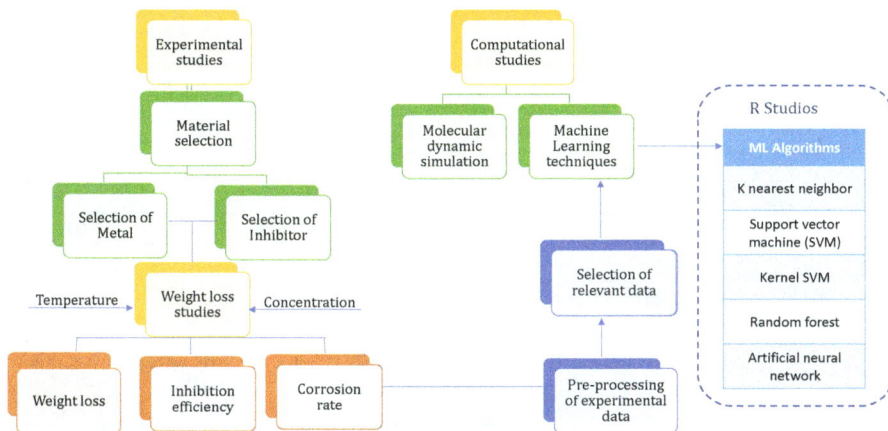

Figure 2. Pictorial representation of the ML process

3.   Results and Discussion

3.1   Weight loss studies

Weight loss studies have been performed for the inhibition of mild steel in the sulphuric acid medium using different concentrations (0.1 to 2 mM) of the inhibitor HCN at different temperatures from 303 to $313 \pm 1K$ after 1 hour of immersion. The concentration of the inhibitor, inhibition efficiency, and corrosion rate are presented in Table 1. The parameters like surface coverage, inhibition efficiency, and corrosion rate were calculated and presented in Table 1. It is observed that the inhibition efficiency is increased and the corrosion rate is decreased when the temperature is increased. This may be due to the strong adsorption of the inhibitor on the metal at low temperature i.e., $303 \pm 1K$. When the temperature is increased, the adsorption of the inhibitor slowly detaches from the metal and diffuses into the solution. The adsorption of the inhibitor is due to the presence of active centers like C≡N, OH, NH_2, O atom, and π electrons[26].

Table 1. Experimental data through weight loss method for the prevention of mild steel in 1M H_2SO_4 at various temperatures

Temperature (K)	Concentration (mM)	Weight (mg/cm^2)	Surface coverage (θ)	Inhibition efficiency (%)	Corrosion rate (g cm^{-2} h^{-1})
303 ± 1K	0	0.0594	0	0	17.66
	0.10	0.0237	0.6010	60.10	7.05
	0.15	0.0223	0.6246	62.46	6.63
	0.20	0.0214	0.6397	63.97	6.36
	0.25	0.0204	0.6566	65.66	6.07
	0.30	0.0201	0.6616	66.16	5.98
	0.35	0.0192	0.6768	67.68	5.71
	0.40	0.0183	0.6919	69.19	5.44
	0.45	0.0181	0.6953	69.53	5.38
	0.50	0.0178	0.7003	70.03	5.29
	0.55	0.0165	0.7222	72.22	4.91
	0.60	0.0159	0.7323	73.23	4.73
	0.65	0.0147	0.7525	75.25	4.37
	0.70	0.0141	0.7626	76.26	4.19
	0.75	0.0131	0.7795	77.95	3.89
	0.80	0.0135	0.7727	77.27	4.01
	0.85	0.0132	0.7778	77.78	3.92
	0.90	0.0129	0.7828	78.28	3.84
	0.95	0.0123	0.7929	79.29	3.66
	1.00	0.0121	0.7963	79.63	3.60
	1.05	0.0118	0.8013	80.13	3.51
	1.10	0.0113	0.8098	80.98	3.36
	1.15	0.0106	0.8215	82.15	3.15
	1.20	0.0103	0.8266	82.66	3.06
	1.25	0.0099	0.8333	83.33	2.94
	1.30	0.0096	0.8384	83.84	2.85
	1.35	0.0093	0.8434	84.34	2.77
	1.40	0.0089	0.8502	85.02	2.65
	1.45	0.0085	0.8569	85.69	2.53
	1.50	0.0084	0.8586	85.86	2.50
	1.55	0.0079	0.8670	86.70	2.35
	1.60	0.0076	0.8721	87.21	2.26
	1.65	0.0068	0.8855	88.55	2.02
	1.70	0.0062	0.8956	89.56	1.84
	1.75	0.0059	0.9007	90.07	1.75

	1.80	0.0054	0.9091	90.91	1.61
	1.85	0.0051	0.9141	91.41	1.52
	1.90	0.0049	0.9175	91.75	1.46
	1.95	0.0046	0.9226	92.26	1.37
	2.00	0.0045	0.9242	92.42	1.34
	0	0.0687	0	0.00	20.43
	0.10	0.0331	0.5182	51.82	9.84
	0.15	0.0312	0.5459	54.59	9.28
	0.20	0.0301	0.5619	56.19	8.95
	0.25	0.0293	0.5735	57.35	8.71
	0.30	0.0281	0.5910	59.10	8.35
	0.35	0.0273	0.6026	60.26	8.12
	0.40	0.0269	0.6084	60.84	8.00
	0.45	0.0254	0.6303	63.03	7.55
	0.50	0.0246	0.6419	64.19	7.31
	0.55	0.0231	0.6638	66.38	6.87
	0.60	0.0227	0.6696	66.96	6.75
	0.65	0.0221	0.6783	67.83	6.57
	0.70	0.0213	0.6900	69.00	6.33
	0.75	0.0209	0.6958	69.58	6.21
	0.80	0.0201	0.7074	70.74	5.98
	0.85	0.0192	0.7205	72.05	5.71
313 ± 1K	0.90	0.0187	0.7278	72.78	5.56
	0.95	0.0183	0.7336	73.36	5.44
	1.00	0.0175	0.7453	74.53	5.20
	1.05	0.0169	0.7540	75.40	5.02
	1.10	0.0163	0.7627	76.27	4.85
	1.15	0.0158	0.7700	77.00	4.70
	1.20	0.0155	0.7744	77.44	4.61
	1.25	0.0151	0.7802	78.02	4.49
	1.30	0.0147	0.7860	78.60	4.37
	1.35	0.0145	0.7889	78.89	4.31
	1.40	0.0142	0.7933	79.33	4.22
	1.45	0.014	0.7962	79.62	4.16
	1.50	0.0137	0.8006	80.06	4.07
	1.55	0.0131	0.8093	80.93	3.89
	1.60	0.0128	0.8137	81.37	3.81
	1.65	0.0119	0.8268	82.68	3.54
	1.70	0.0114	0.8341	83.41	3.39
	1.75	0.0111	0.8384	83.84	3.30

	1.80	0.0109	0.8413	84.13	3.24
	1.85	0.0105	0.8472	84.72	3.12
	1.90	0.0096	0.8603	86.03	2.85
	1.95	0.0091	0.8675	86.75	2.71
	2.00	0.0087	0.8734	87.34	2.59
323 ± 1K	0	0.1414	0	0	42.04
	0.10	0.0764	0.4597	45.97	22.71
	0.15	0.0743	0.4745	47.45	22.09
	0.20	0.0712	0.4965	49.65	21.17
	0.25	0.0694	0.5092	50.92	20.63
	0.30	0.0685	0.5156	51.56	20.37
	0.35	0.0674	0.5233	52.33	20.04
	0.40	0.0634	0.5516	55.16	18.85
	0.45	0.0601	0.5750	57.50	17.87
	0.50	0.0596	0.5785	57.85	17.72
	0.55	0.0573	0.5948	59.48	17.04
	0.60	0.0561	0.6033	60.33	16.68
	0.65	0.0554	0.6082	60.82	16.47
	0.70	0.0541	0.6174	61.74	16.08
	0.75	0.0532	0.6238	62.38	15.82
	0.80	0.0509	0.6400	64.00	15.13
	0.85	0.0483	0.6584	65.84	14.36
	0.90	0.0476	0.6634	66.34	14.15
	0.95	0.0463	0.6726	67.26	13.77
	1.00	0.0459	0.6754	67.54	13.65
	1.05	0.0452	0.6803	68.03	13.44
	1.10	0.0443	0.6867	68.67	13.17
	1.15	0.0437	0.6909	69.09	12.99
	1.20	0.0429	0.6966	69.66	12.75
	1.25	0.0421	0.7023	70.23	12.52
	1.30	0.0414	0.7072	70.72	12.31
	1.35	0.0405	0.7136	71.36	12.04
	1.40	0.0396	0.7199	71.99	11.77
	1.45	0.0391	0.7235	72.35	11.62
	1.50	0.0388	0.7256	72.56	11.54
	1.55	0.0382	0.7298	72.98	11.36
	1.60	0.0375	0.7348	73.48	11.15
	1.65	0.0368	0.7397	73.97	10.94
	1.70	0.0361	0.7447	74.47	10.73
	1.75	0.0349	0.7532	75.32	10.38

	1.80	0.0332	0.7652	76.52	9.87
	1.85	0.0305	0.7843	78.43	9.07
	1.90	0.0289	0.7956	79.56	8.59
	1.95	0.0274	0.8062	80.62	8.15
	2.00	0.0244	0.8274	82.74	7.25
	0	0.1833	0.0000	0.00	54.50
	0.10	0.1094	0.4032	40.32	32.53
	0.15	0.1074	0.4141	41.41	31.93
	0.20	0.1053	0.4255	42.55	31.31
	0.25	0.1024	0.4414	44.14	30.44
	0.30	0.1007	0.4506	45.06	29.94
	0.35	0.0962	0.4752	47.52	28.60
	0.40	0.0913	0.5019	50.19	27.14
	0.45	0.0878	0.5210	52.10	26.10
	0.50	0.0833	0.5456	54.56	24.77
	0.55	0.0821	0.5521	55.21	24.41
	0.60	0.0814	0.5559	55.59	24.20
	0.65	0.0801	0.5630	56.30	23.81
	0.70	0.0783	0.5728	57.28	23.28
	0.75	0.0751	0.5903	59.03	22.33
	0.80	0.0739	0.5968	59.68	21.97
	0.85	0.0711	0.6121	61.21	21.14
$333 \pm 1K$	0.90	0.0684	0.6268	62.68	20.34
	0.95	0.0679	0.6296	62.96	20.19
	1.00	0.0653	0.6438	64.38	19.41
	1.05	0.0645	0.6481	64.81	19.18
	1.10	0.0639	0.6514	65.14	19.00
	1.15	0.0632	0.6552	65.52	18.79
	1.20	0.0625	0.6590	65.90	18.58
	1.25	0.0619	0.6623	66.23	18.40
	1.30	0.0608	0.6683	66.83	18.08
	1.35	0.0599	0.6732	67.32	17.81
	1.40	0.0591	0.6776	67.76	17.57
	1.45	0.0586	0.6803	68.03	17.42
	1.50	0.0572	0.6879	68.79	17.01
	1.55	0.0554	0.6978	69.78	16.47
	1.60	0.0531	0.7103	71.03	15.79
	1.65	0.0508	0.7229	72.29	15.10
	1.70	0.0483	0.7365	73.65	14.36
	1.75	0.0451	0.7540	75.40	13.41

1.80	0.0426	0.7676	76.76	12.67
1.85	0.0391	0.7867	78.67	11.62
1.90	0.0378	0.7938	79.38	11.24
1.95	0.0353	0.8074	80.74	10.50
2.00	0.0334	0.8178	81.78	9.93

3.2   Adsorption isotherm

It is commonly believed that corrosion prevention depends on the mode of inhibitor adsorption and the surface conditions. The inhibitor covered over some part of the metal is only protected whereas the bare area is likely to undergo corrosion, and the inhibitor molecules' blocking effect is merely responsible for the inhibition process. The mechanism of adsorption can be understood well by investigating different types of isotherms like Freundlich, Langmuir, Frumkin, Temkin, etc. Still, the Langmuir isotherm was found to be the suitable one for this studied system because the R^2 value is almost unity. The following equations (3 & 4) are the characteristics of the Langmuir adsorption isotherm equation

$$\frac{C}{\theta} = \frac{1}{K_{ads}} + C \qquad (3)$$

$$K_{ads} = \frac{1}{55.5} \exp\left(\frac{-\Delta G_{ads}^0}{RT}\right) \qquad (4)$$

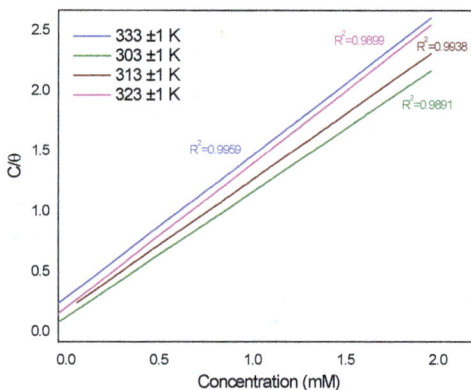

Figure 2. Langmuir adsorption plot

Langmuir plot (C/θ vs. C) for various temperatures studied is shown in Fig. 2. The adsorption isotherm parameters are shown in Table 2 which reveals that the inhibitor molecules adsorbed on the metal surface in the mixed mode (physisorption–chemisorption mechanism) because the values

Advances in Corrosion Science and Surface Engineering   Materials Research Forum LLC
Materials Research Foundations 188 (2026) 208-227   https://doi.org/10.21741/9781644903919-11

of Gibbs standard free energy lie between -20 and -40 kJ/mol. Values around this range are usually caused by mixed mode [27,28] which means both processes occur at the same time. Negative values of Gibbs standard free energy mean the spontaneous reaction of the inhibitor adsorption on the surface of the mild steel.

Table 2. Langmuir adsorption isothermal parameters

Temperature	R^2	Slope	K_{ads} mol lt^{-1}	$-\Delta G^{\circ}_{ads}$ kJ mol^{-1}
303 ±1 K	0.9891	1.031	174.237	23.1218
313 ±1 K	0.9938	1.094	209.008	24.3585
323 ±1 K	0.9899	1.196	232.609	25.4241
333 ±1 K	0.9959	1.184	313.452	27.0372

3.3   Quantum chemical studies

A protonated version of the B3LYP/6-311G (d,p) basis was used to optimize the inhibitor structure. Because of interactions between the acid and the inhibitor in an acidic environment, there may be a coexistence of protonated and non-protonated inhibitor forms. Identifying the preferred species that interact with the metal surface and studying how protonation modifies the structures and molecular characteristics of these forms presents a fascinating area of study. An inhibitor HCN contains heteroatoms, which could be sites for protonation [7]. The calculation results for the various protonation sites reveal that the ethylacetate group's oxygen atoms are the most preferred for protonation in pyran derivatives, while the cyano (-C≡N) group is the most preferred for protonation in cyanopyran derivatives. Since there are no other protons at that position, it is less sterically inhibited. Also, all the chosen compounds share that position, thus it's easy to evaluate protonation's impact on different structures [21].

The frontier molecular orbital theory states that there are two key indicators for determining a chemical species' reactivity i.e., the energy of the highest occupied molecular orbital (E_{HOMO}) and the energy of the lowest unoccupied molecular orbital (E_{LUMO}). So, a higher E_{HOMO} value will improve the transport process via the adsorbed layer, which in turn will make the inhibitor adsorb more efficiently. The converse is also true: a low E_{LUMO} value indicates that a molecule can accept an electron. Another metric that has been discovered to have excellent correlations with experimental inhibition efficiency is the energy gap (ΔE), which is the difference between E_{LUMO} and E_{HOMO}. The hardness or softness of the molecules affects the energy gap. Due to the low energy required to remove electrons from the final occupied orbital, soft molecules are more reactive than strong ones. Table 3 shows the optimized geometry, HOMO, and LUMO. The protonation state quantum chemical characteristics of the HCN inhibitor are displayed in Table 4.

Table 3. Optimized geometry, HOMO, and LUMO of the inhibitor HCN

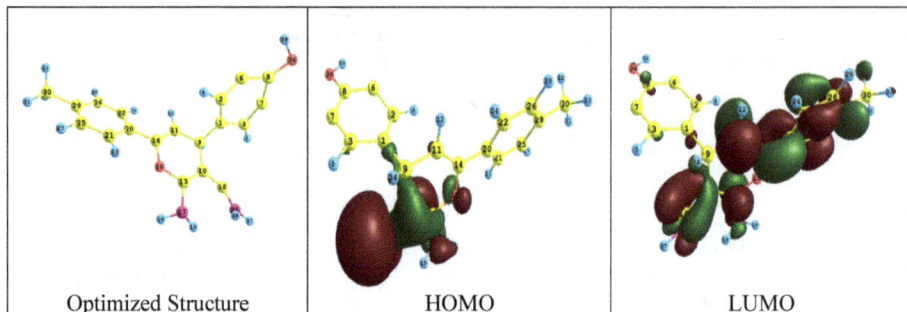

| Optimized Structure | HOMO | LUMO |

Table 4. Quantum chemical parameters for the inhibitor HCN

Quantum chemical parameters	HCN
Total energy (amu)	-993.83
Dipole moment (μ debye)	4.56
E_{HOMO} (eV)	-5.91
E_{LUMO} (eV)	-1.32
Energy gap (ΔE, eV)	4.59
Ionization potential (I, eV)	5.91
Electron affinity (A, eV)	1.32
Hardness (η)	2.30
Softness (σ)	0.44
Electronegativity (χ)	3.62
Electrophilicity(ω)	2.85
Fraction of electrons transferred (ΔN)	0.74

3.4   Machine learning techniques

3.4.1   K-Nearest Neighbors (KNN)

The KNN algorithm is a classification technique for the collection of data based on learning previously categorized data, making it one of the supervised methods in data mining. Included within the supervised learning, where the outcomes of new query instances are categorized to use the majority of the closeness of the present categories in KNN. Many researchers have found that KNN is one of the most widely used pattern recognition algorithms, which achieves exceptionally excellent performance on many data sets [29–31] and provides several key benefits, including

Advances in Corrosion Science and Surface Engineering   Materials Research Forum LLC
Materials Research Foundations 188 (2026) 208-227   https://doi.org/10.21741/9781644903919-11

simplicity and noise resilience [32]. The data set has been labeled, meaning the test sample's chosen neighbors have been accurately identified. This KNN algorithm is used to determine the accuracy of possible inhibition efficiencies that are predictable from the data set. These are experimental values that can be built upon calculating the K member of the nearest values. K is calculated by pre-processing the scale data according to the results of the following data set, taking the K value from the built model. Hence, the algorithm implemented is the five-nearest neighbor algorithm. In this algorithm, the model is trained so that the nearest value distances are calculated for each point to be predicted. The dataset is split in the ratio of 80%:20% as training and test data sets respectively. These training and test datasets are pre-processed and scaled using read and scale functions in R. Next, the KNN function is loaded from the class package.

In the KNN function, the training and test datasets are given including the attribute selection, and the class is set to the training set, the k value is 5, and prob=TRUE where TRUE indicates the inhibition efficiencies of greater than 90%. This 90% is set as optimal inhibition efficiency for successfully preventing corrosion. With this, the KNN algorithm is fitted into the predictions of the training set. Upon the model built on the train set, the predictions are made. A similar plot is simulated and shown in Figure 3b. The accuracy of the model is 86.92% (given(confosion matrix //2)// j-pred

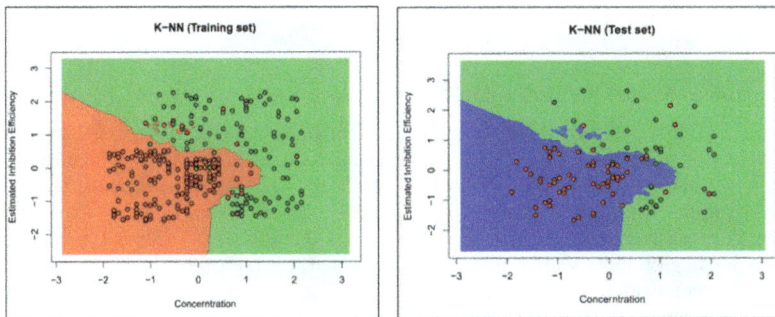

Figure 3: Visualized contour plot of the (a) KNN training set and (b) KNN test set (concentration vs inhibition efficiency)

3.4.2 Random Forests (RF)

Random Forests have been used for the analysis. The dataset taken for this experiment is the same as the data set of KNN. Here the dependent values and independent values are segregated and put into trees then finally these are merged into RF[33]. They have a total of 6 attributes and 160 records; out of those, 128 records are known as Trained sets, i.e., they are used for training the data sets, while 32 are known as test sets. In the train sets, there are 21 trees, and each tree consists of 6 branches. While the test sets have 16 trees in total, as we have mentioned, the test sets are binary trees, each consisting of 2 branches. There are 4 cases in total, meaning there are four different environments. Every environment has the same concentration values. All these four environments consist of multiple concentrations. The leading independent component here is the concentration (mM). The primary dependent component is inhibition efficiency (%). The inhibition efficiency is the percentage value of the surface area i.e. (x100 to surface area). The corrosion rate is given by (g cm^{-2} h^{-1}) [34]. In the Fig. 4, the green line represents the average and error. The black line represents R^2 and the red line represents RMS error. Based on the observation, it is noted that initially, the graph fluctuates high during the prebuild whereas eventually, the graph stabilizes indicating a successful build.

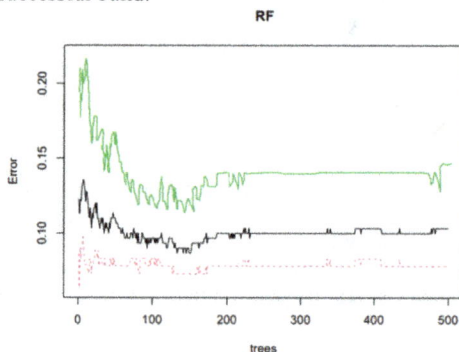

Figure 4: RF plot

3.4.3   Artificial Neural Networks (ANN)

The data set is taken as the input for experimenting. The implementations are based on ANN. The H_2O interface is established, and H_2O is an ANN package[35]. First, the ANN is fit into the training set by taking the number of threads, the model is built using H_2O, and the deep learning package is also used in the R language. The activation denotes the inhibition efficiency (%), the epochs (number of approaches) are taken as 100, the iteration is -2, and the final values are represented as vectors depending on these values the conformation matrix is built [36,37]. Fig. 5 illustrates the architectural framework of an ANN and describes the structure and flow of information within the network. This framework typically consists of three main components: input layer, hidden layers, and output layer.

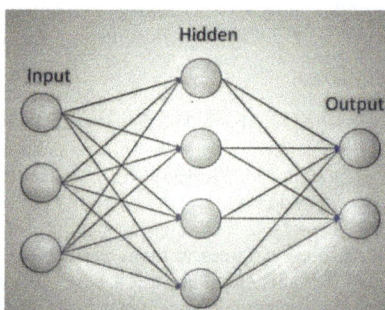

Figure 5. Architectural framework of the ANN process

3.4.4 Support vector machine

SVM creates the best line or decision boundary. The decision boundary or best line is called a hyperplane. The hyperplane is the best line that segregates classes in n-dimensional space. The hyperplane is created by choosing the extreme points or data points by SVM and these extreme points are called support vectors[38]. Fig. 6 explains the contours that visualize the SVM's decision boundaries, highlighting how well the model separates the classes in the given feature space. Fig. 7(a) illustrates the standard SVM, where the decision boundary is linear. Fig. 7(b) demonstrates a Kernel SVM, which uses a non-linear decision boundary to separate classes that are not linearly separable in the original feature space.

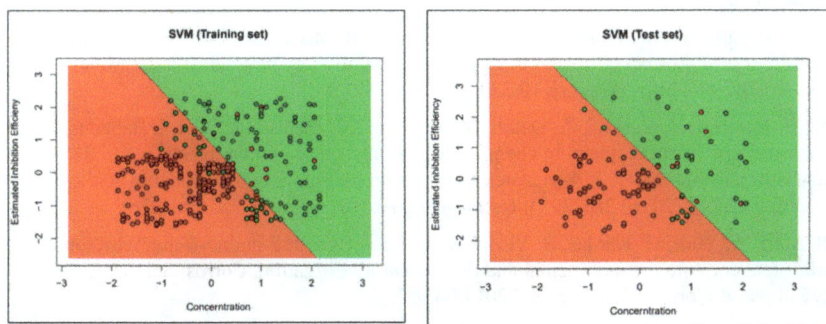

Figure 6. Visualized contour plot of the (a) SVM training set and (b) SVM test set [concentration vs. inhibition efficiency]

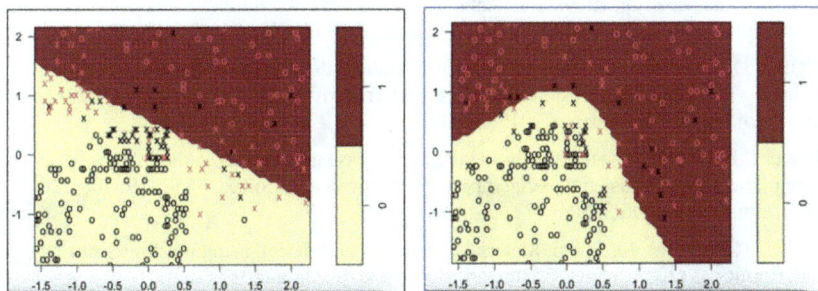

Figure 7.(a) Support vector machine and (b) kernel support vector machine

Conclusion

Computational methods such as DFT and machine learning were effectively employed to predict the corrosion inhibition of mild steel in acidic environments with a cyanopyran derivative. The investigations demonstrated that the inhibitor functioned as a successful component for the protection of mild steel. A comprehensive analysis of all elements associated with the variables is

Advances in Corrosion Science and Surface Engineering   Materials Research Forum LLC
Materials Research Foundations 188 (2026) 208-227   https://doi.org/10.21741/9781644903919-11

impractical due to the multitude of environmental and operational factors that affect the corrosion rate of mild steel in the presence of corrosion inhibitors. Consequently, employing a machine learning model may diminish the necessity for such investigations. Diverse machine learning models, including artificial neural networks, K-nearest neighbors, support vector machines, and random forests, are employed to assess the influence of corrosion inhibitors on the corrosion rates of mild steel based on exposure time, temperature, and concentration. In these experiments, RF was the best algorithm compared to other models based on the mean squared error ranging from 0.005 to 0.093.

References

[1] A.A. Al-Amiery, A.A.H. Kadhum, A. Kadihum, A.B. Mohamad, C.K. How, S. Junaedi, Inhibition of mild steel corrosion in sulfuric acid solution by new schiff base, Mater. 7 (2), (2014a) 787-804 https://doi.org/10.3390/ma7020787

[2] M. El-Belghiti, Y. Karzazi, A. Dafali, B. Hammouti, F. Bentiss, I. Obot, I. Bahadur, E.E. Ebenso. Experimental, quantum chemical and Monte Carlo simulation studies of 3, 5-disubstituted-4-amino-1, 2, 4-triazoles as corrosion inhibitors on mild steel in acidic medium, J. Mol. Liq. 218 (2016) 281-293 https://doi.org/10.1016/j.molliq.2016.01.076

[3] B. Mert, A. Yüce, G. Kardas, B. Yazici, Inhibition effect of 2-amino-4-methylpyridine on mild steel corrosion: experimental and theoretical investigation, Corros. Sci. 85, (2014) 287-295 https://doi.org/10.1016/j.corsci.2014.04.032

[4] H. Huang, F. Bu, Correlations between the inhibition performances and the inhibitor structures of some azoles on the galvanic corrosion of copper coupled with silver in artificial seawater, Corros. Sci. 165 (2020) 108413 https://doi.org/10.1016/j.corsci.2019.108413

[5] G. Gece, The use of quantum chemical methods in corrosion inhibitor studies, Corros. Sci. 50 (2008) 2981-2992 https://doi.org/10.1016/j.corsci.2008.08.043

[6] I. Obot, D. Macdonald, Z. Gasem, Density functional theory (DFT) as a powerful tool for designing new organic corrosion inhibitors. Part 1: an overview, Corros. Sci. 99 (2015) 1-30 https://doi.org/10.1016/j.corsci.2015.01.037

[7] H. Ke, C.D. Taylor, Density functional theory: an essential partner in the integrated computational materials engineering approach to corrosion, Corros. 75 (2019) 708-726 https://doi.org/10.5006/3050

[8] Y. Liu, T. Zhao, W. Ju, S. Shi, Materials discovery and design using machine learning, J. Materiomics. 3 (2017) 159-177 https://doi.org/10.1016/j.jmat.2017.08.002

[9] T.W. Quadri, L.O. Olasunkanmi, O.E. Fayemi, E.D. Akpan, C. Verma, E.S.M. Sherif, K.F. Khaled, E.E. Ebenso, Quantitative structure activity relationship and artificial neural network as vital tools in predicting coordination capabilities of organic compounds with metal surface: A review, Coord. Chem. Rev. 446 (2021) 214101 https://doi.org/10.1016/j.ccr.2021.214101

[10] A.R. Shahmoradi, M. Ranjbarghanei, A.A. Javidparvar, L. Guo, E. Berdimurodov, B. Ramezanzadeh, Theoretical and surface/electrochemical investigations of walnut fruit green husk extract as effective inhibitor for mild-steel corrosion in 1M HCl electrolyte. J. Mol. Liq. 338 (2021) 116550 https://doi.org/10.1016/j.molliq.2021.116550

[11] S.A.X. Stango, U. Vijayalakshmi, Studies on corrosion inhibitory effect and adsorption behavior of waste materials on mild steel in acidic medium, J. Asian Ceram. Soc. 6 (2018) 20-29 https://doi.org/10.1080/21870764.2018.1439608

[12] A. Samide, C. Stoean, R. Stoean, Surface study of inhibitor films formed by polyvinyl alcohol and silver nanoparticles on stainless steel in hydrochloric acid solution using convolutional neural networks, Appl. Surf. Sci. 475 (2019) 1-5 https://doi.org/10.1016/j.apsusc.2018.12.255

[13] V.C. Anadebe, O.D. Onukwuli, F.E. Abeng, N.A. Okafor, J.O. Ezeugo, C.C. Okoye, Electrochemical-kinetics, MD-simulation and multi-input single-output (MISO) modeling using adaptive neuro-fuzzy inference system (ANFIS) prediction for dexamethasone drug as eco-friendly corrosion inhibitor for mild steel in 2 M HCl electrolyte. J. Taiwan Inst. Chem. Eng. 115 (2020) 251-265 https://doi.org/10.1016/j.jtice.2020.10.004

[14] Mohammadreza Aghaaminiha, Ramin Mehrani, Martin Colahan, Bruce Brown, Marc Singer, Srdjan Nesic, Silvia M. Vargas, Sumit Sharma, Machine learning modeling of time-dependent corrosion rates of carbon steel in presence of corrosion inhibitors, Corros. Sci. 193 (2021) 109904. https://doi.org/10.1016/j.corsci.2021.109904

[15] Y. Diao, L. Yan, K. Gao . Improvement of the machine learning-based corrosion rate prediction model through the optimization of input features, Mater. Des. 198 (2021) 109326. https://doi.org/10.1016/j.matdes.2020.109326

[16] C.D. Taylor, B.M. Tossey, High temperature oxidation of corrosion resistant alloys from machine learning. npj Mater. Degrad. 5(1) (2021) 1-10 https://doi.org/10.1038/s41529-021-00184-3

[17] L.B. Coelho, D. Zhang, Y. Van Ingelgem, D. Steckelmacher, A. Nowé, H. Terryn, Reviewing machine learning of corrosion prediction in a data-oriented perspective. npj Mater. Degrad. 6(1) (2022) 1-16 https://doi.org/10.1038/s41529-022-00218-4

[18] A.H. Alamri, N. Alhazmi, Development of data driven machine learning models for the prediction and design of pyrimidine corrosion inhibitors, J. Saudi Chem. Soc. 26 (2022) 101536 https://doi.org/10.1016/j.jscs.2022.101536

[19] V.C. Anadebe, P.C. Nnaji, O.D. Onukwuli, N.A. Okafor, F.E. Abeng, V.I. Chukwuike, C.C. Okoye, I.I. Udoh, M.A. Chidiebere, L. Guo, R.C. Barik, Multidimensional insight into the corrosion inhibition of salbutamol drug molecule on mild steel in oilfield acidizing fluid: Experimental and computer aided modeling approach, J. Mol. Liq. 349 (2022) 118482 https://doi.org/10.1016/j.molliq.2022.118482

[20] A.K. Anees, S.M. Mustafa, B.M. Hameed, Mathematical regression and artificial neural network for prediction of corrosion inhibition process of steel in acidic media, J. Bio Tribo. Corros. 6 (2020) 92 https://doi.org/10.1007/s40735-020-00390-7

[21] J. Saranya, N. Anusuya, F. Benhiba, I. Warad, A. Zarrouk, A Cyanopyran Derivative for Preventing Corrosion of Pipeline Material Used in The Oil and Gas Industry, Anal. Bioanal. Electrochem. 14 (2022) 818-836

[22] M. Irani, R. Chalaturnyk, M. Hajiloo, Application of data mining techniques in building predictive models for oil and gas problems: A case study on casing corrosion prediction. Int. J. Oil Gas Coal Technol. 8 (2014) 369-398 https://doi.org/10.1504/IJOGCT.2014.066304

[23] Y. Liu, T. Zhao, W. Ju, S. Shi, Materials discovery and design using machine learning. J. Mater. 3 (2017) 159-177 https://doi.org/10.1016/j.jmat.2017.08.002

[24] K.T. Butler, D.W. Davies, H. Cartwright, O. Isayev, A. Walsh, Machine learning for molecular and materials science. Nat. Cell Biol. 559 (2018) 547-555 https://doi.org/10.1038/s41586-018-0337-2

[25] Y. Roh, G. Heo, S.E. Whang, A Survey on Data Collection for Machine Learning. IEEE Trans. Knowl. Data Eng. 4347 (2019) 1-20

[26] L.A. Nnanna, B.N. Onwolagba, I.M. Mejeha, K.B. Okeoma, Inhibition effects of some plant extracts on the acid corrosion of aluminium alloy. AJPAC. 4(1) (2010) 11-16

[27] S.A. Haddadi, E. Alibakhshi, G. Bahlakeh, B. Ramezanzadeh, M. Mahdavian, A detailed atomic level computational and electrochemical exploration of the Juglans regia green fruit shell extract as a sustainable and highly efficient green corrosion inhibitor for mild steel in 3.5 wt% NaCl solution. J. Mol. Liq. 284 (2019) 682-699 https://doi.org/10.1016/j.molliq.2019.04.045

[28] E. Ituen, O. Akaranta, A. James, 2017. Evaluation of performance of corrosion inhibitors using adsorption isotherm models: an overview. Chem. Sci. Int. J. 18 (2017) 1-34 https://doi.org/10.9734/CSJI/2017/28976

[29] S. Sanjaya, M.L. Pura, S.K. Gusti, F. Yanto, F. Syafria, K-Nearest Neighbor for Classification of Tomato Maturity Level Based on Hue, Saturation, and Value Colors. IJAIDM. 2 (2019)101-106 https://doi.org/10.24014/ijaidm.v2i2.7975

[30] X. Yan, W. Li, W. Chen, W. Luo, C. Zhang, Q. Wu, H. Liu, Weighted K-Nearest Neighbor Classification Algorithm Based on Genetic Algorithm. TELKOMNIKA Indones. J. Electr. Eng. Comput. Sci. 11 (2013) 6173-6178 https://doi.org/10.11591/telkomnika.v11i10.2534

[31] Y.F. Safri, R. Arifudin, M.A. Muslim, K-Nearest Neighbor and Naive bayes classifier algorithm in determining the classification of healthy card Indonesia giving to the poor. Sci. J. Inform. 5 (2018) 9-18 https://doi.org/10.15294/sji.v5i1.12057

[32] H. Parvin, H. Alizadeh, B. Minati, A Modification on K-Nearest Neighbor Classifier, GJCST. 10 (2010) 37-41

[33] S. Gao, Q. Wu, Z. Zhang, G. Jiang, Simulating active layer temperature based on weather factors on the Qinghai-Tibetan Plateau using ANN and wavelet-ANN models. Cold Reg. Sci. Technol. 177 (2020) 3-10 https://doi.org/10.1016/j.coldregions.2020.103118

[34] M. Ravansalar, T. Rajaee, O. Kisi, Wavelet-linear genetic programming: A new approach for modeling monthly streamflow. J. Hydrol. 549 (2017) 461-475 https://doi.org/10.1016/j.jhydrol.2017.04.018

[35] K.E. Okpalaeke, T.H. Ibrahim, L.M. Latinwo, E. Betiku, Mathematical modelling and optimization studies by an artificial neural network, Genetic algorithm and response surface

Advances in Corrosion Science and Surface Engineering   Materials Research Forum LLC
Materials Research Foundations 188 (2026) 208-227   https://doi.org/10.21741/9781644903919-11

methodology: a case of ferric sulfate-catalyzed esterification of neem (Azadirachta indica) seed oil, Front. Energy Res. 8 (2020) 614621 https://doi.org/10.3389/fenrg.2020.614621

[36] G. Dhande, Z. Shaikh, Analysis of Epochs in Environment based Neural Networks Speech Recognition System, in: 2019 3rd International Conference on Trends in Electronics and Informatics (ICOEI). (2019) 605-608 https://doi.org/10.1109/ICOEI.2019.8862728

[37] S. Afaq, S. Rao, Significance of Epochs on Training a Neural Network. Int. J. Sci. Technol. Res. 19 (2020) 485-488.

[38] N. Guttenberg, Learning to generate classifiers. arXiv. (2018) https://doi.org/10.1016/10.48550/arXiv.1803.11373

Advances in Corrosion Science and Surface Engineering   Materials Research Forum LLC
Materials Research Foundations 188 (2026) 228-248   https://doi.org/10.21741/9781644903919-12

Chapter 12

Corrosion Prevention in Aerospace Industries

Pakanati Siva Prasad[1,2]*, Juan David Matallana Guerrero[2], Anjali Kumari[2]

[1]Department of Metallurgical Engineering and Materials Science, Indian Institute of Technology Bombay, Mumbai, India-400076

[2]Department of Metallurgical and Materials Engineering, Indian Institute of Technology Kharagpur, Kharagpur, India-721302

psiva432@gmail.com

Abstract

Corrosion poses a significant threat to aerospace components, compromising their integrity and safety. This chapter explores corrosion monitoring techniques and prevention strategies in aerospace industries, focusing on aluminum (Al), Al-alloys, manganese (Mn) alloys, and titanium (Ti)-based alloys. Various methods, including coatings, inhibitors, and anodization, are discussed for preventing corrosion-induced damage. Additionally, innovative alloys, advanced protection technologies, and the shift towards composite materials in aerospace component fabrication are highlighted. The integration of these techniques and materials ensures the longevity, security, and dependability of aerospace components, safeguarding against financial losses, environmental disasters, and human life losses.

Keywords

Corrosion Prevention, Aerospace, Corrosion Monitoring, Intergranular Corrosion, Stress Corrosion Cracking, Surface Treatments

Contents

1.   Introduction

Corrosion poses a significant challenge in the aerospace industry, impacting the structural integrity and longevity of aircraft components [1]. Understanding the various forms of corrosion is crucial for developing effective mitigation strategies and ensuring the safety and reliability of aircraft operations. This chapter provides a comprehensive overview of corrosion in the aerospace industry, focusing on its types, causes, and implications.

1.1   Background and Understanding Corrosion in Aerospace Industries

Corrosion, defined as the electrochemical deterioration of a metal due to its chemical reaction with the surrounding environment, poses a significant threat to the structural integrity and safety of aircraft components, particularly those made of Al and steel alloys [2]. The gradual process of corrosion can lead to a reduction in material thickness, and mechanical strength, and can accelerate fatigue crack development and stress corrosion cracking, ultimately compromising the overall structural integrity of the aircraft [1]. If left unchecked, corrosion can result in catastrophic structural failure, highlighting the critical importance of corrosion prevention and control in the aerospace industry [2], [3].

Aircraft structures are exposed to a variety of in-service conditions that promote corrosion. During take-off and landing, aircraft are exposed to dust, gravel, stones, and de-icing salts from runways, which can damage protective coatings and expose bare metals to corrosive elements [2]. Additionally, operational hazards such as oils, fluids, fuel, scratches, abrasions from maintenance actions, battery acid, exhaust gases, and galley spillages can further accelerate corrosion [2]. Maritime aircraft are particularly vulnerable to salt spray, which is known to be highly corrosive and can significantly affect the structural integrity of aircraft components [4].

The Al-alloys, the primary materials used in aircraft construction, exhibit excellent corrosion resistance due to the formation of a barrier oxide film on their surfaces [5], [6]. However, the wide range of environments under which aircraft operate, including urban industrial areas, high humidity, hot deserts, and marine environments, can accelerate corrosion and compromise the integrity of aircraft components [7]. Despite the corrosion resistance of Al-alloys, their susceptibility to corrosion in harsh environments necessitates the development of effective corrosion prevention strategies in the aerospace industry.

Aerospace alloys, including Al-based alloys, magnesium (Mg), nickel, cobalt, and Ti-based alloys, are specifically designed to meet the high strength and corrosion resistance requirements of aircraft components [8]. These alloys are used in various aircraft structures and components, including fuselages, engine blocks, and airframes, where light weight, high strength, and corrosion resistance are essential [9]. As aircraft age and exceed 20 years of service life, corrosion prevention and control become paramount, leading to increased attention to corrosion-related issues in aircraft design, manufacturing, and maintenance [7].

Overall, corrosion poses a significant threat to the safety, reliability, and economic viability of aircraft operations in the aerospace industry. Understanding the mechanisms and factors contributing to corrosion, as well as implementing effective corrosion prevention and control measures, are essential for ensuring the structural integrity and longevity of aircraft components.

2.   Types of Corrosion in Aerospace Industries

In the aerospace industry, various types of corrosion can affect aircraft components, each with its own characteristics and mechanisms. Understanding these types of corrosion is crucial for developing effective mitigation strategies. The following are some common types of corrosion observed in the aerospace industries:

2.1   Uniform Corrosion

Uniform corrosion, also known as surface corrosion, occurs when the entire exposed surface of a metal is corroded at a relatively consistent rate [10]. This type of corrosion is common on aircraft surfaces where the metal is exposed to the atmosphere without protection. It is characterized by a roughened metal surface, etching, and the formation of numerous corrosion spots, often accompanied by powder corrosion products [10].

Poor paintwork or inadequate pre-painting preparations can lead to uniform surface attack corrosion on aircraft surfaces. Inferior paints, salty and changing environmental conditions, and exposure to oxygen in the air can accelerate this type of corrosion, particularly on wing skins or fuselages where the topcoat has delaminated [11].

While uniform corrosion may not penetrate as deeply or rapidly as other types of corrosion, its cumulative effect can significantly weaken the metal over time, compromising the structural integrity of aircraft components [12]. The metal may exhibit a roughened texture, pitting, or etching in severe cases, with a gray-white chalky deposit being a notable sign of surface corrosion [10].

Advances in Corrosion Science and Surface Engineering   Materials Research Forum LLC
Materials Research Foundations 188 (2026) 228-248   https://doi.org/10.21741/9781644903919-12

Prevention and treatment of uniform corrosion involves regular inspections to detect early signs, application of corrosion-resistant coatings or paints, and storing aircraft in controlled environments to reduce exposure to corrosive agents [13]. These measures are essential for maintaining the structural integrity and performance of aircraft components.

2.2   Intergranular Corrosion

Intergranular corrosion, also known as intercrystalline or interdendritic corrosion, occurs along grain boundaries or interdendritic paths within a material [14]. This type of corrosion is particularly problematic in aircraft structures, which often utilize high-strength Al-alloys. The 7xxx series Al-alloys, containing zinc and copper, are especially susceptible to intergranular attack due to the formation of galvanic couples along grain boundaries (ASTM, 1986). This corrosion is most severe in critical aircraft components like stringers and wing spars. Improper heat treatment during alloy production and lack of homogeneity in alloy structures can contribute to this type of corrosion [11].

Detection of intergranular corrosion can be challenging as it may not be visible on the surface of the metal. Microscopic examination is often required to identify corrosion along grain boundaries. In advanced stages, the metal may exhibit flaking or separation along these boundaries, leading to structural weakness and potential component failure [13].

Prevention and treatment of intergranular corrosion involves proper heat treatment of metals to reduce susceptibility, careful selection of alloys known to resist this type of corrosion, and the use of advanced inspection techniques to detect subsurface defects early [13].

2.3   Galvanic Corrosion

Galvanic corrosion, also known as bimetallic corrosion, occurs when two dissimilar metals or alloys come into direct contact with each other in the presence of an electrolyte, such as moisture. The metal with the more negative corrosion potential (anode) corrodes, while the metal with the more positive potential (cathode) remains protected [10]. The Fig. 1 shows the galvanic corrosion between the Mg surface and steel connections. This type of corrosion is particularly concerning in aircraft structures, where various metals and alloys are used in close proximity.

One common example of galvanic corrosion in aircraft is the use of alloy steel fasteners to connect important structures. Although these fasteners are coated with cadmium to protect against corrosion, the cadmium coating can be gradually destroyed by various corrosive media, leading to sacrificial corrosion of the fastener base metal and connected Al alloy components [10].

Preventing galvanic corrosion involves isolating the anode and cathode metals to avoid direct contact and minimizing the accumulation of corrosive media. This can be achieved through the use of insulating barriers, coatings, or sealants between different metals, as well as regular inspections and maintenance to detect and mitigate potential corrosion issues [13], [14], [15].

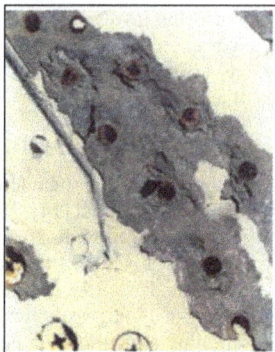

Figure 1. Galvanic corrosion between the magnesium surface and the steel connections [16].

2.4 Crevice Corrosion

Crevice corrosion is a type of localized corrosion that occurs in or around narrow gaps or crevices. It is particularly common in Al structures, where it can affect areas such as joints, seams, fastener heads, under gaskets and seals. This form of corrosion is caused by the limited diffusion of solution in the crevice, which creates a concentration cell and leads to local corrosion [10].

Moisture or other pollutants that are trapped in these crevices can exacerbate the corrosion process. For example, lapped skin joints or rivets on an oil-stained belly are prime spots for crevice corrosion to occur. Additionally, poor drainage holes in long-term parked aircraft can create conditions leading to crevice corrosion in low-lying areas like the cabin door area and fuselage abdomen [10]. The appearance of crevice corrosion on an aircraft is characterized by the trapping of stagnant liquid, which can lead to changes in the local chemistry. As oxygen is consumed in the crevice, a differential aeration cell is created, causing the metal inside the crevice to corrode [13].

To prevent crevice corrosion, it is essential to design aircraft components with minimal tight spaces where moisture can be trapped. Using sealants or coatings to protect vulnerable areas, conducting regular inspections for early detection and intervention, and ensuring proper drainage and ventilation to prevent moisture accumulation are crucial steps in mitigating the risk of crevice corrosion [13], [15].

2.5 Pitting Corrosion

Pitting corrosion is a localized form of aircraft corrosion that results in the formation of small cavities or pits on the metal surface [7]. This type of corrosion is particularly problematic as it can penetrate deep into the metal, making it more serious than other forms of corrosion, such as filiform corrosion. Pitting corrosion is typically caused by trapped moisture, which makes tight crevices particularly vulnerable to this type of damage. If left untreated, pitting corrosion can lead to unsafe stress-corrosion cracking [14].

One important point to note is that pitting corrosion can affect a wide variety of alloys, but it is most commonly associated with Al and Mg. The appearance of pitting corrosion on an aircraft can vary, with pits ranging from shallow to deep (Fig. 2). These pits are often microscopic in size,

Advances in Corrosion Science and Surface Engineering   Materials Research Forum LLC
Materials Research Foundations 188 (2026) 228-248   https://doi.org/10.21741/9781644903919-12

making them challenging to detect in the early stages. They may also be concealed under a dust-like substance, further complicating detection. While the surface may only display a tiny pit, extensive corrosion could be occurring underneath [13].

Pitting corrosion can have serious implications for the structural integrity of aircraft metals. The pits can serve as initiation sites for cracks, potentially leading to catastrophic failures. Additionally, the depth of these pits can be deceptive, with extensive corrosion weakening the structure beneath a seemingly minor surface blemish. Even if the overall material loss is minimal, the structural integrity can be significantly compromised [13].

To prevent and treat pitting corrosion, several measures can be taken. Regular inspections using advanced techniques like ultrasonic testing can help detect even microscopic pits. Applying corrosion-resistant coatings to vulnerable areas can also help protect against pitting corrosion. Additionally, using alloys that are less susceptible to pitting can help mitigate the risk of this type of corrosion [13], [15].

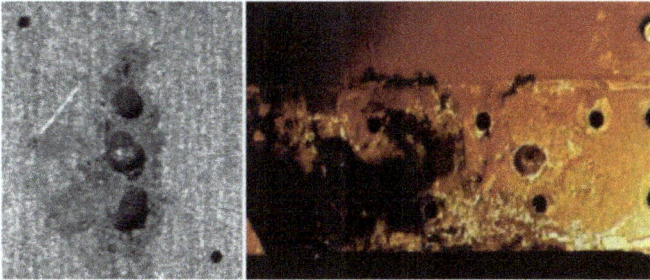

Figure 2. Corrosion pits on the surface [16].

2.6   Stress Corrosion Cracking

Stress corrosion is a type of corrosion that occurs in areas of the aircraft that are under stress, such as the landing gear and engine crankshafts. The stress on these parts can be residual from manufacturing or externally applied through cyclic loading. Furthermore, stress corrosion is often observed in environments with high salt content, such as aircraft flying over seawater or in moist air conditions [11].

The macroscopic and microscopic characteristics of stress corrosion cracking (SCC) have been studied, showing the origin and progression of SCC failure along the flash line of a flap track hinge. Stress corrosion cracking is a significant concern in aircraft structure alloys and can lead to aircraft crashes. In highly stressed parts like landing gear or engine crankshafts, stress corrosion cracking may develop from scratch or surface corrosion. Crankshaft failures are often due to undetected corrosion of this type [11]. The effect of corrosion on the mechanical behaviour of Al-based alloys used in aircraft has been investigated [17]. The results show that corrosion can lead to a reduction in the tensile stress ability of aircraft structures.

Advances in Corrosion Science and Surface Engineering   Materials Research Forum LLC
Materials Research Foundations 188 (2026) 228-248   https://doi.org/10.21741/9781644903919-12

Stress corrosion cracking may first manifest itself in fine ruptures that can be challenging to detect in the early stages. However, these thin ruptures can penetrate deeply into the metal. Certain Al and Ti alloys can be susceptible to stress corrosion cracking when exposed to specific corrosive agents [18], [19]. The corrosion typically starts at localized stress points, such as around fasteners or sharp corners [13], [15].

The macroscopic and microscopic features of stress corrosion cracking (SCC) were attributed by Wanhill and Byrnes [20]. Fig. 3a illustrates the macroscopic genesis and development of SCC failure along a flap track hinge's flash line. Additionally, Figs. 3b and 3c depicted a "woody" surface with elevated grains and a "pancake" microstructure with intergranular failure of the grains, respectively [21]. They observed that stress corrosion cracking (SCC) is an obvious alloy breakdown in aeroplane construction that can lead to an aircraft disaster [20].

To prevent stress corrosion cracking, maintenance often involves non-destructive testing methods to identify potential sites. Proper material selection, protective coatings, and stress-relief treatments are crucial for reducing the risk of stress corrosion cracking. Avoiding designs that introduce unnecessary tensile stresses can also help mitigate the risk [13], [15].

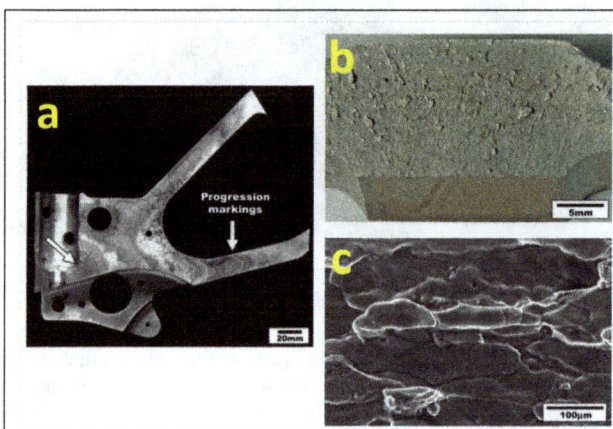

Figure 3. SCC failure of die-forged AA7075-T6 landing gear linkage arm [21].

2.7   Filiform Corrosion

Filiform corrosion is a special form of crevice corrosion that occurs predominantly under protective films. This type of corrosion appears as wire-like grooves that develop unpredictably under the protective layer of metal, often occurring at the heads of fasteners and the edges of skin panels. Filiform corrosion is most prevalent in environments with moderate humidity, typically between 65% and 90% [10].

One of the main causes of filiform corrosion is improper pre-painting processes used for aircraft surfaces made of Al. When the top coat of polyurethane paint does not adhere properly to the

Advances in Corrosion Science and Surface Engineering　　　　　Materials Research Forum LLC
Materials Research Foundations 188 (2026) 228-248　　　　　https://doi.org/10.21741/9781644903919-12

surface due to improper pre-coating, filiform corrosion can occur. This corrosion manifests as minute worm-shaped streaks or in the form of filaments (Fig. 4) beneath the top coat, eventually leading to damage to the coating in the form of flakes and bubbles [11], [22]. Filiform corrosion is unique in that it manifests as thin filaments or threads beneath a metal's coating, such as paint or other protective layers. These filaments spread out, often in a random web-like pattern, causing the coating to lift or bulge. While it jeopardizes the appearance more than the mechanical capabilities of the plane, it is particularly problematic for older aircraft [13].

Preventing filiform corrosion involves ensuring coatings are applied correctly without defects or breaches, reducing exposure to high humidity and acidic conditions, and conducting regular inspections to detect early signs and address any breaches promptly [13].

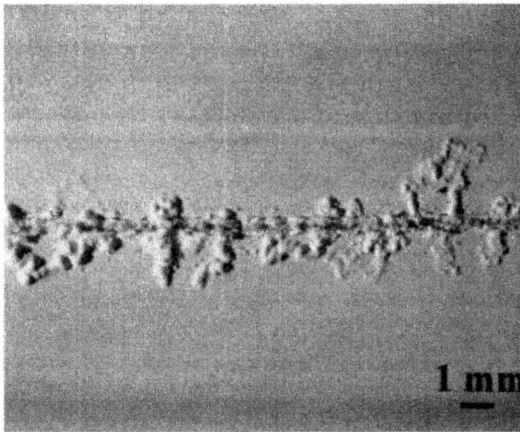

Figure 4. Appearance of filiform corrosion [22].

2.8　Selective Leaching

Selective leaching, also known as de-alloying, is a process in which a less noble metal is selectively removed from alloys under certain conditions. This localized corrosion process typically affects elements such as zinc, Al, nickel, iron, chromium, and cobalt, leaving behind a mechanically weak, porous structure with low ductility [23].

One common example is the preferential dissolution of Al from an Al-alloy, resulting in a process known as de-aluminification. Severe marine environments and acidity can accelerate the depletion of Al from alloys [7]. Preventing selective leaching involves several strategies, including the removal of oxygen from the solution, cathodic protection, careful selection of alloying elements, and reducing grain size in the alloy structure [7]. These measures help to maintain the integrity and performance of the alloy in corrosive environments.

Advances in Corrosion Science and Surface Engineering　　　　Materials Research Forum LLC
Materials Research Foundations 188 (2026) 228-248　　https://doi.org/10.21741/9781644903919-12

3.　　Factors Affecting the Corrosion in Aerospace Industries

3.1.1　Type of Metal

The type of metal used in aircraft parts can significantly impact corrosion. Pure metals are often avoided due to their susceptibility to corrosion. Alloys, which combine different metals, are preferred as they are less prone to rust and corrosion. Active metals like Mg and Al are particularly prone to corrosion and are often alloyed with other metals to reduce this risk [24].

3.1.2　Climatic Conditions

Climatic conditions also play a crucial role in determining corrosion levels. Environments with high moisture content, such as those near seawater or in humid climates, are more corrosive to aircraft parts. Temperature is also a factor, as electrochemical attacks occur more rapidly in hot and moist climates [24].

3.1.3　Faulty Manufacturing Processes

Faulty manufacturing processes can introduce stress into aircraft parts, which can lead to cracking if the threshold for stress corrosion is exceeded. It is essential to maintain cleanliness during manufacturing to prevent such issues [24].

3.1.4　Geographical Location

The geographical location of an aircraft's operation can also affect corrosion levels. Factors such as airborne industrial pollutants, chemicals used on runways, humidity, and wind patterns can all contribute to corrosion. Maps are used to categorize areas based on their corrosion severity, ranging from mild to severe. Identifying these factors can help mitigate corrosion risks in aerospace industries [24].

3.2　　Corrosion Monitoring Techniques

Engineering components, independent of their application in numerous industrial and technological fields, must perform under unique requirements and challenges. These engineered components are often exposed to environmental conditions, such as corrosive environments, high temperatures and thermal cycling, abrasive or erosive environments, mechanical stress and fatigue, and UV radiation. In this case, alloys designed for aerospace applications are often exposed to harsh environmental conditions, in which corrosion occurs, compromising the engineered aerospace component's integrity.

Corrosion monitoring in engineered components constitutes a critical aspect of the maintenance and safety procedures in many industrial sectors for any given engineered component. The prevention of degradation of components by corrosion to retain the reliability and integrity of the components has become a primary tool to avoid failure of engineered components, as their failure constitutes not only financial losses but also possible environmental disasters and human life losses [25]. Regardless of the component application, regular monitoring for prediction and prevention of failures due to corrosion will ensure the longevity of the components due to corrosion-induced damage over time.

Advances in Corrosion Science and Surface Engineering   Materials Research Forum LLC
Materials Research Foundations 188 (2026) 228-248   https://doi.org/10.21741/9781644903919-12

In the aerospace industry, Al, Al-alloys, Mn-alloys, and Ti-based alloys are predominantly employed for aerospace applications. The mechanical and physical properties of aircraft structures are highly compromised by corrosion. Some of the most common corrosion mechanisms are galvanic corrosion between dissimilar metals (GC), pitting corrosion (PC), crevice corrosion (CV), dealloying (i.e., selective dissolution of a metal), intergranular corrosion (IGC), among others [26], [27], [28], [29]. Moreover, aerospace-engineered components are subjected to mechanical stresses, high temperatures, and thermal cycling, leading to corrosion fatigue, stress corrosion cracking (SCC), and hydrogen embrittlement corrosion mechanisms [30].

Some of the monitoring methodologies and techniques can range from visual inspection to non-destructive testing (*NDT*) methods such as ultrasonic NDT, radiography NDT, Eddy current NDT, acoustic emission NDT, weight loss coupons, and electrical resistance (ER) probes, among others [31], [32] and their schematics are shown in Fig.5 Moreover, potentiodynamic polarization techniques (PDP) and electrochemical impedance spectroscopy (EIS) often evaluate the component's corrosion performance. The PDP technique estimates the polarization resistance to assess the corrosion rate at which metal dissolves in a given media, while the EIS techniques give insight into the corrosion mechanisms. Other material techniques such as electrical resistance (ER) probes, gravimetry techniques (TGA), and radioactive tracer methods (RT) allow the acquisition and interpretation of corrosion of real-time data.

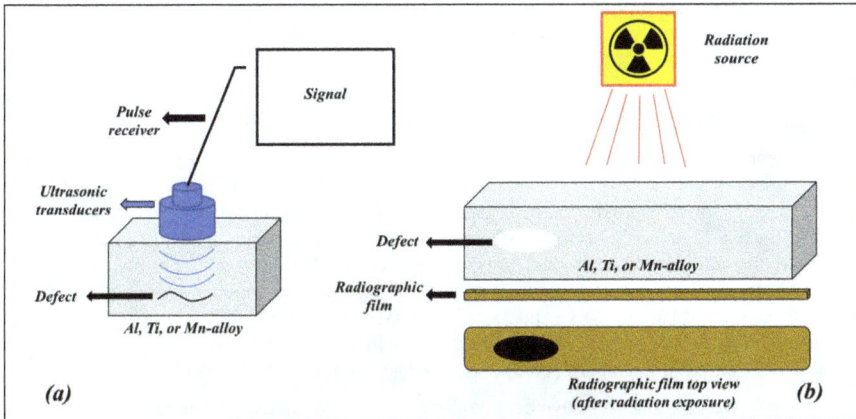

Figure 5. Schematics of some commercially employed NDT methods for corrosion monitoring: (a) ultrasonic NDT and (b) radiography NDT.

Advances in Corrosion Science and Surface Engineering   Materials Research Forum LLC
Materials Research Foundations 188 (2026) 228-248   https://doi.org/10.21741/9781644903919-12

Nowadays, corrosion monitoring has integrated the use of sensors with digital and analytical tools. Smart sensors facilitate real-time monitoring and data analysis of corrosion-related information of a metal system [27]. Traditionally, for aviation structures based on Al alloys, Mg alloys, and Ti-based alloys, corrosion prediction employing analytical methods for prediction of corrosion manifestation and the rate at which it is going to take place is based on traditional corrosion modeling supported by electrochemical mechanisms (PDP, EIS) in aircraft structures [26]. Moreover, corrosion algorithms have been developed for different corrosion growth mechanisms. The author Miyata et al. proposed the general corrosion model, considering relative humidity, temperature, amount of sea particles, and rainfall in real-life applications of aerospace aircraft. *Equ.1* illustrates the general corrosion model proposed by Miyata et al., where A and B are constants, RH is the relative humidity, T is the temperature, SSP is the sea salt particles amount, RF is the amount of rainfall, and f_n is semi-empirical functions [26].

$$c = Af_1(RH)f_2(T)f_3(SSP) + Bf_4(RF)f_2(T)f_5(SSP) \qquad (1)$$

Examining corrosion in engineered components is a complex field that integrates conventional methods with developing technology to protect materials and structures from the harmful consequences of corrosion. The increasing complexity of materials and surroundings necessitates the development of novel monitoring techniques and corrosion-resistant materials. By adopting these technological developments, various industries may effectively guarantee the designed component's longevity, security, and dependability, safeguarding the environment, economies, and human lives.

4.   Corrosion Prevention Strategies used in Aerospace industries

Over the years, aircraft components have been manufactured from lightweight Al and Al-alloys, which possess good strength-to-weight ratios and corrosion resistance. The Al-alloys are characterized by forming passive oxide layers (10-80 nm), increasing their resistance to corrosion. However, these alloys do not perform well in high and low-pH environments [33]. Moreover, it is well known that alloying with a second metal matrix will lead to micro-galvanic coupling, inducing corrosion between the two metals present in the alloy. The most commonly employed Al-alloys in the aerospace industry are the 7xxx and 2xxx series, which constitute the alloying of the Al matrix with zinc and copper, respectively. This chapter will discuss some of the methods used for corrosion prevention in engineered components for aerospace applications [34].

Corrosion prevention strategies have been implemented in the industry to increase the service life of aerospace-engineered components and ensure their structural integrity. The most commonly used methods are coatings, corrosion inhibitors, and anodization.

4.1   Coatings

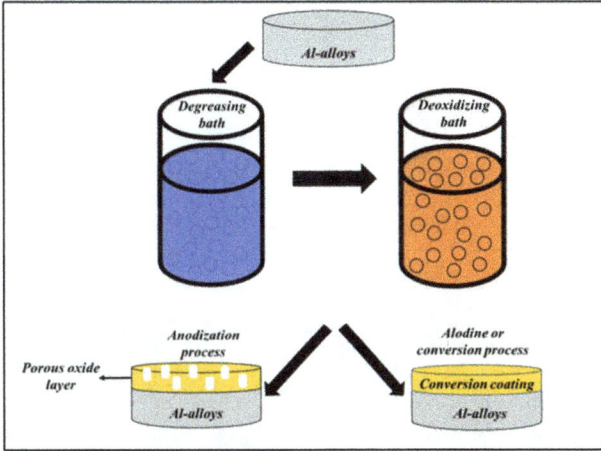

Figure 6. Schematic of surface treatment preparation for Al-based alloys and conventional anodization and chromium conversion coating processes.

Protective coatings to protect the component metal surface against environmental attack are one of the most employed methods in the industry. The aerospace industry requires high-quality coatings to paint or repaint the existing aircraft fleet. Each component relies on many layers of coatings to improve adhesion, protect against environmental elements and corrosion, enhance visual appeal, and fulfil other specific duties. Fig. 6 illustrates the simplified Al-based alloy surface preparation schematic before the anodization process and chromium conversion coatings. For Al-based alloys, a three-layer protective coating is generally employed. Alodine coating (Chromium conversion coating), primer coating, and a polymeric-based top coating [28]. The first layer is called an Alodine coating, often employed to passivate the industry's Al, Zn, Mn, and steel. The Alodine process is a surface treatment in which the Al surface gets chemically treated to produce a very thin passive oxide layer. This method differs from the conventional anodization process, as, in this case, an electrolytic process forms the passive oxide layer [35]. After the chromium conversion coating takes place, surface preparation by applying a primer coating (pigmented organic resin matrix) is performed to ensure the adhesion between the prior surface and the polymeric-based coating. The top layer coating consists of epoxy, polyurethane, polyamide, or polyester base and is employed to guard the component against environmental conditions such as UV light, dirt particles, and radiation [33], [36]. Fig. 7 shows the schematic of the coating protection process for Al-based alloys.

Materials Research Forum LLC
https://doi.org/10.21741/9781644903919-12

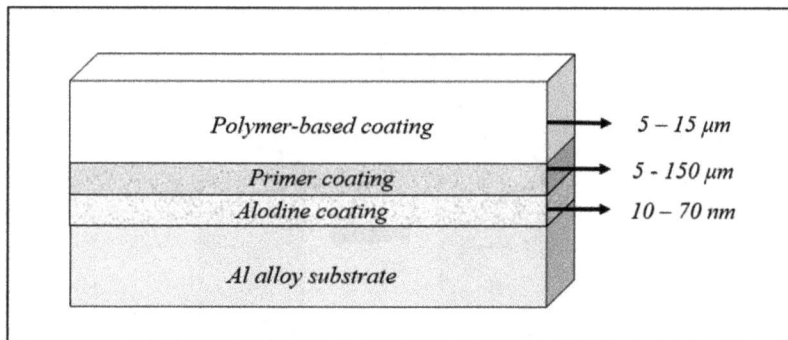

Figure 7. Three-step process for coating protection of Al-based alloys for aerospace applications.

4.2 Corrosion inhibitors

Generally, metallic coatings protect and impart other functionalities to the engineered component, as discussed in this chapter. Moreover, particular applications of metallic coatings can be either single-layer or multi-layered coatings. These coatings employ inhibiting compounds such as hexavalent chromium, which is the most efficient corrosion inhibitor for Al and Mg alloys [33], [36], [37], [38]. Hexavalent chromium has been observed to inhibit both anodic and cathodic corrosion sites and restrict chlorine absorption, thereby helping to stabilize the protective oxide film. However, although it provides substantial advantages for corrosion prevention in Al and Mg alloys, it is crucial to recognize the potential environmental and health risks associated with its use, as it is carcinogenic [28], [36]. Moreover, it is worth mentioning that in the case of Al and Al-based alloys, the use of other types of organic compounds, such as Fluorescein, 1,5-Dihydroxynapthalene, N-4-phenthiazolyl-2-cyanoacetamide,2-Hydroxy-3-(a-hydroxyphenyl)-1-phenylpropan, sodium benzonate, glutaric acid among others have been employed as corrosion inhibitors by many authors in chloride and alkaline media [35], [39], [40], [41].

Furthermore, with increasing legislation and a shift towards sustainable practices, other practices for corrosion inhibitors that do not contain Hexavalent chromium are adopted. Nevertheless, chromate conversion coatings continue to be a popular option for corrosion prevention in many sectors as long as appropriate procedures for handling and disposing are implemented to minimise environmental and health hazards. However, as more viable alternatives are widely embraced, the use of chromate conversion coatings may decline.

4.3 Corrosion-resistant aerospace designs

Aerospace materials are crucial for the construction of spacecraft components, supporting the aircraft's structural integrity during flight operations. Components such as wings, landing gear, tail booms, rotor blades, and the structural airframe are vital, where safety design standards cannot be compromised [42]. Lightweight airframe constructions with improved mechanical properties are essential for the aviation industry, enhancing payload capacity and fuel efficiency [43]. The Al and its derivatives are widely accepted due to their admirable strength, high fracture and fatigue performance, and toughness, meeting the demands for reduced overall structural weight. Similarly, Ti is also highly demanded in the aircraft industry [44].

Various alloying metals like zinc, copper, and magnesium are often added to Al to improve its mechanical strength, but this can also increase the alloy's susceptibility to corrosion [6], [45]. The material property requirements for aviation materials vary depending on the specific component being considered. The materials selected for aircraft components are directly influenced by design requirements such as loading conditions, ease of manufacturing, geometrical limitations, environmental influences, and maintenance requirements [46].

4.3.1 Design criteria for fuselages and wings

Design criteria for fuselages and wings are critical, requiring an exceptional blend of high mechanical strength and low weight, along with the ability to withstand adverse environments of high moisture, extreme temperatures (-30 °C to 370 °C), UV light, and thunder disturbances. These materials must have low densities and high formability to easily shape aerodynamically. The inclusion of novel materials like carbon nanotubes (CNTs) has proven to improve the mechanical characteristics of Al and Ti matrix composites [47].

4.3.2 Design criteria for engine and engine parts

For engine and engine parts, high mechanical strength and low density are essential. Aerospace engines experience massive heat and mass transfer, friction, and extreme temperatures ranging from 300°C to 600°C, leading to hot corrosion and mechanical failure. Materials for engine parts must withstand this extreme environment with their mechanical strength intact. Nickel-based superalloys, for example, can retain their strength even at high working temperatures, making them ideal for turbine and combustion chamber components, where temperatures may range from 1100 to 1250 °C [48].

Corrosion and friction tend to increase with relative movement between two sections. Therefore, limiting these effects is crucial to addressing problems caused by doping and relative motion. Grease (lubricating material) is applied to aircraft connecting surfaces to reduce friction and corrosion from the relative motion of parts [49].

In the case of the space shuttle orbiter, corrosion control is based on a service life of 15-20 years, making material selection and design critical [50]. Dissimilar material contacts are avoided to prevent galvanic corrosion. Contacts are often sealed with vulcanized rubber or epoxy resin to combat water inclusion in the systems [51]. Launchpads at satellite centers also face different working environments with exhaust impingement, extreme temperatures, and acid accumulation, which require standard corrosion protection coatings for launch structures [52].

4.3.3 Innovative Alloys and Materials

The Al-alloys are highly valued for their specific density, corrosion resistance, damage resistance, and ability to withstand high temperatures, making them one of the most promising materials for high-performance aircraft parts [53]. Al-Li, Al-Mg, Al-Zn, and Al-Cu are some of the strongest and most corrosion-resistant materials, with enhanced fatigue strength and fracture toughness due to their chemical composition and effective thermal treatment [54]. Al-based metal Matrix Composites (MMCs) consist of a secondary material, such as ceramics or fiber reinforcements (carbon fibers), with Al alloys (Al-Si, Al–Cu, Al-Si–Mg) as matrix materials. These composites

Advances in Corrosion Science and Surface Engineering   Materials Research Forum LLC
Materials Research Foundations 188 (2026) 228-248   https://doi.org/10.21741/9781644903919-12

offer better strength and toughness, hot corrosion resistance, improved wear resistance, and recyclability [55]. Their usage is rapidly growing in airframe structures, as seen in commercial aircraft like the Airbus A350XWB, A380, and Boeing 787. These composites not only reduce weight but also lower overall maintenance costs. Materials like SiC, Al_2O_3, C, B, B_4C, AlN, SiO_2, and BN exhibit excellent wear resistance and high specific modulus [56].

Magnesium sheets have become popular due to their low density compared to steel and Al, with Mg-based alloys achieving tensile strengths of up to 610 MPa. These alloys also offer high stiffness and damping capabilities [57]. The Mg-based alloys, particularly Mg-Al systems, are ideal for designing lightweight structures for military and commercial aircraft. Complex aircraft structures with high mechanical strength and refined microstructure use Mg-based alloys such as Mg-Y, Mg-Zn-Zr, and Mg-Al-Zn due to their high formability [58]. Reinforcements like B_4C, Al_2O_3, and SiC, along with thermally stable reinforcements, are often added to improve tribological properties [59], [60], [61]. However, due to their low corrosion resistance, the use of Mg alloys in the aircraft industry is limited.

The Ti-based alloys, including Ti-10V-2Fe-3Al, B120VCA, and Ti-6Al-4V, have lower density and higher mechanical strength even at very high temperatures [62]. Titanium Matrix Composites (TMCs) consist of materials like BN, Boron, BC alumina, SiC, and silica as reinforcements added to Ti alloys as the matrix material. These ceramics can withstand temperatures as high as 1500°C [63]. Ti alloys retain their mechanical strength even at elevated temperatures compared to Al, making them more suitable for aircraft and missile structures operating at high temperatures. TMCs reinforced with B4C have shown better friction and wear behaviour with a 20% reinforcement content. By systematically controlling the microstructure, volume fraction, and dispersion of these reinforcements, the tribological and mechanical strength of the materials can be tailored [64].

4.3.4 Advanced Corrosion Protection Technologies

Various cutting-edge methods are being developed to protect aerospace-manufactured components from corrosion. The goal is to enhance performance, longevity, and sustainability while reducing environmental effects. These technologies utilize advanced materials, coatings, surface treatments, and monitoring systems to improve resistance against corrosion and prolong the lifespan of crucial aeronautical components.

Some of the new technologies that are worth mentioning are nano-technology assisted techniques, where nano-scale conversion coatings, nano-scaled ceramic coatings, sol-gel coatings, nano-composite coatings, and layer-by-layer coatings have found themselves within a wide range of applications whose objective is to improve the life duty, reliability, and performance of aerospace engineered components [28]. For this particular case, nanostructured coatings involving nickel, zinc, chromium, and molybdenum layers increase the polarization resistance of the system, decreasing the corrosion current and corrosion potential, thus reducing the component corrosion rates. These layers offer enhanced surface barrier properties, galvanic protection, and self-healing abilities of microscopic defects.

Throughout the chapter, we have discussed the most commonly used methods for corrosion protection for Al, Mg, and Ti-based alloys. However, discussing new trends in using composite materials to fabricate engineered components in the aerospace industry is appropriate. Modern aeroplanes constructed by Boeing and Airbus are expanding their horizons and changing from

Advances in Corrosion Science and Surface Engineering   Materials Research Forum LLC
Materials Research Foundations 188 (2026) 228-248   https://doi.org/10.21741/9781644903919-12

metal alloys to composite materials. Some of these materials are polymer-based, known as polymer matrix composites (PMC), thermosets, and thermoplastics. Moreover, metal matrix composites (MMC), ceramic matrix composites (CMC), shape memory composite materials (SMCM), biocomposite materials (BCM), and nanocomposite materials have also been considered [65]. Different coating manufacturing methods are not limited to anodization and chromium conversion. Methods such as the sol-gel route, layer-by-layer (LBL) nanofilms, and surface modification techniques such as laser surface texturing, plasma electrolytic oxidation (PEO), and ion implantation have been explored to improve the corrosion resistance of aircraft components.

Nowadays, the aircraft industry manufactures most aircraft components from composite materials, either traditional metals, ceramics, or polymers. A greener alternative has been presented with fiber-reinforced polymers, which possess a good weight-to-strength ratio. These composites are a mixture of natural or synthetic fibers in a polymer matrix (epoxy resins) [66], [67]. The properties of natural reinforced fiber composites (NFRPC) have drawn attention as they offer good mechanical properties, good tensile strength, and low acquisition cost. Natural fibers such as jute, hemp, flax, rubberwood, etc., have been considered. On the other hand, petroleum-based fibers such as fiberglass, carbon, aramid, and basalt are employed to manufacture synthetic fiber-reinforced polymer composites (SFRPC) [68], [69].

However, although SFRPC and NFRPC offer unique characteristics with good mechanical, tribological, and corrosion-resistant properties, they have limitations and drawbacks. In the case of natural fibers (NF), they are prone to water absorption, low thermal stability, and biodegradability over time, while synthetic fibers (SF), even though they have high thermal stability, are expensive, they are not environmentally friendly, and are more brittle than their counterpart. Therefore, their application in the aerospace industry has been limited to small components such as bridge decks, reinforcement bars, girders, handrails, or staying cables [65].

Conclusion

In conclusion, aerospace materials are crucial for ensuring the safety, efficiency, and performance of aircraft. Al-alloys, Ti-alloys, and innovative materials like MMCs and Mg-based alloys are extensively used in aircraft structures due to their unique properties. These materials offer a combination of strength, lightweight, and corrosion resistance, making them ideal for aerospace applications. Corrosion monitoring and prevention are critical aspects of maintaining aerospace component integrity and safety. The utilization of coatings, inhibitors, anodization, and advanced technologies ensures the longevity and reliability of aerospace components. The shift towards composite materials further enhances corrosion resistance, mechanical strength, and overall performance. Continued research and development in corrosion prevention technologies will be essential for ensuring the safety and efficiency of aerospace operations.

References

[1] H. Zhu and J. Li, "Advancements in corrosion protection for aerospace aluminum alloys through surface treatment," *Int J Electrochem Sci*, vol. 19, no. 2, p. 100487, Feb. 2024. https://doi.org/10.1016/J.IJOES.2024.100487

[2] L. Li, M. Chakik, and R. Prakash, "A Review of Corrosion in Aircraft Structures and Graphene-Based Sensors for Advanced Corrosion Monitoring," *Sensors 2021, Vol. 21, Page 2908*, vol. 21, no. 9, p. 2908, Apr. 2021. https://doi.org/10.3390/S21092908

[3] J. Demo and F. Friedersdorf, "Aircraft corrosion monitoring and data visualization techniques for condition based maintenance," *IEEE Aerospace Conference Proceedings*, vol. 2015-June, Jun. 2015. https://doi.org/10.1109/AERO.2015.7119048

[4] M. E. Hoffman and P. C. Hoffman, "Corrosion and fatigue research — structural issues and relevance to naval aviation," *Int J Fatigue*, vol. 23, pp. 1–10, Jan. 2001. https://doi.org/10.1016/S0142-1123(01)00115-3

[5] M. P. Martínez-Viademonte, S. T. Abrahami, T. Hack, M. Burchardt, and H. Terryn, "A Review on Anodizing of Aerospace Aluminum Alloys for Corrosion Protection," *Coatings 2020, Vol. 10, Page 1106*, vol. 10, no. 11, p. 1106, Nov. 2020. https://doi.org/10.3390/COATINGS10111106

[6] T. Dursun and C. Soutis, "Recent developments in advanced aircraft aluminium alloys," *Materials & Design (1980-2015)*, vol. 56, pp. 862–871, Apr. 2014. https://doi.org/10.1016/J.MATDES.2013.12.002

[7] R. Asmatulu, "Nanocoatings for corrosion protection of aerospace alloys," *Corrosion Protection and Control Using Nanomaterials*, pp. 357–374, Jan. 2012. https://doi.org/10.1533/9780857095800.2.357

[8] S. Gialanella and A. Malandruccolo, "Alloys for Aircraft Structures," *Topics in Mining, Metallurgy and Materials Engineering*, pp. 41–127, 2020. https://doi.org/10.1007/978-3-030-24440-8_3/FIGURES/50

[9] J. R. Davis, "ASM Specialty Handbook: Al and Al alloys," p. 784, 1993, Accessed: Mar. 26, 2024. [Online]. Available: https://www.asminternational.org/asm-specialty-handbook-aluminum-and-aluminum-alloys/results/-/journal_content/56/06610G/PUBLICATION/

[10] Y. Chen and R. Xing, "Corrosion protection of aluminum alloy skin for long-term parking aircraft," in *IOP Conference Series: Earth and Environmental Science*, IOP Publishing Ltd, Jan. 2021. https://doi.org/10.1088/1755-1315/631/1/012035

[11] G. S. Malhi *et al.*, "Corrosion in Aircraft Components: Types, Impacts and Protection Measures," *Corrosion in Aircraft Components: Types, Impacts and Protection Measures Article in International Journal of Advanced Science and Technology*, vol. 29, no. 10S, pp. 4891–4896, 2020, [Online]. Available: https://www.researchgate.net/publication/342153745

[12] S. J. Findlay and N. D. Harrison, "Why aircraft fail," *Materials Today*, vol. 5, no. 11, pp. 18–25, Nov. 2002. https://doi.org/10.1016/S1369-7021(02)01138-0

[13] "Aircraft Corrosion - Filiform, Pitting & Corrosion Types." Accessed: Mar. 26, 2024. [Online]. Available: https://blog.eplane.com/aircraft-corrosion-filiform-pitting-other-corrosion-types/

[14] Zaki. Ahmad, "Principles of corrosion engineering and corrosion control," p. 656, 2006.

[15]   "Complete Guide to Aircraft Corrosion Types | Mid-America Areotech." Accessed: Mar. 26, 2024. [Online]. Available: https://www.maaero.com/a-complete-guide-to-the-types-of-corrosion-in-aircraft/

[16]   M. Czaban, "Aircraft corrosion - Review of corrosion processes and its effects in selected cases," *Fatigue of Aircraft Structures*, vol. 2018, no. 10, pp. 5–20, Dec. 2018. https://doi.org/10.2478/FAS-2018-0001

[17]   S. G. Pantelakis, P. G. Daglaras, and C. A. Apostolopoulos, "Tensile and energy density properties of 2024, 6013, 8090 and 2091 aircraft aluminum alloy after corrosion exposure," *Theoretical and Applied Fracture Mechanics*, vol. 33, no. 2, pp. 117–134, May 2000. https://doi.org/10.1016/S0167-8442(00)00007-0

[18]   "Stress-Corrosion Cracking of Titanium Alloys," *Stress-Corrosion Cracking*, pp. 271–302, Dec. 2017. https://doi.org/10.31399/ASM.TB.SCCMPE2.T55090271

[19]   E. L. Colvin, "Aluminum Alloys: Corrosion," *Encyclopedia of Materials: Science and Technology*, pp. 107–110, Jan. 2001. https://doi.org/10.1016/B0-08-043152-6/00022-X

[20]   R. J. H. Wanhill and R. T. Byrnes, "Stress Corrosion Cracking in Aircraft Structures," in *Aerospace Materials and Material Technologies* , Springer, Singapore, 2017, pp. 387–410. https://doi.org/10.1007/978-981-10-2143-5_19

[21]   R. J. H. Wanhill, R. T. Byrnes, and C. L. Smith, "Stress corrosion cracking (SCC) in aerospace vehicles," *Stress corrosion cracking: Theory and practice*, pp. 608–650, Jan. 2011. https://doi.org/10.1533/9780857093769.4.608

[22]   C. Vargel, "Filiform corrosion," *Corrosion of Aluminium*, pp. 247–265, Jan. 2020. https://doi.org/10.1016/B978-0-08-099925-8.00019-3

[23]   Einer Bardal, *Corrosion and Protection*. London: Springer, 2004. https://doi.org/https://doi.org/10.1007/b97510

[24]   "Factors That Affect Your Aircrafts Corrosion." Accessed: Mar. 26, 2024. [Online]. Available: https://www.acornwelding.com/blog/post/factors-affect-aircrafts-corrosion/

[25]   P. R. Roberge, *Corrosion Inspection and Monitoring*. 2006. https://doi.org/10.1002/9780470099766

[26]   S. Benavides, "Corrosion control in the aerospace industry," *Corrosion control in the aerospace industry*, pp. 1–312, 2009. https://doi.org/10.1533/9781845695538

[27]   L. Yang, *Techniques for corrosion monitoring*, no. september 2016. 2008.

[28]   R. Asmatulu, *Nanocoatings for corrosion protection of aerospace alloys*. Woodhead Publishing Limited, 2012. https://doi.org/10.1533/9780857095800.2.357

[29]   K. R. Larsen, "A new approach to corrosion-resistant aerospace designs," *Mater Perform*, vol. 56, no. 4, pp. 17–21, 2017.

[30]   J. W. Gooch and J. K. Daher, *Electromagnetic shielding and corrosion protection for aerospace vehicles*. 2007. https://doi.org/10.1007/978-0-387-46096-3

[31]   R. S. Dwyer-Joyce, "The Application of Ultrasonic NDT Techniques in Tribology," *http://dx.doi.org/10.1243/135065005X9763*, vol. 219, no. 5, pp. 347–366, May 2005. https://doi.org/10.1243/135065005X9763

[32]   C. H. Chen, *Ultrasonic and advanced methods for nondestructive testing and material characterization.* 2007. https://doi.org/10.1142/6327

[33]   M. P. Martínez-Viademonte, S. T. Abrahami, T. Hack, M. Burchardt, and H. Terryn, "A review on anodizing of aerospace aluminum alloys for corrosion protection," *Coatings*, vol. 10, no. 11, pp. 1–30, 2020. https://doi.org/10.3390/coatings10111106

[34]   S. S. Li *et al.*, "Development and applications of aluminum alloys for aerospace industry," *Journal of Materials Research and Technology*, vol. 27, pp. 944–983, Nov. 2023. https://doi.org/10.1016/J.JMRT.2023.09.274

[35]   K. Xhanari and M. Finšgar, "Organic corrosion inhibitors for aluminum and its alloys in chloride and alkaline solutions: A review," *Arabian Journal of Chemistry*, vol. 12, no. 8, pp. 4646–4663, Dec. 2019. https://doi.org/10.1016/J.ARABJC.2016.08.009

[36]   R. L. Twite and G. P. Bierwagen, "Review of alternatives to chromate for corrosion protection of aluminum aerospace alloys," *Prog Org Coat*, vol. 33, no. 2, pp. 91–100, 1998. https://doi.org/10.1016/S0300-9440(98)00015-0

[37]   W. Kaysser, "Surface modifications in aerospace applications," *Surface Engineering*, vol. 17, no. 4, pp. 305–312, 2001. https://doi.org/10.1179/026708401101517926

[38]   F. Andreatta and L. Fedrizzi, *Corrosion inhibitors*, vol. 233. 2016. https://doi.org/10.1007/978-94-017-7540-3_4

[39]   M. Abdallah, O. A. Hazazi, A. Fawzy, S. El-Shafei, and A. S. Fouda, "Influence of N-thiazolyl-2-cyanoacetamide derivatives on the corrosion of aluminum in 0.01 M sodium hydroxide," *Protection of Metals and Physical Chemistry of Surfaces*, vol. 50, no. 5, pp. 659–666, Sep. 2014. https://doi.org/10.1134/S2070205114050025/METRICS

[40]   S. Rajendran, C. Thangavelu, A. Angamuthu, and S. Jayakumar, "Inhibition of corrosion of aluminium in alkaline medium by glutaric acid in conjunction with zinc sulphate and diethylene triamine penta (Methylene phosphonic acid)," *Arch Appl Sci Res*, 2013.

[41]   V. V. Dhayabaran, J. P. Merlin, I. S. Lydia, R. Shanthi, and R. Sivaraj, "Inhibition of corrosion of aluminium in presence of fluorescein in basic medium," *Ionics (Kiel)*, vol. 10, no. 3–4, pp. 288–290, 2004. https://doi.org/10.1007/BF02382831/METRICS

[42]   R. Soni, R. Verma, R. Kumar Garg, and V. Sharma, "A critical review of recent advances in the aerospace materials," *Mater Today Proc*, no. August, 2023. https://doi.org/10.1016/j.matpr.2023.08.108

[43]   X. Zhang, Y. Chen, and J. Hu, "Recent advances in the development of aerospace materials," *Progress in Aerospace Sciences*, vol. 97, no. August 2017, pp. 22–34, 2018. https://doi.org/10.1016/j.paerosci.2018.01.001

[44]   J. C. Williams and R. R. Boyer, "Opportunities and Issues in the Application of Titanium Alloys for Aerospace Components," *Metals 2020, Vol. 10, Page 705*, vol. 10, no. 6, p. 705, May 2020. https://doi.org/10.3390/MET10060705

[45]   M. Chandrasekaran and Y. M. S. John, "Effect of materials and temperature on the forward extrusion of magnesium alloys," *Materials Science and Engineering: A*, vol. 381, no. 1–2, pp. 308–319, Sep. 2004. https://doi.org/10.1016/J.MSEA.2004.04.057

[46]   A. S. Warren, "Developments and challenges for aluminum--A boeing perspective," *Materials forum*, vol. 28, pp. 24–31, 2004, Accessed: Apr. 09, 2024. [Online]. Available: http://www.icaa-conference.net/ICAA9/data/papers/INV%203.pdf

[47]   S. C. Tjong, "Recent progress in the development and properties of novel metal matrix nanocomposites reinforced with carbon nanotubes and graphene nanosheets," *Materials Science and Engineering R: Reports*, vol. 74, no. 10, pp. 281–350, 2013. https://doi.org/10.1016/j.mser.2013.08.001

[48]   S. L. Soo, R. Hood, D. K. Aspinwall, W. E. Voice, and C. Sage, "Machinability and surface integrity of RR1000 nickel based superalloy," *CIRP Ann Manuf Technol*, vol. 60, no. 1, pp. 89–92, 2011. https://doi.org/10.1016/j.cirp.2011.03.094

[49]   S. Q. A. Rizvi, "Chapter 12 | Additives and Additive Chemistry," *Fuels and Lubricants Handbook: Technology, Properties, Performance, and Testing, 2nd Edition*, pp. 351–512, Nov. 2019. https://doi.org/10.1520/MNL3720150036

[50]   S. Benavides, "Corrosion in the aerospace industry," *Corrosion control in the aerospace industry*, pp. 1–14, Jan. 2009. https://doi.org/10.1533/9781845695538.1

[51]   J. K. Lomness and L. M. Calle, "Comparison of the Chromium Distribution in New Super Koropon Primer to 30 Year Old Super Koropon Using Focused Ion Beam/Scanning Electron Microscopy," 2020.

[52]   Protective coating of carbon steel, stainless steel, and aluminium on launch structures, facilities, and ground support equipment. 2016.

[53]   S. G. Pantelakis, A. N. Chamos, and A. T. Kermanidis, "A critical consideration for the use of Al-cladding for protecting aircraft aluminum alloy 2024 against corrosion," *Theoretical and Applied Fracture Mechanics*, vol. 57, no. 1, pp. 36–42, 2012. https://doi.org/10.1016/j.tafmec.2011.12.006

[54]   J. Yu and X. Li, "Modelling of the precipitated phases and properties of Al-Zn-Mg-Cu alloys," *J Phase Equilibria Diffus*, vol. 32, no. 4, pp. 350–360, 2011. https://doi.org/10.1007/s11669-011-9911-0

[55]   T. Dursun and C. Soutis, "Recent developments in advanced aircraft aluminium alloys," *Mater Des*, vol. 56, pp. 862–871, 2014. https://doi.org/10.1016/j.matdes.2013.12.002

[56]   D. K. Koli, G. Agnihotri, and R. Purohit, "Advanced Aluminium Matrix Composites: The Critical Need of Automotive and Aerospace Engineering Fields," *Mater Today Proc*, vol. 2, no. 4–5, pp. 3032–3041, 2015. https://doi.org/10.1016/j.matpr.2015.07.290

[57]   W. W. Jian *et al.*, "Ultrastrong Mg Alloy via Nano-spaced Stacking Faults," *Mater Res Lett*, vol. 1, no. 2, pp. 61–66, 2013. https://doi.org/10.1080/21663831.2013.765927

[58]   Y. Chen, Z. Xu, C. Smith, and J. Sankar, "Recent advances on the development of magnesium alloys for biodegradable implants," *Acta Biomater*, vol. 10, no. 11, pp. 4561–4573, Nov. 2014. https://doi.org/10.1016/J.ACTBIO.2014.07.005

[59]   S. Jayalakshmi, S. V. Kailas, and S. Seshan, "Tensile behaviour of squeeze cast AM100 magnesium alloy and its Al2O3 fibre reinforced composites," *Compos Part A Appl Sci Manuf*, vol. 33, no. 8, pp. 1135–1140, Aug. 2002. https://doi.org/10.1016/S1359-835X(02)00049-0

[60]   C. Y. H. Lim, S. C. Lim, and M. Gupta, "Wear behaviour of SiCp-reinforced magnesium matrix composites," *Wear*, vol. 255, no. 1–6, pp. 629–637, Aug. 2003. https://doi.org/10.1016/S0043-1648(03)00121-2

[61]   Q. C. Jiang, H. Y. Wang, B. X. Ma, Y. Wang, and F. Zhao, "Fabrication of B4C particulate reinforced magnesium matrix composite by powder metallurgy," *J Alloys Compd*, vol. 386, no. 1–2, pp. 177–181, Jan. 2005. https://doi.org/10.1016/J.JALLCOM.2004.06.015

[62]   Ikuhiro INAGAKI, Yoshihisa SHIRAI, Tsutomu TAKECHI, and Nozomu ARIYASU, "Application and Features of Titanium for the Aerospace Industry," *Nippon Steel & Sumitomo Metal*, vol. 106, no. 106, pp. 22–27, 2014.

[63]   K. Zhu, Y. J. Xu, T. Jing, and H. L. Hou, "Fracture behavior of a composite composed by Ti-aluminide multi-layered and continuous-SiCf-reinforced Ti-matrix," *Rare Metals*, vol. 36, no. 12, pp. 925–933, 2017. https://doi.org/10.1007/s12598-017-0883-z

[64]   R. Chaudhari and R. Bauri, "A novel functionally gradient Ti/TiB/TiC hybrid composite with wear resistant surface layer," *J Alloys Compd*, vol. 744, pp. 438–444, 2018. https://doi.org/10.1016/j.jallcom.2018.02.058

[65]   N. Mazlan, S.M.Sapuan, and R. A. Ilyas, *Advanced Composites in Aerospace Engineering Applications*, Springer Cham, 2022. https://doi.org/10.1007/978-3-030-88192-4

[66]   S. Vigneshwaran *et al.*, "Recent advancement in the natural fiber polymer composites: A comprehensive review," *J Clean Prod*, vol. 277, p. 124109, 2020. https://doi.org/10.1016/j.jclepro.2020.124109

[67]   S. Waghmare, S. Shelare, K. Aglawe, and P. Khope, "A mini review on fibre reinforced polymer composites," *Mater Today Proc*, vol. 54, pp. 682–689, 2022. https://doi.org/10.1016/j.matpr.2021.10.379

[68]   F. G. Alabtah, E. Mahdi, and F. F. Eliyan, "The use of fiber reinforced polymeric composites in pipelines: A review," *Compos Struct*, vol. 276, no. August, p. 114595, 2021. https://doi.org/10.1016/j.compstruct.2021.114595

[69]   S. Navaratnam, K. Selvaranjan, D. Jayasooriya, P. Rajeev, and J. Sanjayan, "Applications of natural and synthetic fiber reinforced polymer in infrastructure: A suitability assessment," *Journal of Building Engineering*, vol. 66, no. January, p. 105835, 2023. https://doi.org/10.1016/j.jobe.2023.105835

Advances in Corrosion Science and Surface Engineering   Materials Research Forum LLC
Materials Research Foundations 188 (2026) 249-260   https://doi.org/10.21741/9781644903919-13

Chapter 13

Advances in Corrosion Research

P. Selvakumar[1], R. Jagadeeswari[2*], G. Kausalya Sasikumar[3], Rajender Boddula[4]

[1]Department of Humanities and Sciences, Gokaraju Rangaraju Institute of Engineering and Technology, Hyderabad, 500090, Telangana, India

[2]Department of Chemistry, KPR Institute of Engineering and Technology, Coimbatore -641407, Tamil Nadu India

[3]Centre of Research and Development, KPR Institute of Engineering and Technology, Coimbatore 641407, Tamil Nadu, India

[4]School of Sciences, Woxsen University, Telangana, Hyderabad, 502345, India

jagadeeswarichem@gmail.com

Abstract

The widespread and expensive problem of corrosion has substantial economic and environmental consequences. Protective coatings designed to combat corrosion are essential in shielding various industrial sectors from its harmful effects. The environmental toxicity associated with organic and inorganic compounds utilized as corrosion inhibitors has emerged as a significant issue in recent decades. The control of metal corrosion rates through nanomaterials represents a novel approach and discovery in the field of nanotechnology. Nanomaterials exhibit enhanced corrosion inhibition properties due to their increased surface-to-volume ratio. Numerous studies have shown the effectiveness of nanomaterials in inhibiting corrosion. This section provides an overview of current technological approaches, with a particular focus on nanotechnology. It also discusses some initial research findings regarding the corrosion-resistant properties of nanostructures and their practical applications.

Keywords

Nanotechnology, Nanocoatings, Nanoparticles, Nanoalloys, Corrosion Inhibitors

Contents

1.   Introduction

The field of nanotechnology is experiencing rapid growth, with wide-ranging applications in scientific research. Particles at the nanoscale, measuring between 1-100 nm, have shown versatile uses across multiple sectors, including industrial, biomedical, environmental, food and agricultural industries. This broad applicability stems from their exceptional and superior fundamental chemical and physical characteristics, such as enhanced surface area, mechanical durability, optical properties, and chemical responsiveness [1]. A significant industrial application of nanoparticles is their exceptional capacity to protect metals from corrosion in diverse environments [2]. Metals possess several desirable qualities, including electrical and thermal conductivity, high melting and boiling points, superior tensile strength, high mass-to-volume ratio, and ductility, rendering them essential as base materials or raw materials in industrial and other applications. The interaction between metals and their surroundings triggers chemical or electrochemical reactions at the interface of the metal and its environment. These processes occur as a result of the metal's exposure to the elements around it. Consequently, refined metals transform into more chemically stable forms such as oxides, hydroxides, or sulfides, a process known as corrosion. This interaction affects the structures and properties of various materials.

Corrosion represents an irreversible form of material damage or destruction and is a highly costly phenomenon, constituting a significant global challenge. The economic impact of corrosion is estimated to be approximately 4% of the gross national product, or equivalent to one-fifth of global steel production designed to replace losses caused by corrosion [3]. Corrosion is influenced by numerous physical and chemical environmental factors, including pH variations, gases, humidity, contaminants, salts, temperature, and electrolyte types present on the metal surface [4]. To reduce corrosion rates, several strategies are implemented, including applying protective coatings to metal surfaces, modifying the surrounding environment, utilizing corrosion-inhibiting substances, and adjusting pH levels and electrical potential through cathodic or anodic processes [5]. The use of inhibitors to protect metals from corrosion is a prominent strategy. These inhibiting agents create a defensive layer on the metal's surface, aiding in the regulation of corrosion rates. A wide array of organic and inorganic inhibitors is employed to mitigate corrosion in metalworking industries.

Nevertheless, these substances are frequently expensive and potentially harmful to both the environment and human well-being, which restricts their usage. Nanomaterials and their components function as efficient corrosion inhibitors owing to their greater surface area-to-volume ratio in comparison to traditional macroscopic substances [6]. Nanocompounds serve to inhibit surface reactions and regulate corrosion rates by blocking active locations on metallic surfaces. These compounds also enhance various material properties, including durability, hardness, optical qualities, strength, and resistance to thermal changes [7]. These substances are eco-friendly and decompose naturally. This chapter aims to investigate innovative methods for employing nanostructures as more efficient inhibitors of corrosion.

Advances in Corrosion Science and Surface Engineering   Materials Research Forum LLC
Materials Research Foundations 188 (2026) 249-260   https://doi.org/10.21741/9781644903919-13

2.   Nanoparticle-Based Inhibitors

Metal and their oxide nanoparticles demonstrate efficacy in corrosion control. The application of these nanoparticles as effective corrosion inhibitors on metal surfaces has been demonstrated. Multiple research studies have shed light on the mechanisms of corrosion prevention through the process of metal nanoparticles and their oxides adhering to metallic surfaces. Recent literature has reported the utilization of metal and metal oxide nanoparticles to enhance corrosion resistance, including Ag, TiO_2, Cu_2O, ZnO, ZrO_2, and SiO_2 [8]. The inhibitory effect of silver nanoparticles on the corrosion of carbon steel in acidic solution has been documented [9]. In aqueous acidic environments, silver nanoparticles display enhanced reactivity. Researchers employed polarization techniques and EIS (electrochemical impedance spectroscopy) to assess corrosion inhibition effectiveness. Results showed that Ag nanoparticles formed self-assembled monolayers on the metal surface, with Ag-poly (ethylene glycol) thiol acting as a mixed-type inhibitor. Another investigation examined the corrosion rate of silver nanoparticles combined with surfactants on the same metal, revealing a notable reduction in carbon steel corrosion. The silver nanoparticles exhibited effective performance and protective qualities against corrosion [10]. The corrosion resistance of TiO_2 nanoparticle coatings on centrifugal pump blades was evaluated using atomic absorption spectrometry, given the superior mechanical characteristics of TiO_2 (Fig. 1) [11].

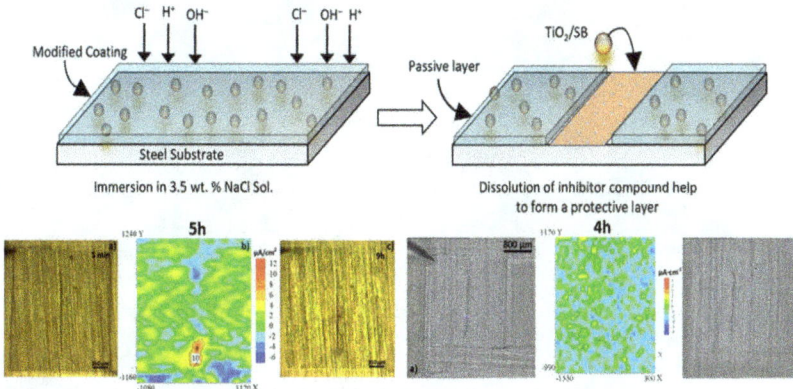

Figure 1. Nanoparticle based corrosion inhibitor.

Research has shown that the effectiveness of titania nanoparticles as inhibitors is influenced by the thickness of the protective layer on the surface. Scientists investigated the corrosion-resistant properties of titania (TiO_2) nanoparticles embedded in a nickel matrix when exposed to alkaline environments. Samples with nickel matrices containing TiO_2 nanoparticle coatings demonstrated superior hardness compared to pure nickel coatings. Additionally, these matrix-form samples exhibited improved resistance to corrosion [12]. The corrosion resistance of copper was examined using an anodization method, which involved coating copper with nanocopper oxide. This technique exhibited promising efficiency in inhibiting corrosion [13]. Studies have demonstrated the effectiveness of ZnO nanoparticles in inhibiting corrosion of mild steel across various solutions (HCl, NaCl, H_2O, NaOH). The corrosion rate was determined using the weight loss technique [14].

Advances in Corrosion Science and Surface Engineering   Materials Research Forum LLC
Materials Research Foundations 188 (2026) 249-260   https://doi.org/10.21741/9781644903919-13

3. Nanotubes as Inhibitors

Nanotubes possess hollow tubular nanostructures and have demonstrated several applications due to their greater physicochemical properties [15]. These formations can serve as protective coatings to prevent corrosion [16]. On metal surfaces, nanotubes create a protective layer that connects metal and polymer-based composites. They enhance sacrificial protection in polymer coatings and serve as corrosion-inhibiting additives (Fig. 2) [17].

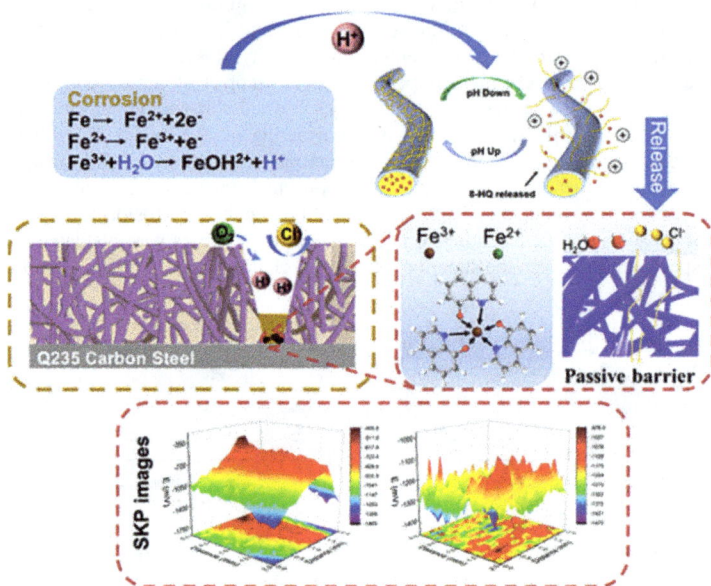

Figure 2. Nanocarbon applied as corrosion inhibitor.

The incorporation of benzotriazole-loaded nanotubes into paint can improve its resistance to corrosion and enhance the coating's tensile strength [18]. The time required for the formation of a protective metal layer through copper complexation is related to the release kinetics of benzotriazole. Research has been published on the anticorrosive properties of a hybrid film composed of MWCNT (multi-wall carbon nanotube) with BTESPT (bis-[triethoxysilylpropyl] tetra sulfide) silane, and a pure BTESPT silane hybrid film, applied to stainless steel [19]. The study found that the nanotube silane film, which has been modified, demonstrates superior effectiveness in regulating the corrosion rate compared to a film consisting solely of BTESPT.

4. Nanofiber as Inhibitor

Nanofibers are typically defined as filaments with dimensions smaller than 100 nm in diameter. These microscopic structures find use in a wide range of applications, including their incorporation into self-repairing coatings designed to safeguard metals against corrosive processes (Fig. 3) [20].

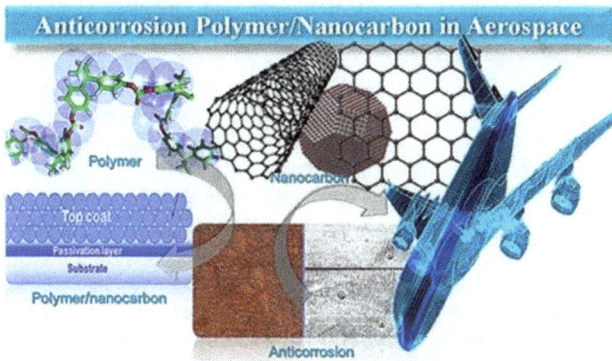

Figure 3. Nanofiber used as corrosion inhibitor.

Aluminum substrates have been coated with nanofibers composed of polyvinyl chloride doped with ceria. This nanofiber coating helps to reduce the cathodic reaction on aluminum substrates in acidic environments, thus decreasing electrochemical corrosion [21]. Metal surfaces have also been protected from corrosion using polyamide nanofiber coatings. Additionally, cellulose nanofibers have been employed to manage the corrosion of carbon steel in alkaline solutions. Electrochemical impedance spectroscopy was used to measure the corrosion rate of carbon steel, revealing the outstanding self-healing properties of cellulose nanofibers [22].

5. Nanocomposites

Nanocomposites offer an efficient strategy to reduce corrosion-related expenses. The fabrication of these nanocomposites involves the use of polymers and nanomaterials (Fig. 4) [23].

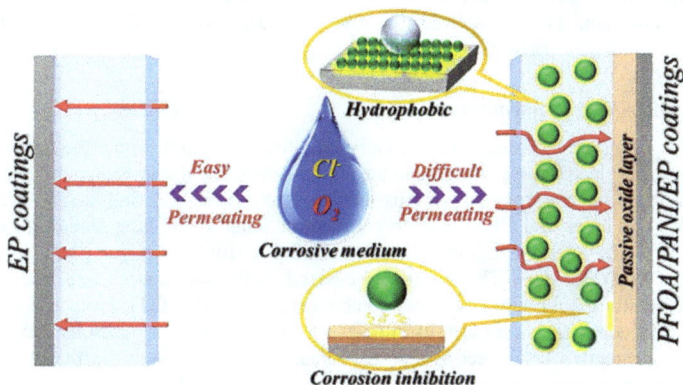

Figure 4. Nanocomposite used as corrosion inhibitor.

Advances in Corrosion Science and Surface Engineering   Materials Research Forum LLC
Materials Research Foundations 188 (2026) 249-260   https://doi.org/10.21741/9781644903919-13

Nanocomposites are generally composed of organic and inorganic elements. The inorganic components contribute to enhanced adhesion, increased ductility, and improved mechanical strength in these materials. Conversely, the organic constituents provide flexibility, better compatibility, and a reduction in porosity and defects. In the preparation of nanocomposite coatings, several organic polymers are frequently utilized. These include, but are not limited to, polyurethane, epoxy, PEG (polyethylene glycol), PANI (polyaniline), polyacrylic, polystyrene, polyvinyl alcohol (PVA), polyamide and polypyrrole [24]. Metal nanoparticles, along with their carbides, phosphates, and oxides, typically make up the inorganic components [25].

The effectiveness of these nanoparticles in inhibiting corrosion is directly related to their concentration and the temperature. The nanocomposite molecules form a protective layer on the metal surface, shielding it from corrosive agents in harsh environments. Incorporating inert nanoparticles (such as carbides, oxides, polytetrafluoroethylene, carbon nanotubes) into the metallic matrix through galvanic plating has been shown to enhance both anticorrosive and mechanical properties compared to pure metal coatings [6]. The composite layers that result demonstrate enhanced resistance to corrosion and wear, greater hardness, improved tribological characteristics, and better paint adhesion. Research has been conducted on the corrosion inhibition effects of zinc oxide nanocomposite, incorporating specific polymers (polyvinylpyrrolidone (PVP), polyethylene glycol (PEG), polyacrylonitrile (PAN)), on mild steel when exposed to HCl solution [26]. Images of the surface at a microscopic level reveal the degree of damage inflicted by corrosive ions in various corrosive environments. The incorporation of nanocomposites enhances the ability to inhibit corrosion, with greater effectiveness observed as the concentration of polymer inhibitors increases in both nanocomposite and composite coatings. These coatings also serve to hinder the breakdown of mild steel when exposed to acidic conditions [27]. According to experimental findings, the rate of corrosion for conducting polyaniline nanocomposite is roughly 34 times less than that of unprotected low-carbon steel. Additionally, these composite coatings offer improved thermal resistance to the underlying material [28]. A superior inhibitor exhibits several desirable qualities, such as strong inhibitory performance, affordability, minimal toxicity, and simple manufacturing processes [29]. Nanocomposites offer similar advantages while also reducing environmental impact. The use of nanocomposites to slow down corrosion is particularly important. These materials improve the adhesion of polymers and decrease the porousness of metal nanoparticles.

6.   Nano alloys and Nanocoatings

The formation of an oxide film enables nanocrystal alloys to exhibit properties such as hardness, resistance to wear, electrical resistivity, and the ability to withstand corrosion at high temperatures [30]. Researchers have employed various methods to investigate and assess the corrosion characteristics of nanocrystal alloys in different environments. Using laser cladding technology, a Fe-based amorphous–nanocrystalline coating containing in situ synthesized particles was applied to the surface of Q235 steel. The resulting Fe-based composite coating exhibited uniformity, density, and metallurgical bonding without notable crack formation [31]. Protective coatings play a crucial role in combating the harmful impacts of corrosion on various substances, especially metals. These protective layers act as a shield, creating a barrier between the material and its surrounding environment (Fig. 5) [32].

Advances in Corrosion Science and Surface Engineering   Materials Research Forum LLC
Materials Research Foundations 188 (2026) 249-260   https://doi.org/10.21741/9781644903919-13

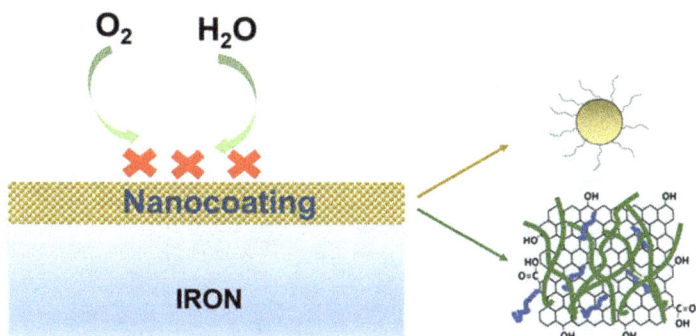

Figure 5. Nanocoating prevents metal from environmental condition.

Effective corrosion management relies on a thorough grasp of various corrosion protection coatings and their protective mechanisms. These coatings fall into three main categories: organic, inorganic, and metallic, each with its own characteristics, modes of action, and uses, providing specific benefits for different corrosion issues. Among these, organic coatings are the most widely employed. They consist of a mixture of resins, binders, pigments, and additives, creating a continuous layer on the substrate that acts as a physical barrier against corrosive elements. The thickness and composition of this film can be adjusted to meet specific needs, allowing for versatile application. Moreover, organic coatings often demonstrate strong adhesion to the underlying material, which enhances their protective capabilities [33]. Ceramic coatings and other non-organic protective layers are renowned for their ability to withstand extreme temperatures and their long-lasting nature [34]. The inorganic nature of these coatings provides exceptional resistance to chemical corrosion and allows them to endure severe environmental conditions, making them ideal for use in extreme settings. The protective qualities of inorganic coatings stem from their chemical stability and their capacity to create stable oxide barriers that shield against corrosive substances [35]. Protection against corrosion can be achieved through metallic coatings that function via sacrificial or barrier methods. Sacrificial metallic coatings, typically made of zinc or aluminum, are applied to the base metal as a protective layer. These coatings possess a higher electrochemical potential compared to the underlying metal, causing them to corrode first. This process allows the sacrificial coating to degrade while shielding the substrate, effectively slowing or preventing corrosion. In contrast, barrier metallic coatings operate by forming a physical obstruction between the substrate and corrosive elements. Often composed of metals like stainless steel or nickel alloys, these coatings impede the movement of corrosive agents to the substrate, thereby decreasing the rate of corrosion [36].

7.   Nanotechnology-Enhanced Methods

Self-healing coatings have revolutionized the field of organic coatings, representing a significant advancement in corrosion protection. These innovative coatings possess the remarkable ability to

Advances in Corrosion Science and Surface Engineering   Materials Research Forum LLC
Materials Research Foundations 188 (2026) 249-260   https://doi.org/10.21741/9781644903919-13

repair themselves autonomously when damaged, thereby prolonging their functional life and providing continuous defense against corrosion.

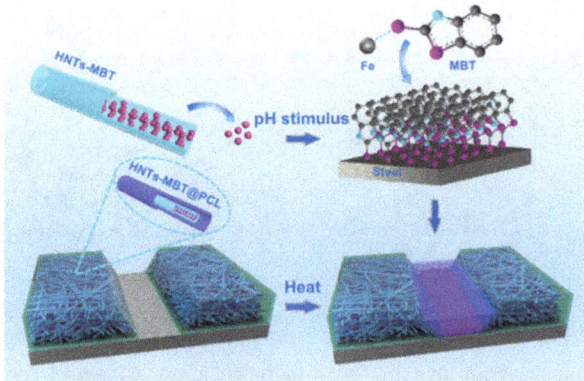

Figure 6. Nanotech displaying self-healing coating.

Various techniques have been developed to imbue organic coatings with self-healing properties, each contributing to this extraordinary capability. One notable method involves embedding microcapsules or micro/nanocontainers within the coating matrix. These containers are filled with specialized healing agents, such as corrosion inhibitors or polymer precursors, which are crucial for the self-repair process [37]. External forces that harm the coating cause these microcapsules to break open, releasing healing substances into the damaged area. This coordinated process helps to seal and repair cracks or imperfections, effectively restoring the coating's functionality. This method has received considerable interest due to its proven ability to impart self-healing capabilities to coatings in harsh environments (Fig. 6) [38].

In this scenario, it is essential to delve into specific instances that highlight nanotechnology's effectiveness in enhancing coating performance, while thoroughly evaluating the protective capabilities of these nanoparticle-enhanced coatings across diverse environments and metal surfaces. The integration of nanoparticles into organic coatings has brought about numerous advantages, resulting in significant improvements in mechanical durability, barrier functionality, and resistance to ultraviolet (UV) light [39]. As an example, coatings enhanced with advanced ceramics like alumina or zirconia have shown exceptional resistance to high-temperature corrosion effects. Likewise, ceramic coatings infused with specialized compounds, such as yttria-stabilized zirconia, have proven to be outstanding protective materials in aerospace applications. The harsh and fast-moving environment within aircraft engines requires coatings with superior wear resistance. These infused ceramic coatings have proven their effectiveness by strengthening metallic components against abrasive forces and high-temperature erosive elements [39].

Conclusion

Natural corrosion is a highly destructive and expensive phenomenon. The investigation of cutting-edge materials, including nanomaterials and biologically-inspired coatings, shows great promise for improving resistance to corrosion and longevity. Researchers have examined and analyzed various nanofillers and nanostructured materials for corrosion protection purposes, such as nanocrystal alloys, metal-based nanoparticles, nanocontainers, nanocomposites, nanotubes, and nanofibers. The primary discoveries and improvements outlined in this chapter showcase the advancements made in comprehending and tackling corrosion-related issues. Nevertheless, there remains substantial work to be done in exploring and creating solutions that meet the changing requirements of industries and the increasing demand for sustainable alternatives. The advancement of corrosion protection coatings will be driven by continuous scientific studies, collaboration between stakeholders, and the adoption of cutting-edge technologies. These efforts will contribute to enhancing the longevity, security, and environmental sustainability of various industrial sectors.

References

[1] K.A. Altammar, A review on nanoparticles: characteristics, synthesis, applications, and challenges, Frontiers in Microbiology 14 (2023) 1155622.

[2] S.M. Ramteke, M. Walczak, M. De Stefano, A. Ruggiero, A. Rosenkranz, M. Marian, 2D materials for Tribo-corrosion and-oxidation protection: A review, Advances in Colloid and Interface Science 331 (2024) 103243.

[3] R.E. Lobo, B. Guzmán, P.A. Orrillo, C.C. Domínguez, L.E. Jimenez, M.I. Torino, Corrosion: Basics, Adverse Effects and Its Mitigation, in: R. Aslam, M. Mobin, J. Aslam (Eds.), Sustainable Food Waste Management, Springer Nature Singapore, Singapore, 2024: pp. 3–22. https://doi.org/10.1007/978-981-97-1160-4_1.

[4] N. Haridharan, R.V. Shiva Kumar, Anticorrosion and Antifouling Coating Materials, in: H. Murthy, V.J. Pillai, K.P. Kumar, M. Cowan (Eds.), Novel Anti-Corrosion and Anti-Fouling Coatings and Thin Films, 1st ed., Wiley, 2024: pp. 353–398. https://doi.org/10.1002/9781394234318.ch12.

[5] K.M. Shwetha, B.M. Praveen, B.K. Devendra, A review on corrosion inhibitors: types, mechanisms, electrochemical analysis, corrosion rate and efficiency of corrosion inhibitors on mild steel in an acidic environment, Results in Surfaces and Interfaces (2024) 100258.

[6] A.O. Alao, A.P. Popoola, O. Sanni, The Influence of Nanoparticle Inhibitors on the Corrosion Protection of Some Industrial Metals: A Review, J Bio Tribo Corros 8 (2022) 68. https://doi.org/10.1007/s40735-022-00665-1.

[7] N. Sharma, Ferrite Nanoparticles for Corrosion Protection Applications, in: P. Sharma, G.K. Bhargava, S. Bhardwaj, I. Sharma (Eds.), Engineered Ferrites and Their Applications, Springer Nature Singapore, Singapore, 2023: pp. 227–240. https://doi.org/10.1007/978-981-99-2583-4_12.

[8] S. Silviana, C. Lukmilayani, Metal Coatings Derived from Modified Silica as Anti-Corrosion, in: Defect and Diffusion Forum, Trans Tech Publ, 2024: pp. 77–95. https://www.scientific.net/DDF.431.77 (accessed January 25, 2025).

[9] S.R. Al-Mhyawi, Green synthesis of silver nanoparticles and their inhibitory efficacy on corrosion of carbon steel in hydrochloric acid solution, International Journal of Electrochemical Science 18 (2023) 100210.

[10]   E. Ituen, A. Singh, L. Yuanhua, O. Akaranta, Biomass-mediated synthesis of silver nanoparticles composite and application as green corrosion inhibitor in oilfield acidic cleaning fluid, Cleaner Engineering and Technology 3 (2021) 100119.

[11]   M. Nawaz, S. Ahmad, M.G. Taryba, M.F. Montemor, R. Kahraman, R.A. Shakoor, Improvement in inhibition performance of anti-corrosion coatings using polyolefin matrix embedded with modified TiO2 nanoparticles, Progress in Organic Coatings 195 (2024) 108659.

[12]   J. Li, S. Wang, B. Chen, G. Pan, G. Zhang, Z.-B. Ding, Y. Liu, TiO$_2$ Nanoparticle/Carboxymethyl Cellulose Coatings for Photocatalytic Dye Degradation in Simulated Seawater, ACS Appl. Nano Mater. 8 (2025) 315–328. https://doi.org/10.1021/acsanm.4c05676

[13]   F.A. Ansari, D.S. Chauhan, Anticorrosive properties of copper oxide nanomaterials in aggressive media, in: Smart Anticorrosive Materials, Elsevier, 2023: pp. 229–249. https://www.sciencedirect.com/science/article/pii/B9780323951586000163 (accessed January 25, 2025).

[14]   S. Ramamoorthy, S. Surendhiran, D. Senthil Kumar, G. Murugesan, M. Kalaiselvi, S. Kavisree, S. Muthulingam, S. Murugesan, Evaluation of photocatalytic and corrosion properties of green synthesized zinc oxide nanoparticles, J Mater Sci: Mater Electron 33 (2022) 9722–9731. https://doi.org/10.1007/s10854-022-07776-y

[15]   S. Rathinavel, K. Priyadharshini, D. Panda, A review on carbon nanotube: An overview of synthesis, properties, functionalization, characterization, and the application, Materials Science and Engineering: B 268 (2021) 115095.

[16]   D. Yan, Z. Zhang, W. Zhang, Y. Wang, M. Zhang, T. Zhang, J. Wang, Smart self-healing coating based on the highly dispersed silica/carbon nanotube nanomaterial for corrosion protection of steel, Progress in Organic Coatings 164 (2022) 106694.

[17]   A. Kausar, I. Ahmad, T. Zhao, Corrosion-resisting nanocarbon nanocomposites for aerospace application: an up-to-date account, Applied Nano 4 (2023) 138–158.

[18]   L.M. Muresan, Nanocomposite coatings for anti-corrosion properties of metallic substrates, Materials 16 (2023) 5092.

[19]   S.E. Elugoke, T.W. Quadri, L.O. Olasunkanmi, O.E. Fayemi, A.S. Adekunle, E.E. Ebenso, Production and corrosion protection properties of carbon nanotubes, in: Smart Anticorrosive Materials, Elsevier, 2023: pp. 63–90. https://www.sciencedirect.com/science/article/pii/B9780323951586000126 (accessed January 25, 2025).

[20]   R. Wang, L. Cao, W. Wang, Z. Mao, D. Han, Y. Pei, Y. Chen, W. Fan, W. Li, S. Chen, Construction of Smart Coatings Containing Core–Shell Nanofibers with Self-Healing and Active Corrosion Protection, ACS Appl. Mater. Interfaces 16 (2024) 42748–42761. https://doi.org/10.1021/acsami.4c09260

[21]   D.D. Thiruvoth, M. Ananthkumar, Evaluation of cerium oxide nanoparticle coating as corrosion inhibitor for mild steel, Materials Today: Proceedings 49 (2022) 2007–2012.

[22]   A. Yabuki, C. Nishikawa, I.W. Fathona, Synergistic effect by release of corrosion
inhibitors via cellulose nanofibers in self-healing polymer coatings to prevent corrosion of
carbon steel, Journal of Industrial and Engineering Chemistry (2024).
https://www.sciencedirect.com/science/article/pii/S1226086X24007354 (accessed January 25,
2025).

[23]   A. Kausar, I. Ahmad, P. Bocchetta, High-performance corrosion-resistant
polymer/graphene nanomaterials for biomedical relevance, Journal of Composites Science 6
(2022) 362.

[24]   N. Dhiman, N. Singla, Smart Nanocoating- an Innovative Solution to Create Intelligent
Functionality on Surface, ChemistrySelect 9 (2024) e202403038.
https://doi.org/10.1002/slct.202403038

[25]   A.J. Kora, Applications of inorganic metal oxide and metal phosphate-based
nanoceramics in dentistry, Industrial Applications of Nanoceramics (2024) 63–77.

[26]   T.W. Quadri, L.O. Olasunkanmi, O.E. Fayemi, E.E. Ebenso, Utilization of ZnO-based
materials as anticorrosive agents: A review, Inorganic Anticorrosive Materials (2022) 161–
182.

[27]   M. Hasanin, S.A. Al Kiey, Development of ecofriendly high performance anti-corrosive
chitosan nanocomposite material for mild steel corrosion in acid medium, Biomass Conv.
Bioref. 13 (2023) 12235–12248. https://doi.org/10.1007/s13399-021-02059-8

[28]   F.R. Rangel-Olivares, E.M. Arce-Estrada, R. Cabrera-Sierra, Synthesis and
characterization of polyaniline-based polymer nanocomposites as anti-corrosion coatings,
Coatings 11 (2021) 653.

[29]   S.H. Alrefaee, K.Y. Rhee, C. Verma, M.A. Quraishi, E.E. Ebenso, Challenges and
advantages of using plant extract as inhibitors in modern corrosion inhibition systems: Recent
advancements, Journal of Molecular Liquids 321 (2021) 114666.

[30]   M. Udhayakumar, N. Radhika, K.L. Arun, A Comprehensive Review on Nanocrystalline
Coatings: Properties, Challenges and Applications, J Bio Tribo Corros 8 (2022) 83.
https://doi.org/10.1007/s40735-022-00683-z

[31]   X. Shang, C. Zhang, M. Shan, Q. Liu, H. Cui, Effect of In situ Synthesis on the
Microstructure, Corrosion, and Wear Resistance of Fe-Based Amorphous–Nanocrystalline
Coatings, J Therm Spray Tech 32 (2023) 259–272. https://doi.org/10.1007/s11666-022-
01480-3

[32]   L.T. Nhiem, D.T.Y. Oanh, N.H. Hieu, Nanocoating toward anti-corrosion: A review,
Vietnam Journal of Chemistry 61 (2023) 284–293. https://doi.org/10.1002/vjch.202300025

[33]   M.H. Nazari, Y. Zhang, A. Mahmoodi, G. Xu, J. Yu, J. Wu, X. Shi, Nanocomposite
organic coatings for corrosion protection of metals: A review of recent advances, Progress in
Organic Coatings 162 (2022) 106573.

[34]   S. Trevisan, W. Wang, B. Laumert, A high-temperature thermal stability and optical
property study of inorganic coatings on ceramic particles for potential thermal energy storage
applications, Solar Energy Materials and Solar Cells 239 (2022) 111679.

[35] D. Zhang, F. Peng, X. Liu, Protection of magnesium alloys: From physical barrier coating to smart self-healing coating, Journal of Alloys and Compounds 853 (2021) 157010.

[36] R. Aslam, M. Mobin, S. Zehra, J. Aslam, A comprehensive review of corrosion inhibitors employed to mitigate stainless steel corrosion in different environments, Journal of Molecular Liquids 364 (2022) 119992.

[37] A. Thakur, S. Kaya, A. Kumar, Recent Innovations in Nano Container-Based Self-Healing Coatings in the Construction Industry, CNANO 18 (2022) 203–216. https://doi.org/10.2174/1573413717666210216120741

[38] X. Fu, W. Du, H. Dou, Y. Fan, J. Xu, L. Tian, J. Zhao, L. Ren, Nanofiber Composite Coating with Self-Healing and Active Anticorrosive Performances, ACS Appl. Mater. Interfaces 13 (2021) 57880–57892. https://doi.org/10.1021/acsami.1c16052

[39] H.S. Aljibori, A. Alamiery, A.A.H. Kadhum, Advances in corrosion protection coatings: A comprehensive review, Int. J. Corros. Scale Inhib 12 (2023) 1476–1520.

Keyword Index

About the Editors

Dr. Rajender Boddula is a highly accomplished researcher and academic with a strong focus on materials science, energy, and environmental sustainability. He currently serves as an Associate Professor at the School of Sciences, Woxsen University, Hyderabad, India, and also holds an adjunct professorship at Graphic Era Hill Deemed University, India.

Dr. Boddula earned his Master of Science in Organic Chemistry from Kakatiya University, Warangal, India, in 2008. He completed his Ph.D. in Chemistry with the highest honors in 2014 from CSIR-Indian Institute of Chemical Technology (CSIR-IICT) and Kakatiya University, with his doctoral research focusing on "Synthesis and Characterization of Polyanilines for Supercapacitor and Catalytic Applications."

His extensive postdoctoral career includes prestigious fellowships and research positions globally: as a Chinese Academy of Sciences-President's International Fellowship Initiative (CAS-PIFI) fellow at the Chinese Academy of Science-National Center for Nanoscience and Technology (CAS-NCNST, Beijing), a Postdoctoral Researcher at National Tsing-Hua University (NTHU, Taiwan), and a Postdoctoral Researcher at the Center for Advanced Materials (CAM), Qatar University. His diverse research engagements have contributed significantly to fields such as functional materials synthesis, biofuels, water splitting, and biomass/CO_2 valorization.

Dr. Boddula's research expertise spans sustainable nanomaterials, graphene and polymer composites, heterogeneous catalysis, environmental remediation technologies, energy conversion and storage systems (including supercapacitors, and batteries). He is a prolific author with over 96 research articles, 42 book chapters, and 66 edited books published with renowned international publishers like Springer, Elsevier, Wiley-Scrivener, and CRC Press.

Dr. Boddula is the recipient of several academic honors including the Australian Endeavour Research Fellowship, CSIR-UGC National Fellowships, and CAS-PIFI. Dr. Boddula also holds significant editorial roles, serving on the editorial boards of "Bioinorganic Chemistry and Applications" (Hindawi Publishers) and "Hydrogen" (MDPI) journals, Associate Editorial Board Member of "Current Science, Engineering and Technology (CSET)" (Bentham Publishers), and Review Editor for "Frontiers in Chemistry" and reviewer for various leading international journals, in addition to guest-editing for several other international journals.

G. Kausalya Sasikumar holds a Bachelor of Engineering degree in Electronics and Communication Engineering and a Master of Technology with a specialization in Nanoscience and Technology from Anna University, India. She achieved a first-class distinction in both degrees and secured a university rank for her bachelor's degree.

During her academic career, she undertook projects at renowned Indian institutes such as NIT and IICT, working under the supervision of esteemed scientists. Additionally, she received the Korean Government Scholarship and the Brain-pool fellowship from 2015-2018 for her research program. She is currently pursuing her PhD in the Faculty of Technology stream at Anna University, India. She also works as a Women Scientist in the DST-sponsored project under the Women Scientist Scheme (WOS-A). Her research interests primarily focus on Carbon nanostructured materials, Ferrites, and Polymers for various applications such as Biophotonics, Biosensing, Bioimaging, Nanophosphors, Energy Storage, and Energy Harvesters. She has published research articles in peer-reviewed journals such as Materials Letters, Catalysis Communication, RSC Advances, ChemElectroChem, Microchemical Journal and Current Chromatography among others. Her academic and research credentials demonstrate her expertise and commitment to excellence in her field. Her impressive achievements highlight her ability to work collaboratively and independently, making her an asset to any academic or research institution.

Ramyakrishna Pothu submitted her PhD thesis in the chemistry under the supervision of Prof. Jianmin Ma in the Hunan University (China). She obtained her Bachelor degree from Satavahana University, India in 2013, and her Postgraduate degree from Osmania University, India in 2015, respectively. She has published several scientific articles in peer-reviewed international journals and co-authored more than thirty book chapters by various publishers and she has co-edited book with Wiley publishers. *Her main research interests focus on the functional nanomaterials and its composites for energy and environmental science.*

Dr. Noora Al-Qahtani obtained a bachelor's degree with honors in Physics and Biomedical Sciences from the College of Arts and Sciences at Qatar University in 2008. She continued her graduate studies and obtained a master's degree in Materials Science and Engineering from the University of Sheffield in 2015, then a Ph.D. in Materials Science and Engineering in corrosion from Imperial College London in 2021. She is the first Qatari female in this field who hold a doctorate degree. Dr. Noora Al-Qahtani is currently a Research Projects Manager &Research Assistant Professor in the Center for Advanced Materials at QU and Section Head of Student and Mentor Training at QUYSC. During Dr. Al-Qahtani's academic career, she chaired and was involved in many projects; She also was a member of various committees at the university. In addition to her work in research, she is teaching undergraduate courses. She has an active research agenda with collaborative projects with colleagues at QU and other international universities. Her research interest focused on sustainability and applied electrochemistry, corrosion, and STEM education research for young students. Her interests encompassed archaeology from a scientific aspect and concentrated on linking the gap between fundamental corrosion studies and practical engineering solutions. Dr. Al-Qahtani has been awarded the following grants as lead Principal Investigator in National Capacity Building Program Grant from Qatar University and is currently a primary investigator and participating in 3 NPRP projects that are worth of 1.9 million USD and funded by the Qatar National Research Fund (NPRP). Dr. Al-Qahtani has a noticeable record of publications in academic journals, peer-reviewed conference proceedings, and academic conference presentations. She published one patent and over ten articles in internationally renowned journals and conference proceedings related to the corrosion application of different coatings, inhibitors, and synthesis of various nanomaterials. Also, she has been an active member of prominent organizations in science and administration. She participated in several training sessions and workshops in strategic management, research administration, international relations, and education strategies.

www.ingramcontent.com/pod-product-compliance
Lightning Source LLC
Chambersburg PA
CBHW071338210326
41597CB00015B/1489